Silicon Based Polymers

François Ganachaud · Sylvie Boileau ·
Bruno Boury
Editors

Silicon Based Polymers

Advances in Synthesis and Supramolecular Organization

Springer

Editors

Dr François Ganachaud
Ecole Nationale Supérieure de
Chimie de Montpellier
UMR 5253 - Institut Charles
Gerhardt
Ingénierie et Architectures Macromoléculaires
8 rue de l'Ecole Normale
34296 Montpellier CX 05
France

Dr Sylvie Boileau
Institut de Chimie et des
Matériaux
Paris-Est (ICMPE)
SPC-CNRS
2 à 8 rue Henri Dunant
94320 Thiais
France

Pr Bruno Boury
Université Montpellier II
Institut Charles Gerhardt
Chimie Moléculaire et organisation
du solide
place Eugène Bataillon
34095 Montpellier CX 05
France

ISBN: 978-1-4020-8527-7 e-ISBN: 978-1-4020-8528-4

Library of Congress Control Number: 2008931415

Printed on acid-free paper

9 8 7 6 5 4 3 2 1

springer.com

Preface

Interest in silicon-based materials has established an international community whose roots initially dug in organic and silicon chemistry, polymer and sol-gel chemistry, and material science. This fact led Prof. Interrante and Prof. Kawakami to launch a series of workshops that began in Tatsunokuchi, Japan, in 1999 as a satellite workshop to the XIIth International Symposium on Organosilicon Chemistry in Sendai. The first workshop was attended by over 100 participants from Asia, Europe, and the USA and its success emphasized the perceived need for a continuing forum for organosilicon polymer research; a decision was then taken to maintain these workshops on a bi-annual basis. The second and third workshops were held in Canterbury, UK, in 2001, and Troy, USA, in 2003, both workshops achieved a similarly strong international participation.

The fifth meeting was held in France. The city of Montpellier was chosen as organosilicon chemistry plays an important role in the academic field of the University and Ecole de Chimie (The XI International Symposium on Organosilicon Chemistry was organized in Montpellier in 1996, the III Sol-Gel Symposium was organized in Montpellier in 1985 and the XIV edition was also held in September 2007, in the same city). The next ISPO 08 was held in Busan, Korea, 6–8 June 2008, just after the 15th International Symposium on Organosilicon Chemistry.

Those international workshops are designed to highlight advanced research and technological innovations by bringing together scientists with common interests in macromolecular organosilicon compounds and systems. Indeed, silicon-containing materials and polymers are used all over the world in industry, domestic products and high technology applications. Among them, silicones are certainly the most well-known materials, still renewed in their properties and preparation processes and still full of potential. Lesser known (but on the brink of future exploitation) silicon containing-polymers are now close to maturity and some of them are already on the market. Indeed, the chemists have in their grasp polysilsesquioxanes, polycarbosilanes, polysilanes, and polysilazanes, and others that combine silicon and different elements with different linkages. We have learnt how to build chains, dendrimers, hyperbranched and cross-linked networks, physical and chemical gels. The result is a source of materials with applications in materials for optics, electronics, ionic electrolytes, liquid crystals, bio-materials, ceramics and concrete, paintings and coatings etc. all needed to face a myriad of environmental, energy and technological issues.

This research interest is international in nature with strong participation from scientists in Asia, Europe and North America and especially students and young scientists.

The contributions in this book reflect the diversity of silicon-based materials and the new developments in the strategies and concepts at work.

The editors are especially grateful to the authors for their contributions and cooperation which made this book possible. We thank all of them for their participation.

François Ganachaud
Sylvie Boileau
Bruno Boury

Contents

Contributors

Noureddine Amdouni
Laboratoire de Physico-Chimie des Matériaux Solides, Faculté des Sciences de Tunis, Manar II, 2092 Tunis, Tunisia

Maryse Bacquet
Laboratoire de Chimie Organique et Macromoléculaire, UMR CNRS 8009, Université des Sciences et Technologies de Lille, 59655 Villeneuve d'Ascq France

Bharathi Balagam
Engineered Nanomaterials Laboratory, Department of Chemistry and Physics, William Paterson University, 300, Pompton Road, Wayne, New Jersey 07470-2103

Eric Besson
Institut Charles Gerhardt, UMR 5253 CNRS, Chimie Moléculaire et Organisation du Solide, Université Montpellier II, Place E. Bataillon, 34095, Montpellier Cedex 5, France

Corinne Binet
Lab. de Dynamique et Structure des Matériaux Moléculaires, UMR CNRS 8024, Université des Sciences et Technologies de Lille, 59655 Villeneuve d'Ascq, France

Sylvie Boileau
Institut de Chimie et des Matériaux, Paris-Est (ICMPE), SPC-CNRS, 2 à 8 rue Henri Dunant, 94320 Thiais, France, boileau@glvt-cnrs.fr

Bruno Boury
Institut Charles Gerhardt Montpellier, UMR 5253 CNRS-UM2-ENSCM-UM1, CMOS, Place E. Bataillon, 34 095 Montpellier, France, boury@univ-montp2.fr

Bernard Boutevin
Institut Charles Gerhardt – UMR 5253 CNRS, Ingénierie et Architectures Macromoléculaires, Ecole Nationale Supérieure de Chimie de Montpellier, 8, rue de l'école normale, 34296 Montpellier cedex 5

Michael A. Brook
Department of Chemistry, McMaster University, 1280 Main Street West, Hamilton, ON, Canada, L8S 4M1, mabrook@mcmaster.ca

Bhanu P. S. Chauhan
Engineered Nanomaterials Laboratory, Department of Chemistry and Physics, William Paterson University, 300, Pompton Road, Wayne, New Jersey 07470-2103, chauhanbps@wpunj.edu or nano.chem@wpunj.edu

Yves Chevalier
University of Lyon, Laboratoire d'Automatique et de Génie des Procédés (LAGEP), UMR 5007 CNRS – Université Claude Bernard Lyon 1, 69622 Villeurbanne, France

Robert J. P. Corriu
Institut Charles Gerhardt, UMR 5253 CNRS, Chimie Moléculaire et Organisation du Solide, Université Montpellier II, Place E. Bataillon, 34095, Montpellier Cedex 5, France

Mihaela Cristea
"P.Poni" Institute of Macromolecular Chemistry, Aleea Gr. Ghica Voda 41A, Iasi, Romania, mcristea@icmpp.ro

Anna Czech
GE Silicones, 771 Old Sawmill River Road, Tarrytown, New York

Vincent Darras
Laboratoire de Physicochimie des Polymères et des Interfaces (LPPI) – Université de Cergy-Pontoise – 5, mail Gay-Lussac, Neuville-sur-Oise – 95031 Cergy-Pontoise Cedex – France; Institut de Chimie des Matériaux Paris Est, SPC-CNRS UMR 7582, 2 rue H. Dunant, 94320 Thiais, France, vincent.darras@polymtl.ca

Bogumiła Delczyk
Centre of Molecular and Macromolecular Studies, Polish Academy of Sciences, Sienkiewicza 112, 90–363 Łódź, Poland

Wojciech Duczmal
Department of Chemical Technology and Engineering, University of Technology and Life Sciences, Seminaryjna 3, 85–326 Bydgoszcz

Matthieu Dumont
Lab. des Interactions Moléculaires et Réactivité Chimique et Photochimique, UMR 5623 CNRS, Université de Toulouse, 118 Rte de Narbonne, 31062 Toulouse cedex 9, France

Ramona Enea
Technical University "Gh. Asachi" Iasi, Department of Natural and Synthetic Polymers, Bd. Mangeron 71, 700050-Iasi, Romania, ramon_enea@yahoo.com

Amanda S. Fawcett
Department of Chemistry, McMaster University, 1280 Main Street West, Hamilton, ON, Canada, L8S 4M1

Odile Fichet
Laboratoire de Physicochimie des Polymères et des Interfaces (LPPI) – Université de Cergy-Pontoise – 5, mail Gay-Lussac, Neuville-sur-Oise – 95031 Cergy-Pontoise Cedex – France, odile.fichet@u-cergy.fr

Juliette Fitremann
Laboratoire des Interactions Moléculaires et Réactivité Chimique et Photochimique, Université de Toulouse, UMR 5623 CNRS – Université Paul Sabatier, Bâtiment 2R1, 118 Route de Narbonne, F-31062 Toulouse Cedex 9, France, fitremann@chimie.ups-tlse.fr

Elke Fritz-Langhals
Wacker Chemie AG, Consortium für elektrochemische Industrie, Zielstattstraße 20–22, D-81379 München, Germany, elke.fritz-langhals@wacker.com

François Ganachaud
Institut Charles Gerhardt – UMR 5253 CNRS, Ingénierie et Architectures Macromoléculaires, Ecole Nationale Supérieure de Chimie de Montpellier, 8, rue de l'école normale, 34296 Montpellier cedex 5, ganachau@enscm.fr

Tomasz Ganicz
Centre of Molecular and Macromolecular Studies, Polish Academy of Sciences, Sienkiewicza 112, 90–363 Łódź, Poland

Alain Gibaud
Laboratoire de Physique de l'Etat Condensé, Université du Maine, UMR 6087 CNRS, 72085 Le Mans Cedex 09, France

Stéphane Gineste
CRPP, CNRS, Av. A. Schweitzer, 33600 Pessac, France

Takayuki Hattori
Goi research center, Chisso Petrochemical Corporation, Ichihara, Chiba 290–8551, Japan

Souhaira Hbaieb
Laboratoire de Physico-Chimie des Matériaux Solides, Faculté des Sciences de Tunis, Manar II, 2092 Tunis, Tunisia

Simon J. Holder
Functional Materials Group, School of Physical Sciences, University of Kent, Canterbury, Kent. CT2 7NH. UK, S.J.Holder@kent.ac.uk

Nicolae Hurduc
Technical University "Gh. Asachi" Iasi, Department of Natural and Synthetic Polymers, Bd. Mangeron 71, 700050-Iasi, Romania, nhurduc@ch.tuiasi.ro

Richard G. Jones
Functional Materials Group, School of Physical Sciences, University of Kent, Canterbury, Kent. CT2 7NH. UK, R.G.Jones@kent.ac.uk

Rafik Kalfat
Laboratoire de Physico-Chimie des Matériaux Solides, Faculté des Sciences de Tunis, Manar II, 2092 Tunis, Tunisia; University of Lyon, Institut National de Recherche et d'Analyse Physico-chimique, Pôle Technologique Sidi-Thabet, 2020 Sidi-Thabet, Tunisia

Anna Kowalewska
Centre of Molecular and Macromolecular Studies, Polish Academy of Sciences, Sienkiewicza 112, 90–363 Łódź, Poland, anko@cbmm.lodz.pl

Heinz Langhals
Department of Chemistry, LMU University of Munich, Butenandtstr. 13, D-81377 Munich, Germany, Langhals@lrz.uni-muenchen.de

Elisabeth Laurent
Lab. des Interactions Moléculaires et Réactivité Chimique et Photochimique, UMR 5623 CNRS, Université de Toulouse, 118 Rte de Narbonne, 31062 Toulouse cedex 9, France

Claire Longuet
Institut Charles Gerhardt – UMR 5253 CNRS, Ingénierie et Architectures Macromoléculaires, Ecole Nationale Supérieure de Chimie de Montpellier, 8, rue de l'école normale, 34296 Montpellier cedex 5, France

Jean-Daniel Marty
Lab. des Interactions Moléculaires et Réactivité Chimique et Photochimique, UMR 5623 CNRS, Université de Toulouse, 118 Rte de Narbonne, 31062 Toulouse cedex 9, France

Hideyuki Matsumoto
Faculty of Engineering, Gunma University, Kiryu, Gunma 376–8515, Japan.

Monique Mauzac
Laboratoire des Interactions Moléculaires et Réactivité Chimique et Photochimique, Université de Toulouse, UMR 5623 CNRS – Université Paul Sabatier, Bâtiment 2R1, 118 Route de Narbonne, F-31062 Toulouse Cedex 9, France

Ahmad Mehdi
Institut Charles Gerhardt, UMR 5253 CNRS, Chimie Moléculaire et Organisation du Solide, Université Montpellier II, Place E. Bataillon, 34095, Montpellier Cedex 5, France, ahmad.mehdi@univ-montp2.fr

Anne-Françoise Mingotaud
Lab. des Interactions Moléculaires et Réactivité Chimique et Photochimique, UMR 5623 CNRS, Université de Toulouse, 118 Rte de Narbonne, 31062 Toulouse cedex 9, France, afmingo@chimie.ups-tlse.fr

Ioana Moleavin
Technical University "Gh. Asachi" Iasi, Department of Natural and Synthetic Polymers, Bd. Mangeron 71, 700050-Iasi, Romania, ioana_moleavin@yahoo.com

Michel Morcellet
Laboratoire de Chimie Organique et Macromoléculaire, UMR CNRS 8009, Université des Sciences et Technologies de Lille, 59655 Villeneuve d'Ascq France

Waêl Moukarzel
Laboratoire des Interactions Moléculaires et Réactivité Chimique et Photochimique, Université de Toulouse, UMR 5623 CNRS – Université Paul Sabatier, Bâtiment 2R1, 118 Route de Narbonne, F-31062 Toulouse Cedex 9, France

Ahlem Noomen
University of Lyon, Laboratoire d'Automatique et de Génie des Procédés (LAGEP), UMR 5007 CNRS – Université Claude Bernard Lyon 1, 69622 Villeurbanne, France; Laboratoire de Physico-Chimie des Matériaux Solides, Faculté des Sciences de Tunis, Manar II, 2092 Tunis, Tunisia; University of Lyon, Institut de Chimie et Biochimie Moléculaires et Supramoléculaires (ICBMS), UMR 5246 CNRS – Université Claude Bernard Lyon 1, 69622 Villeurbanne, France

Małgorzata Noskowska
Hollister, Poleczki 47, 02–822 Warszawa, Poland

Nobumasa Ootake
Goi research center, Chisso Petrochemical Corporation, Ichihara, Chiba 290–8551, Japan

Guillaume Palaprat
Lab. des Interactions Moléculaires et Réactivité Chimique et Photochimique, UMR 5623 CNRS, Université de Toulouse, 118 Rte de Narbonne, 31062 Toulouse cedex 9, France

Hélène Parrot-Lopez
Institut de Chimie et Biochimie Moléculaires et Supramoléculaires (ICBMS), UMR 5246 CNRS – Université Claude Bernard Lyon 1, 69622 Villeurbanne, France

Alexandra Penciu
University of Lyon, Laboratoire d'Automatique et de Génie des Procédés (LAGEP), UMR 5007 CNRS – Université Claude Bernard Lyon 1, 69622 Villeurbanne, France

Emmanuel Pouget
Institut Charles Gerhardt – UMR 5253 CNRS, Ingénierie et Architectures Macromoléculaires, Ecole Nationale Supérieure de Chimie de Montpellier, 8, rue de l'école normale,34296 Montpellier cedex 5

Jitendra S. Rathore
Engineered Nanomaterials Laboratory, Department of Chemistry and Physics, William Paterson University, 300, Pompton Road, Wayne, New Jersey 07470-2103

Catherine Réyé
Institut Charles Gerhardt, UMR 5253 CNRS, Chimie Moléculaire et Organisation du Solide, Université Montpellier II, Place E. Bataillon, 34095, Montpellier Cedex 5, France

Ana Maria Resmerita
Technical University "Gh. Asachi" Iasi, Department of Natural and Synthetic Polymers, Bd. Mangeron 71, 700050-Iasi, Romania, resmerita_anamaria@yahoo.com

Alok Sarkar
Engineered Nanomaterials Laboratory, Department of Chemistry and Physics, William Paterson University, 300, Pompton Road, Wayne, New Jersey 07470-2103, USA

Dan Scutaru
Technical University "Gh. Asachi" Iasi, Department of Natural and Synthetic Polymers, Bd. Mangeron 71, 700050-Iasi, Romania, dscutaru@ch.tuiasi.ro

Włodzimierz A. Stańczyk
Centre of Molecular and Macromolecular Studies, Polish Academy of Sciences, Sienkiewicza 112, 90–363 Łódź, Poland, was@bilbo.cbmm.lodz.pl

Anna Szeląg
Centre of Molecular and Macromolecular Studies, Polish Academy of Sciences, Sienkiewicza 112, 90–363 Łódź, Poland

Ryouji Tanaka
Department of Applied Chemistry, Gunma University, Kiryu, Gunma 376–8515, Japan

Dominique Teyssié
Laboratoire de Physicochimie des Polymères et des Interfaces (LPPI) – Université de Cergy-Pontoise – 5, mail Gay-Lussac, Neuville-sur-Oise – 95031 Cergy-Pontoise Cedex – France, dominique.teyssie@u-cergy.fr

David B. Thompson
Department of Chemistry, McMaster University, 1280 Main Street West, Hamilton, ON, Canada, L8S 4M1

Frédéric Vidal
Laboratoire de Physicochimie des Polymères et des Interfaces (LPPI) – Université de Cergy-Pontoise – 5, mail Gay-Lussac, Neuville-sur-Oise – 95031 Cergy-Pontoise Cedex – France, frederic.vidal@u-cergy.fr

Stéphanie Willai
Laboratoire de Chimie Organique et Macromoléculaire, UMR CNRS 8009, Université des Sciences et Technologies de Lille, 59655 Villeneuve d'Ascq France

Kazuhiro Yoshida
Goi research center, Chisso Petrochemical Corporation, Ichihara, Chiba 290–8551, Japan, k.yoshida@chisso.co.jp

Lacramioara Zadoina
Lab. des Interactions Moléculaires et Réactivité Chimique et Photochimique, UMR 5623 CNRS, Université de Toulouse, 118 Rte de Narbonne, 31062 Toulouse cedex 9, France

Part 1
Functional Polysiloxanes

New Avenues, New Outcomes: Nanoparticle Catalysis for Polymer Makeovers

Bhanu P. S. Chauhan, Bharathi Balagam, Jitendra S. Rathore, and Alok Sarkar

Abstract In this chapter, nanocluster catalyzed modifications of organic and silicon based polymers are described. The tailoring of the polymeric templates was carried out under mild conditions and led to hybrid polymers in quantitative yields. Detailed characterization studies indicated that the integrity of the polymeric templates was not compromised during the functionalization process. The nanoparticle catalysis was found to be quite effective and highly selective. In most cases exclusive ß-hydrosilylation products were obtained without any rearrangement or isomerization reactions. Detailed characterization and property profiling of the new hybrid polymers is also presented.

1 Introduction

Modern applications of polymeric materials desire and in certain instances require various functions in one family of polymeric architectures. To achieve such a goal most of the time one needs to devise new functional monomers with desired functionalities and study their polymerization to generate new polymeric materials. This approach is time tested and finally leads to desired materials. Though, one drawback to this approach is that for each new polymer a new functional monomer has to be synthesized and in certain instances new processes have to be developed to convert them to useful polymeric materials. In order to expedite the discovery of new functional polymers, one strategy could be to use a template polymer and investigate strategies to modify the property profile of such templates to achieve desired polymeric materials with required functionalities. This strategy allows a fast and efficient way to obtain functional polymeric materials in an economical fashion (Scheme 1).

B.P.S. Chauhan
Engineered Nanomaterials Laboratory, Department of Chemistry,
William Paterson University, 300, Pompton Road, Wayne, New Jersey 07470-2103
e-mail: chauhanbps@wpunj.edu

F. Ganachaud et al. (eds.), *Silicon Based Polymers*,

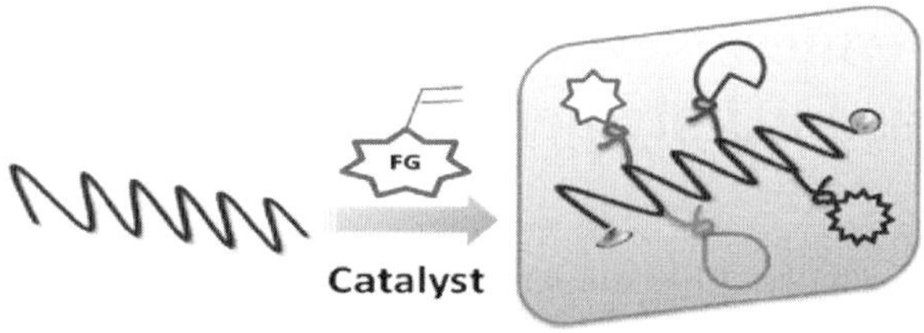

Scheme 1 Cartoon representation of polymer makeover strategy

The synthesis of mixed functionality polymers can be achieved via two types of strategies. One in which, an organic functionality is introduced into an inorganic backbone and another type could be where, inorganic or organometallic functional groups are attached to organic backbones. This type of functional group tailoring can lead to diverse family of structures, which can have desirable features and advantageous properties of both classes (organic and inorganic) of polymers. Owing to these attributes, throughout this chapter, we will use term "hybrid" to such type of polymers.

The synthesis of hybrid polymers can be regarded as an attractive avenue to develop new materials with wide-ranging desired property profiles [1, 2–7]. Recent disclosures from our laboratory have introduced new strategies for the generation of catalytically active metal nanoclusters and their utility as potent catalysts for silaesterification [8] chemo-selective hydrogenation [9] and hydrosilylation [10–13] reactions. In this chapter, we describe the strategy to generate families of new inorganic/organic hybrid polymer systems derived from hydrosilylation of polymeric templates such as polybutadienes and polyhydrosiloxane (PMHS) using various organic and inorganic functional groups.

The unsaturated polymers such as polyisoprenes and polybutadienes represent an ideal class of polymers for polymer tailoring reactions since these polymers contain one active double bond per each monomer unit after polymerization. Among them the polybutadienes are well known, least expensive and are readily available with a wide range of molecular weight with different degree of unsaturation. Moreover, their unique elastomeric properties make these polymers indispensable for the production of synthetic rubber. Various catalytic and non-catalytic reactions across the double bonds are known which can be performed to achieve new organo-functional polymers. The most common catalytic methods to modify these polymeric double bonds can be classified as hydrogenation, hydroformylation, hydrocarboxylation, epoxidation, oxidation, hydrosilylation and vinyl coupling reactions [1, 14–22]. In addition, a considerable number of patents have been granted on different studies of functionalization of polybutadienes. Although a large number of functional materials have been generated based on various catalytic reactions on these polymers, very little is known about these polymers covalently linked with silicon functionality.

2 Results and Discussion

2.1 Tailoring of Polybutadiene Based Polymers

In recent years, we have been investigating easy and economical functionalization of widely used carbon based polymers such as polybutadienes. The preliminary results of these studies have led our group to discover a highly selective and mild synthetic route to silyl-functionalization of 1,2-polybutadienes (PBD) via Pt-nanocluster catalyzed hydrosilylation of olefin bonds. Unlike other catalytic systems, our system was found to be equally effective with all varieties of functional silanes such as halo-, alkyl-, aryl- and alkoxy- silanes affording high yields and selectivities. In addition, all the hydrosilylation reactions were found to be very clean with the ease of product separation and purifications (Scheme 2).

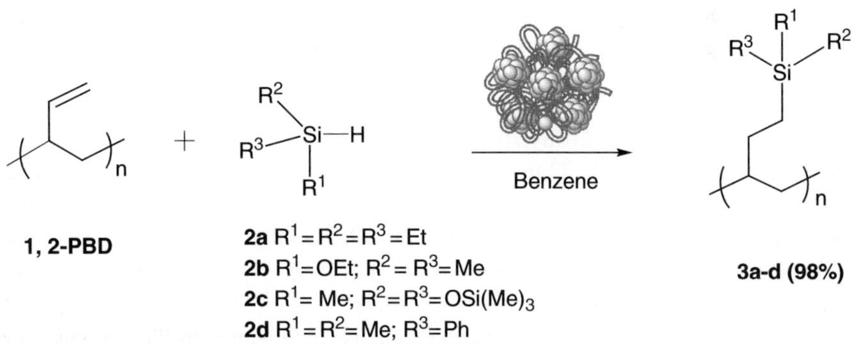

Scheme 2 General strategy to incorporate organosilicon moiety into polybutadiene

Size exclusion chromatography (SEC) was used to determine the molecular weight and chain length properties of functionalized polymers **3a–3d** with reference to polystyrene standards (Fig. 1). The SEC chromatograms of the products have

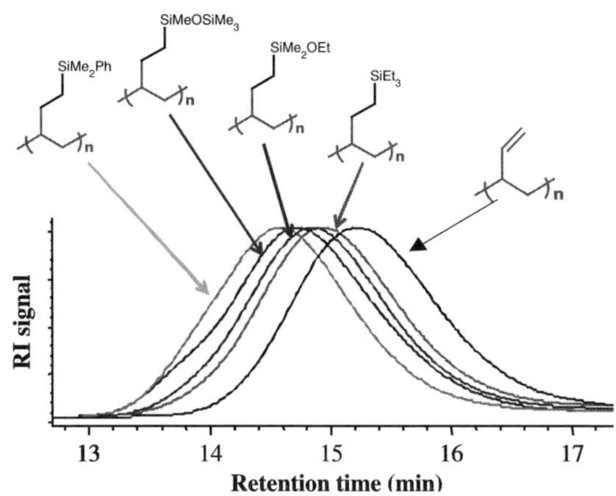

Fig. 1 SEC traces of the PBD-1 and Polymers **3a, 3b, 3c** and **3d**; mobile phase: THF (1mL/min). Taken from Ref [13]. (See also Plate 1 in the Colour Plate Section)

clearly been shifted towards the high molecular weight region, while retaining a fair molecular weight distribution (M_w / M_n ~ 1.4–1.5). This analysis confirmed that no other side reactions such as chain scission, cross-linking etc. occurred during the course of the hydrosilylation reaction leaving large-scale molecular structure intact.

The structure and regioselectivity of these functionalized polymers **3a–3d** were determined by ^{1}H, ^{13}C, DEPT and ^{29}Si NMR studies. ^{1}H and ^{13}C NMR studies of the products suggested that the hydrosilylation of PBD-1 occurs selectively via an anti-Markovnikov addition i.e., the Si atom being attached at the terminal position of the 1,2-PBD unit (β-product). Distortionless Enhancement by Polarization Transfer (DEPT) of ^{13}C NMR was used to differentiate between α and β product of hydrosilylated polymer. The selectivity towards β-product was verified for all the products by DEPT technique. For instance, the DEPT spectra of **3d** (Fig. 2) has shown one upward resonance at 33.70 which can be assigned to the methine carbon (CH) and three downward signals at 10.76, 26.72 and 37.85, were attributed to the three methylene (CH_2) carbons of polymer backbone of β-product. The opposite regioselectivity (β) would have generated two methine and two methylene carbons, lacking of which, clearly indicates the exclusive formation of α-product.

The regioselectivity of the product was further established by comparing the spectroscopic results with the hydrosilylation reaction of 1-heptene using four model silanes, **2a–2d** (Scheme 2). The ^{1}H, ^{13}C and ^{29}Si NMR results of hydrosilylation of 1-heptene (ß-product) were in good agreement with those of hydrosilylated polymers. For example, the hydrosilylation of **PBD-1** using silanes **2a, 2b, 2c, 2d** gave single ^{29}Si resonances at δ(ppm) 7.06, 18.01, 7.02/–20.36 and –1.7 respectively, which are comparable to the Si shifts observed for the hydrosilylated 1-heptene (7.7,17.46, 6.94/–21.05, –2.66). The ^{29}Si NMR, with only one detectable ^{29}Si resonance indicates the high selectivity of these reactions with exclusive formation of ß-adduct. In particular, the hydrosilylation reaction of the silane **2d** is noteworthy because this organosilane is known to favor Markovnikov product [15]. In the present catalysis under standard conditions, even silane **2d** led to formation of exclusive ß-product in quantitative yields.

To widen the scope of hydrosilylation of polybutadienes, we have screened a variety of cholorofunctional silanes, **2e–2i** to generate functional butadienes, which

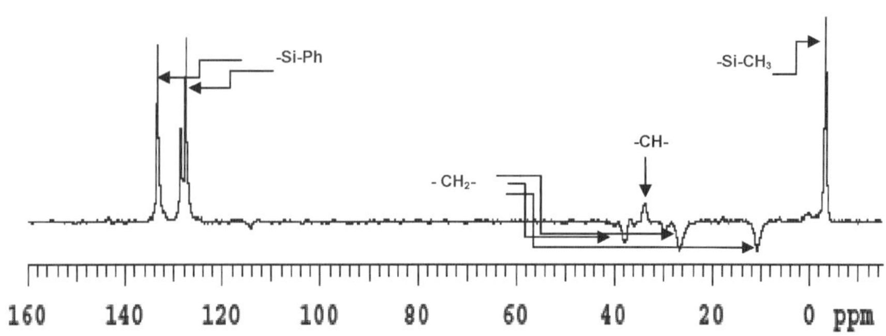

Fig. 2 ^{13}C NMR (DEPT) spectra of **3d** indicating aromatic, CH, CH_2 and CH_3 carbons

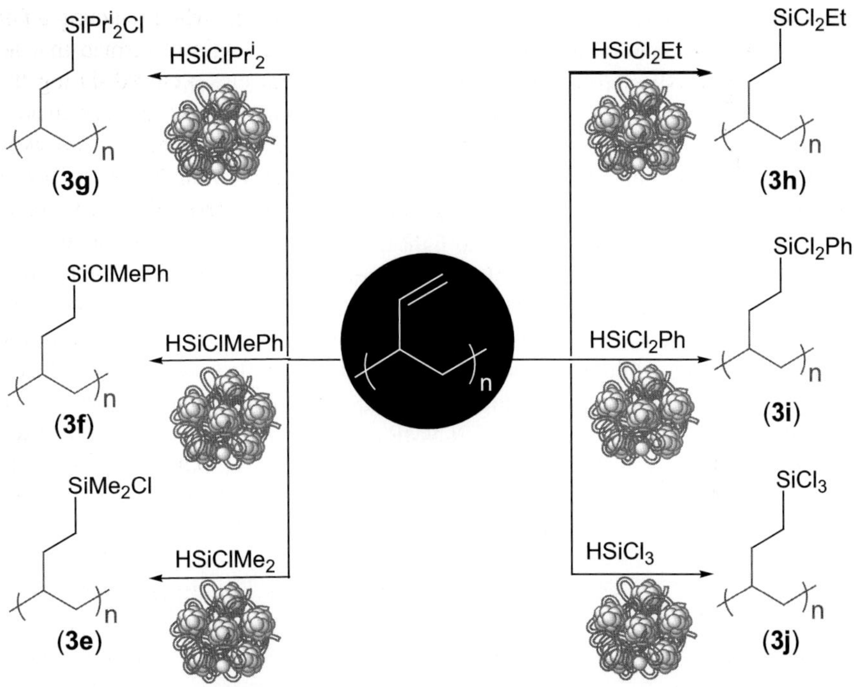

Scheme 3 Pt-nanoparticle catalyzed reaction of polybutadiene with various chlorosilanes

Table 1 Hydrosilylation of 1,2 Polybutadiene using Pt-nanoclusters

Silane Substrates	^{29}Si -NMR of silanes	Conditions [a]	Product (Yield %) [b]	^{29}Si-NMR of products
$HSiMe_2Cl$	12.2	25°C/16h	**3e** (98)	32.9
HSiPhMeCl	3.5	70°C/24h	**3f** (98)	22.1
$HSiPr^i_2Cl$	24.9	70°C/24h	**3g** (98)	36.8
$HSiCl_2Et$	13.1	45°C/24h	**3h** (98)	35.8
$HSiCl_2Ph$	−1.92	80°C/24h	**3i** (98)	20.2
$HSiCl_3$	−9.8	25°C/24h	**3j** (98)	13.5

[a]Conditions: [PBD-1] = 1.0 mmol; [silane] = 1.2 mmol; Pt-nanocluster = (0.010 gm); Solvent: Benzene, [b]Yields are based on isolated products. The isolation was performed under nitrogen atmosphere.

after classical reactions may lead to novel polymers **3e–3i** (Scheme 3, Table 1). Almost complete functionalization (95–98%) of 1,2 vinyl groups of the polymer was achieved with variety of chloroalkyl or chloroalkylphenyl silanes. The results are summarized in Table 1. All the products have shown high yields and selectivity, regardless of the nature of the silane, as evidenced by ^{1}H, ^{13}C, ^{29}Si NMR spectroscopy (Table 1). This general trend of selectivity is very encouraging and bodes well for the commercial applications of nanoparticle catalysts.

Cyclic siloxanes are one of the important class of inorganic-organic hybrid compounds, which have often been used to improve the properties of the materials [23–25].

To investigate the catalytic efficiency and selectivity of Pt-nanoclusters in hydrosilylation reaction of PBD with oligosiloxanes, the reaction of (D_3DH) with PBD was studied in presence of Pt-nanoclusters (Scheme 4). In a typical procedure, a Schlenk tube (10 mL), equipped with magnetic stirrer and oil bath was charged with Pt-nanoclusters (0.005 g, 0.001 mmol), degassed and flushed with nitrogen. The 1,2 PBD (0.056 g, 1.0 mmol) dissolved in dry benzene (2 ml) was added to the Schlenk tube followed by the addition of D_3DH (1.0 mmol) under the constant flow of nitrogen. After a few minutes of stirring, the reaction mixture turned into light brown homogeneous solution, indicating the formation of soluble nanoclusters. The reaction was monitored by ^1H NMR spectroscopy. After the completion of reaction, catalyst particles (fine powdery precipitate) were visible again, which were very conveniently separated by centrifugation (1 h). After separation of the catalysts the solvent was evaporated to obtain the product polymer **4** (1.0 mmol) as a pale yellow viscous liquid. The product thus obtained was analyzed by GPC and ^1H, ^{13}C, DEPT, ^{29}Si NMR spectroscopic techniques (Scheme 4).

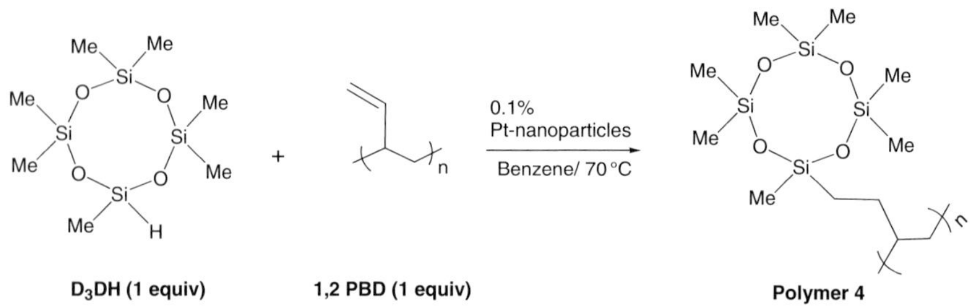

Scheme 4 Pt-Nanoparticle Catalyzed Hydrosilylation of 1, 2-polybutadiene with D_3DH

Size exclusion chromatography (SEC) was used to determine the molecular weight and chain length properties of functionalized polymers with reference to polystyrene standards. The SEC chromatograms of the products have clearly been shifted toward the high molecular weight region, while retaining a quite narrow molecular weight distribution ($Mw/Mn = 1.4–1.5$). This analysis confirmed that no other side reactions such as chain scission, cross-linking, etc. occurred during the course of the hydrosilylation reaction, leaving the large-scale molecular structure intact.

^1H, ^{13}C and ^{29}Si NMR studies of the products suggested that the hydrosilylation of PBD occurs selectively via an anti-Markovnikov addition, i.e., the Si-atom being attached at the terminal position of olefin bonds (β-product) (Fig. 3).

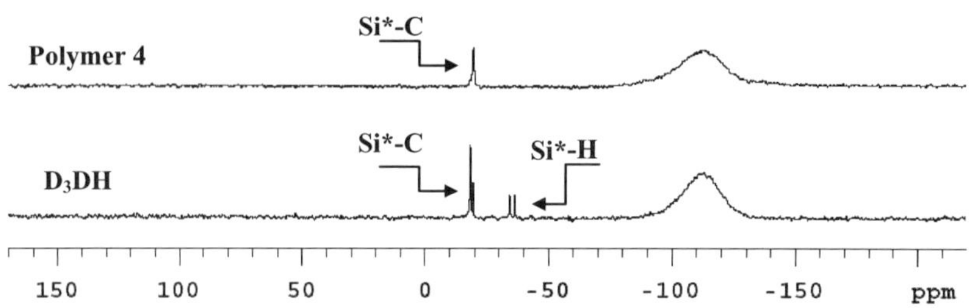

Fig. 3 Coupled ^{29}Si NMR of D_3DH and Polymer 4 showing disappearance of Si signals associated with Si centers containing Si—H bonds

2.2 *Tailoring of Poly(siloxane) Based Polymers*

The other type of hybrid system could be where organic functional groups are substituted on a main chain organometallic/inorganic polymer template. One of the well-established strategies is to substitute functionality via catalytic hydrosilylation reaction of alkenes. Generally, organic moieties can be covalently attached either to the siloxane polymer backbone [26] or via end grafting to polymer end groups to generate hybrid siloxane-polymers [27–30]. The choice of studying such systems is based on the utility of such polymers in applications ranging from coatings to highly sophisticated electronic materials. New selective methods to modify such polymers will lead to generation of new materials with various property profiles. Our selection of siloxane polymer for the main-chain grafting is polymethylhydrosiloxane (PMHS) because of the presence of Si-H bonds in the main chain. Moreover, commercial availability of PMHS in large scale with varying molecular weights and solubility in common organic solvent makes PMHS an ideal precursor to the functionalized organosilicon polymers (Scheme 5).

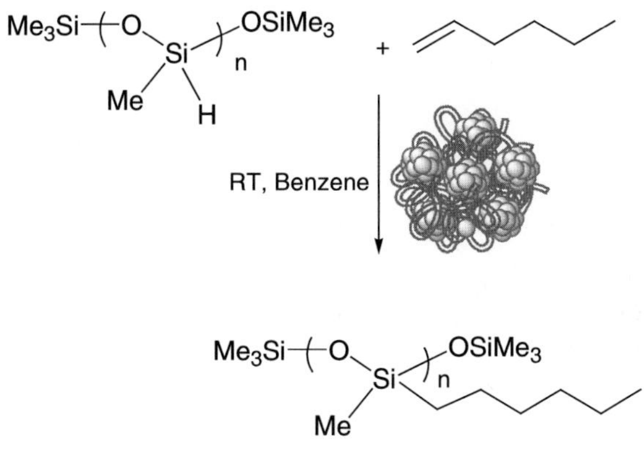

Scheme 5 Pt-nanocluster catalyzed reaction of 1-hexene with PMHS

To functionalize polymethylhydrosiloxane with organic functional groups, PMHS 1 (0.06 mL, 1.0 mmol; Mw ~ 2000 g.mol^{-1}), Pt-nanoclusters (0.01 g, 0.001 mmol Pt), and 1-hexene (0.13 mL, 1.0 mmol) were mixed together in 4-mL benzene and stirred at room temperature (Scheme 4). After 1-h of stirring, colorless reaction mixture gradually turned to homogenized brownish-yellow solution suggesting the dispersion of Pt-nanoclusters.

The progress of the reaction was monitored with ^{1}H NMR spectroscopy, which manifested gradual disappearance of peaks associated with Si-H (δ 4.58) and olefinic-protons (δ 4.8–5.9) with concomitant appearance of new peaks in the methylene region (δ 0.4–1.2). Complete disappearance of peaks associated with starting materials was observed after 20-h of the reaction. Isolation of the product was carried out by high-speed centrifugation (20 min) of reaction mixture, which led to the precipitation of Pt-nanoclusters. After separation of the catalyst, evaporation of the solvent furnished a polymer **2** (Entry 2, Table 2) as a viscous liquid. The molecular weight analysis by SEC of **2** (Mw ~ 4600 g.mol^{-1} and Mw/Mn = 1.2) was consistent with the proposed structure and in good agreement with calculated value (Mw ~ 4500 g.mol^{-1}). ^{13}C and ^{29}Si NMR analysis of the polymer **2** was carried out to elucidate the regioselectivity of the addition reaction.

Addition of Si-H bonds to alkene can take place either in an anti-Markovnikov (β-addition) fashion resulting in silicon being attached at the terminal position or in Markovnikov (α-addition) fashion resulting in silicon bonded to the iso-position of the 1,2 alkene. Moreover, as stated in the introduction, regioselectivity of organic incorporations into the silicones was reported to play a crucial role in

Table 2 Pt-nanocluster catalyzed regioselective grafting of PMHS

Entry	Alkene	Reaction Conditions	Product	Yield (%)
1		18h/RT	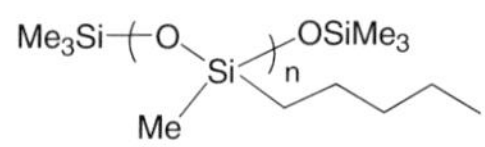	98
2	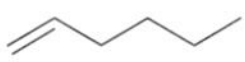	18h/RT	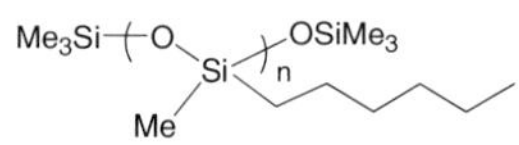	98
3		18h/RT	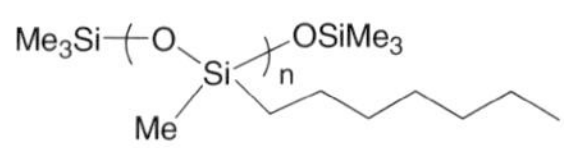	97
4	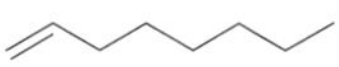	18h/RT	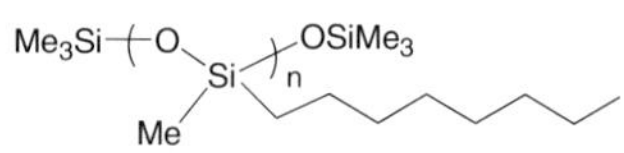	99
5	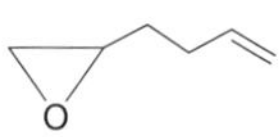	32h/RT	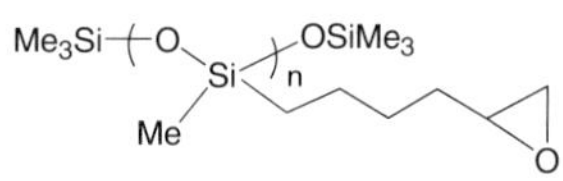	97
6	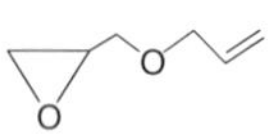	32h/RT	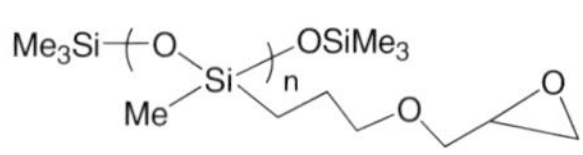	97
7	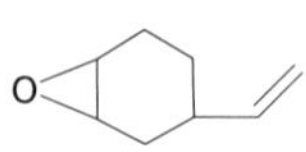	34h/RT	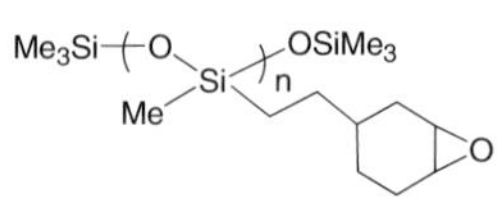	90
8	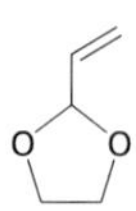	34h/RT	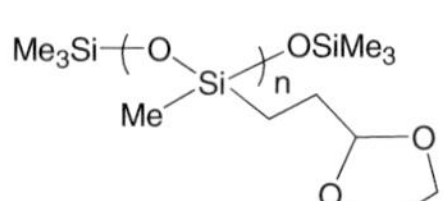	90

defining the property profile in terms of durability, flexibility and film toughness of the composites. ^{13}C NMR spectra of the solution revealed two new peaks at δ 23.31 (-CH_2Si) and δ 14.31 (-CH_2CH_2Si) respectively, indicating exclusive β-silylation of 1-hexene at terminal carbons. Furthermore, ^{29}Si NMR analysis of the reaction mixture showed only a single peak at δ –22.40 (-CH_2*Si*Me) corresponding to hexyl-substituted Silicon centers besides the peak at δ –7.28 originating from the terminal -*Si*Me_3 groups. The reaction was found to be quite general for other linear alkenes (entry **1, 3, 4**, and **5**; Table **2**) and provided corresponding β-adducts very selectively.

Hydrosilylation of functionalized olefins is often challenging due to the side reactions. For example, epoxy-containing olefins are susceptible to ring-opening polymerization and carbonyl-bearing olefins often yield a mixture of C-and O-silylated products. Using Pt nanocluster as the catalyst, epoxy and carbonyl containing olefins were silylated with PMHS without any side reactions. A high degree of regioselectivity was achieved by Pt-nanoclusters yielding >95% anti-Markovnikov adduct during the hydrosilylation of epoxy-functionalized linear olefins. Regioselectivity of the product remained unaffected (i.e. >95% anti-Markovnikov) by varying the functional groups (carbonyl, ether, and epoxide) on the olefins (Table **2**). Functionalization of PMHS with allylacetate (entry **9**, Table **2**) was also achieved selectively (>95% anti-Markovnikov) without any side reactions. In all of the cases, the substitution was either quantitative or near quantitative.

Pt-nanocluster catalyzed hydrosilylation has also been explored with aromatic alkenes using PMHS. As shown in the results summarized in Table 3, substrate dependent (alkene) regioselectivity pattern was displayed for the hydrosilylation

Table 3 Pt-nanocluster catalyzed regioselective hydrosilylation of aromatic olefins

Entry	Alkene	Reaction Conditions	Major Product	(β/α)	Yield (%)
1	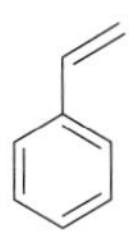	48h/RT	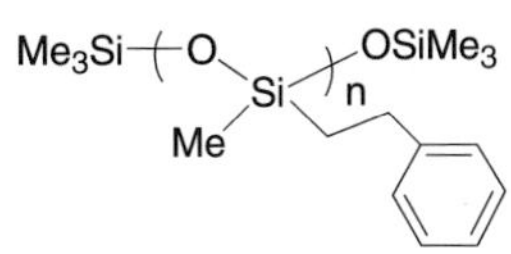	55/45	98
2	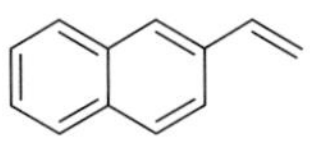	72h/RT	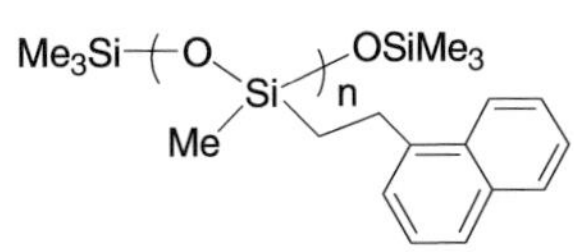	65/35	98
3	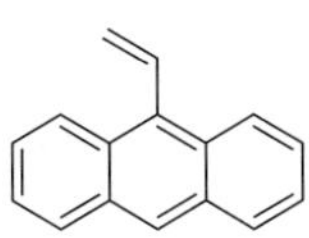	78h/RT	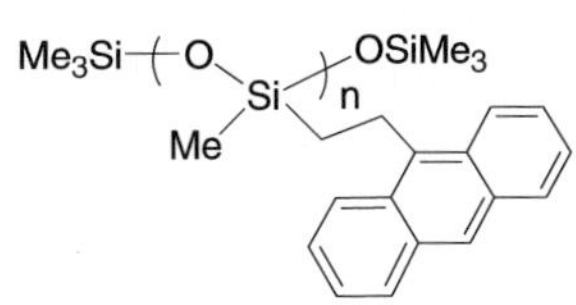	95/5	98

of olefins containing aromatic rings (entry **1, 2, 3**; Table **3**). Hydrosilylation of styrene yielded polymer, which was a mixture of regioisomers (β/α, 55/45). An improvement in the selectivity towards β-silylated product (β/α: 65/35) was observed with 2-vinylnaphthalene (entry **2**, Table **3**). A further increase in the number of aromatic rings (entry **3**, Table **3**; 9-vinylanthracene) led to almost exclusive β-silylated polymer (entry **3**; Table **3**). Increased steric hindrance in conjunction with electronic effects of aromatic rings seems to affect the above regioselectivity pattern in the hydrosilylated product.

Metal containing polymers are a very interesting class of polymers due to their potential applications in various fields. In order to explore the possibility of incorporating metallocenes into the polymer main chain reactivity nanoparticle catalyzed hydrosilylation of PMHS was also investigated. Reaction of vinylferrocene with PMHS yielded corresponding addition product regioselectively in near quantitative yields and under moderate conditions and without any side reaction involving metallocene-ring (Scheme 6). Detailed studies are underway to compare the physical and chemical properties of metal containing polymers to investigate if regeoselective attachment of metallic functionality impacts the assembly as well as the morphology of the resulting polymers.

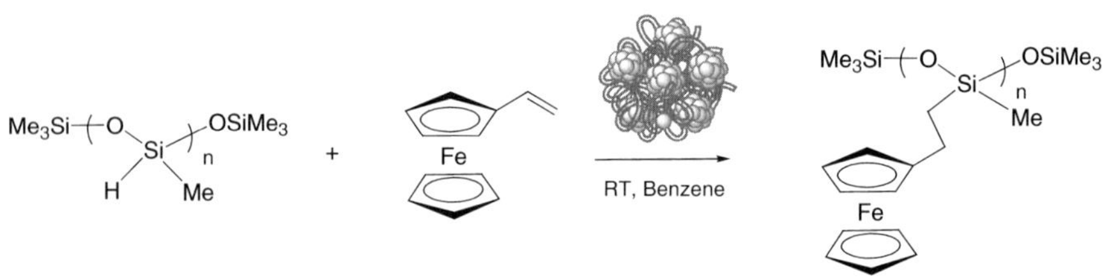

Scheme 6 Regiospecific addition of vinylferrocene to PMHS

3 Conclusions

In conclusion in this chapter we have described polymer makeover by reacting organic or inorganic/organometallic polymer templates with variety of functional alkenes in presence of Pt-nanoparticle catalyst. In both cases nanoparticle catalysts were very active and more importantly highly selective to provide almost exclusively anti-Markovnikov products. Thorough characterization of the resulting products indicated that the integrity of the main chain of the template polymer was unaffected during and after functionalization process.

These results also demonstrate that the template based polymer modification can be quite fruitful strategy and can lead to libraries of materials for commodity applications as well as for various sophisticated functions. Currently, we are studying comparative catalysis not only to improve the new functional group compatibility but also to explore reversal of regionselectivities and efficiencies.

4 Experimental Section

All of the experiments and manipulations were performed under nitrogen using standard Schlenk line techniques. Solvents were purchased from EM science (Merck) and distilled over sodium/benzophenone before use. PMHS (Mw ≈ 2000 g.mol^{-1}), $Me_2Pt(COD)$, polybutadienes and silanes were purchased from Aldrich Chemical Co., and Gelest Chemical Co. and used without further purification. ^{1}H NMR, ^{13}C NMR and ^{29}Si NMR spectra were recorded on 200 MHz and 600 MHz Varian Unity NMR instruments with $CDCl_3$ as an internal standard. SEC analysis was carried out on Alliance GPCV 200 (Water) instrument, equipped with two silica columns, HRSE and HR-1 with the pore size range of 100–5000 Å and $2000-4\times10^{6}$ Å respectively. This instrument was calibrated using polystyrene standards. THF was used as an eluent at the flow rate of 1 ml/min at 40°C. A third order calibration curve was used to measure the molecular weight of unknown samples. Philips CM 100 transmission electron microscope (TEM) was employed to examine the reaction mixture for the presence of Pt-nanoclusters. Scanning electron microscope Amray 1910 (SEM) was used to analyze solid Pt-nanocluster. The amount of Pt present in the cluster was determined on the basis of SEM data, which is ~0.001 mmol per 0.01 gm of the solid.

4.1 *Synthesis of Pt-Nanoclusters*

Me_2Pt (COD) (cod: 1,5 cyclo-octadiene) (0.25 mmol, 0.084 g), and PMHS (10.00 mmol, 0.60 mL) were added in 50 mL toluene in a 200-mL RB. Reaction mixture was stirred at 80°C for 24 h under positive pressure of nitrogen. Reduction of Me_2Pt (COD) was monitored with UV-vis spectroscopy, which revealed a gradual disappearance of peaks associated with the Me_2Pt (COD) leading to a featureless spectrum. After verifying the presence of "Pt"-nanoclusters in the solution with EM-spectroscopy, flow of nitrogen was stopped and the flask was exposed to air. Further stirring of the solution in air yielded a black gummy solid. The solid was washed thoroughly with toluene/benzene (100 mL) to remove organic moieties. Air-drying of the gummy solid furnished a black powder, which was characterized with various spectroscopic techniques and used as hydrosilylation catalyst. ^{29}Si CP/MAS δ −36.15 (D, Si-H), −81.46(T), −102.58(Q, SiO_2), −112.26(Q, SiO_2). ^{13}C CP/MAS δ −3.35 ($-SiCH_3$). FT-IR spectra (KBr, cm^{-1}) 2175 (Si-H), 1120 (Si-O-R), 1050 (Si-O-Si), 893 (Si-OH).

4.2 *General Procedure For Hydrosilylation of Polybutadienes*

A Schlenk tube (10 ml), equipped with magnetic stirrer and oil bath was charged with Pt-nanoclusters (0.01 g, 0.001 mmol Pt), degassed and flushed with dry Nitrogen.

The PBD-1 (0.054 g, 1 mmol) dissolved in freshly prepared dry benzene (2 ml) was added to the Schlenk tube, followed by the addition of **2a** (0.13 ml, 1.2 mmol) under the constant flow of nitrogen. After few minutes of stirring, the reaction mixture turned into light brown homogeneous solution, indicating the formation of soluble nanoclusters.[12] The reaction progress was monitored by ^{1}H NMR spectroscopy. On completion of the reaction, the solid catalyst was separated by centrifugation. The filtrate was evacuated to obtain the crude product. The product thus obtained was passed through the silica gel column (hexane) and analyzed by using SEC and ^{1}H, ^{13}C, DEPT, ^{29}Si NMR techniques.

4.3 Characterization of the Silylated Polybutadienes

Polymer, 3a (Scheme 2): ^{1}HNMR ($CDCl_3$, 200 MHz): δ (ppm) 0.44 (t, 9H), 0.85 (q, 6H), 0.0–2.0 (br, 7H). ^{13}CNMR/DEPT ($CDCl_3$, 200 MHz): δ (ppm) 3.29 (-Si*CH*$_2$CH$_3$), 7.52 (-SiCH$_2$*CH*$_3$), 27.0 (-*CH*$_2$CH$_2$Si), 34.7 (-*CH*CH$_2$-), 38.1 (-CH*CH*$_2$-). ^{29}SiNMR ($CDCl_3$, 600 MHz): δ (ppm) 7.76.

Polymer, 3b (Scheme 2): ^{1}HNMR ($CDCl_3$, 200 MHz): δ (ppm) –0.04 (s, 6H), 0.2–2.0 (br, 7H), 1.05 (t, 3H), 3.52 (q, 2H). ^{13}CNMR/DEPT ($CDCl_3$, 200 MHz): δ (ppm) –2.14 (-Si*CH*$_3$), 11.29 (-CH$_2$*CH*$_2$Si-), 18.305(-OCH$_2$*CH*$_3$), 25.79 (-SiCH$_2$*CH*$_2$-), 33.57(-*CH*CH$_2$-), 37.68 (-CH*CH*$_2$-), 58.16 (-O*CH*$_2$CH$_3$). ^{29}SiNMR ($CDCl_3$, 600 MHz): δ (ppm) 18.01

Polymer, 3c (Scheme 2): ^{1}HNMR ($CDCl_3$, 200 MHz): δ (ppm) 0.27 (s, 3H), 0.4–2.0 (br, 7H), 7.34 (m, 3H), 7.50 (m, 2H). ^{13}CNMR/DEPT ($CDCl_3$, 200 MHz): δ (ppm) –3.22 (-Si*CH*$_3$Ph), 10.76 (-CH$_2$*CH*$_2$Si-), 26.72(-SiCH$_2$*CH*$_2$-), 33.7(-CH*CH*$_2$-), 37.85 (-*CH*CH$_2$-), 127.49 (-Si*Ph*), 128.12 (-Si*Ph*), 128.54 (-Si*Ph*), 133.3 (-Si*Ph*). ^{29}SiNMR ($CDCl_3$, 600 MHz): δ (ppm) –1.795.

Polymer, 3d (Scheme 2): ^{1}HNMR ($CDCl_3$, 200 MHz): δ (ppm) –0.10 (s, 3H), –0.02 (s, 6H), 0.0–2.0 (br, 7H). ^{13}CNMR/DEPT ($CDCl_3$, 200 MHz): δ (ppm) –0.72 (-Si*CH*$_3$OSiCH$_3$), 1.62 (-SiCH$_3$OSi*CH*$_3$), 12.93 (-CH$_2$*CH*$_2$Si-), 25.95 (-*CH*$_2$CH$_2$Si-), 33.7(-*CH*CH$_2$-), 37.96 (-CH*CH*$_2$-).^{29}SiNMR ($CDCl_3$, 600 MHz): δ (ppm) –20.36, 7.02.

Polymer, 3e (Table 1): ^{1}HNMR ($CDCl_3$, 200 MHz): δ (ppm) 0.07 (s, 6H), 0.0–2 (br, 7H). ^{13}CNMR/DEPT ($CDCl_3$, 200 MHz): δ (ppm) 1.62 (-Si*CH*$_3$), 14.37 (-CH$_2$*CH*$_2$Si-), 26.1(-*CH*$_2$CH$_2$Si-), 33.6(-*CH*CH$_2$-), 38.1 (-CH*CH*$_2$-). ^{29}SiNMR ($CDCl_3$, 600 MHz): δ (ppm) 32.97.

Polymer, 3f (Table 1): ^{1}HNMR ($CDCl_3$, 200 MHz): δ (ppm) 0.49 (s, 3H), 0.0–2.0 (br, 7H), 7.23 (m, 3H), 7.45 (m, 2H). ^{13}CNMR/DEPT ($CDCl_3$, 200 MHz): δ (ppm) 0.26 (-Si*CH*$_3$), 13.52 (-CH$_2$*CH*$_2$Si-), 25.83 (-*CH*$_2$CH$_2$Si-), 33.43 (-*CH*CH$_2$-), 37.86 (-CH*CH*$_2$-), 128.06 (-Si*Ph*), 130.27 (-Si*Ph*), 133.27 (-Si*Ph*), ^{29}SiNMR ($CDCl_3$, 600 MHz): δ (ppm) 22.17.

Polymer, 3g (Table 1): ^{1}HNMR ($CDCl_3$, 200 MHz): δ (ppm) 0.0–2 (br, 21H). ^{13}CNMR/DEPT ($CDCl_3$, 200 MHz): δ (ppm) 8.05 (-CH_2*CH_2*Si-), 13.73 (-Si*CH*(CH_3)$_2$), 16.99 (-SiCH(*CH_3*)$_2$), 25.8 (-*CH_2*CH_2Si-), 33.93(-*CH*CH_2-), 37.61(-CH*CH_2*-). ^{29}SiNMR ($CDCl_3$, 600 MHz): δ (ppm) 36.8.

Polymer, 3h (Table 1): ^{1}HNMR ($CDCl_3$, 600 MHz): δ (ppm) 0.0–2 (br, 12 H). ^{13}CNMR/DEPT ($CDCl_3$, 200 MHz): δ (ppm) 6.16 (-Si*CH_2*CH_3), 12.52 (-SiCH_2*CH_3*), 15.52 (-CH_2*CH_2*Si-), 25.37 (-*CH_2*CH_2Si-), 33.0 (-*CH*CH_2-), 37.7(-CH*CH_2*-). ^{29}SiNMR ($CDCl_3$, 600 MHz): δ (ppm) 35.84.

Polymer 3i (Table 1): ^{1}HNMR ($CDCl_3$, 200 MHz): δ (ppm) 0.0–2.0 (br, 7H), 7.39 (m, 3H), 7.62 (m, 2H). ^{13}CNMR/DEPT ($CDCl_3$, 200 MHz): δ (ppm) 16.58(-CH_2*CH_2*Si-), 25.47 (-*CH_2*CH_2Si-), 33.18 (-*CH*CH_2-), 37.86 (-CH*CH_2*-), 128.34 (-Si*Ph*), 131.66 (-Si*Ph*), 132.28 (-Si*Ph*), 133.29 (-Si*Ph*). ^{29}SiNMR ($CDCl_3$, 600 MHz): δ (ppm) 20.24.

Polymer, 3j (Table 1): ^{1}HNMR ($CDCl_3$, 200 MHz): δ (ppm) 0.4–2.0 (br, 7H). ^{13}CNMR/DEPT ($CDCl_3$, 200 MHz): δ (ppm) 20.2 (-CH_2*CH_2*Si-), 25.76(-*CH_2*CH_2Si-), 32.97 (-*CH*CH_2-), 37.83 (-CH*CH_2*-). ^{29}SiNMR ($CDCl_3$, 600 MHz): δ (ppm) 13.59.

Polymer 4 (Scheme 4): ^{1}HNMR ($CDCl_3$, 200 MHz): δ (ppm) –0.036 (s, 21H), 0.5–1.9 (br, 7H). ^{13}CNMR/DEPT ($CDCl_3$, 200 MHz): δ (ppm) –4.83 (-Si*CH_3*), 12.42 (-CH_2*CH_2*Si-), 25.6 (-*CH_2*CH_2Si-), 33.66 (-*CH*CH_2-), 37.9 (-CH*CH_2*-). ^{29}SiNMR ($CDCl_3$, 600 MHz): δ (ppm) –20.096 (s).

4.4 Typical Procedure of Hydrosilylation of Olefins with PMHS

Pt-nanoclusters (0.010 g, 0.001 mmol Pt), PMHS (1.0 mmol, 0.06 mL), and 1-hexene (1.00 mmol, 0.13 mL) were mixed together in 4 mL of freshly distilled benzene in a 50-mL Schlenk-tube. Schlenk-tube was degassed three-time using standard Schlenk-tube-technique followed by stirring at room temperature. ^{1}H NMR spectroscopy was used to monitor the progress of catalytic transformation. After confirming the total substitution, product separation was carried out by high-speed centrifugation of reaction mixture (20 min), which led to precipitation of nanoclusters. Evaporation of solvent yielded pure poly(methylhexyl)siloxane. Isolation of the product was followed by multinuclear NMR spectroscopy characterization of the product.

Poly(methylpentyl)siloxane (Entry 1, Table 2): ^{1}H NMR ($CDCl_3$, 200 MHz) δ 0.063, 0.51(m), 0.88(m), 1.29(m); ^{13}C NMR ($CDCl_3$, 200 MHz) δ 1.69, 14.20, 17.90, 22.65, 22.99, 35.85; ^{29}Si NMR ($CDCl_3$, 600 MHz) δ –21.83, 7.24.

Poly(methylhexyl)siloxane) (Entry 2, Table 2): ^{1}H NMR ($CDCl_3$, 200 MHz) δ –0.003, 0.46(m), 0.81(m), 1.22(m); ^{13}C NMR ($CDCl_3$, 200 MHz) δ –0.094, 2.047,14.31, 17.99, 22.90, 23.31, 31.94, 33.41; ^{29}Si NMR ($CDCl_3$, 600 MHz) δ –22.40, 7.28.

Poly(methylheptylsiloxane) (Entry 3, Table 2): ^{1}H NMR ($CDCl_3$, 200 MHz) δ 0.19, 0.67(m), 0.92(m), 1.33(m); ^{13}C NMR ($CDCl_3$, 200 MHz) δ 1.28, 3.03, 15.62, 19.43, 23.54, 24.40, 30.77, 33.50, 35.11; ^{13}C NMR ($CDCl_3$, 200 MHz) δ, –21.86, 7.34

Poly(methyloctylsiloxane) (Entry 4, Table 2): ^{1}H NMR ($CDCl_3$, 200 MHz) δ 0.11, 0.48(m), 0.90(m), 1.30(m); ^{13}C NMR ($CDCl_3$, 200 MHz) δ –0.32, 1.90, 14.16, 17.79, 22.86, 23.25, 29.45, 29.54, 32.14, 33.42; ^{29}Si NMR ($CDCl_3$, 600 MHz) δ –20.98, 7.00

Poly[(methyl (1,2 epoxy-5-hexane)siloxane] (Entry 5, Table 2): ^{1}H NMR ($CDCl_3$, 200 MHz) δ 0.11, 0.56(m), 1.44(m), 2.31(m), 2.57(m), 2.78(m); ^{13}C NMR ($CDCl_3$, 200 MHz) δ –0.52, 17.42, 22.76, 29.38, 32.03, 46.38, 51.64; ^{29}Si NMR ($CDCl_3$, 600 MHz) δ –22.16, 7.18.

Poly[methyl(propylglycidylether)siloxane] (Entry 6, Table 2): ^{1}H NMR ($CDCl_3$, 200 MHz) δ 0.07, 0.55(m), 0.88(m), 1.57(m), 2.41(m), 2.98(m), 3.24(m), 3.45(m), ^{13}C NMR ($CDCl_3$, 200 MHz) δ –0.22, 13.89, 23.72, 43.24, 50.63, 69.81, 73.92, ^{29}Si NMR ($CDCl_3$, 600 MHz) δ –21.60, 7.78.

Poly[methyl (4-ethylcyclohexyl-1, 2-epoxide)siloxane] (Entry 7, Table 2): ^{1}H NMR ($CDCl_3$, 200 MHz) δ 0.25, 0.30, 0.28, 0.67, 1.22(m), 1.38 (m), 1.41(m), 1.59(m), 1.75(m), 1.85(m), 1.97(m), 2.08(m), 2.27(m), 2.34(m), 2.59(m), 3.04(m), 3.14(m), ^{13}C NMR ($CDCl_3$, 200 MHz) δ 0.28, 2.39, 13.02, 13.15, 15.06, 15.10, 21.46, 23.68, 24.23, 24.76, 25.78, 25.90, 26.65, 27.65, 30.12, 31.07, 32.35, 32.86, ^{29}Si NMR ($CDCl_3$, 600 MHz) δ –21.49.

Poly[methyl(2-ethyl-1,3-dioxolane)siloxane] (Entry 8, Table 2): ^{1}H NMR ($CDCl_3$, 200 MHz) δ –0.06, 0.06, 0.87(t), 1.07(t), 1.62(m), 3.78(m), ^{13}C NMR ($CDCl_3$, 200 MHz) δ –0.45, 8.25, 27.26, 66.65, 73.16, 106.08, ^{29}Si NMR ($CDCl_3$, 600 MHz) δ –21.94, 7.80.

Poly[methy(ethylphenyl)siloxane] (mixture of β:α products) (Entry 1, Table 3): ^{1}H NMR ($CDCl_3$, 200 MHz) δ –0.07(m), 0.76(m), 1.28(m), 2.56(m), 6.6(m), 7.2(m), 7.56(m); ^{13}C NMR ($CDCl_3$, 200 MHz) δ 1.2, 1.8, 14.8, 19.5, 29.1, 30.7, 113, 124, 125, 126, 127, 128, 137, 144; ^{29}Si NMR ($CDCl_3$, 600 MHz) δ –26.1, –22.2.

Poly[methyl (2-ethylnapthalene)siloxane] (Entry 2, Table 3): ^{1}H NMR ($CDCl_3$, 200 MHz) δ –0.008, 0.84(m), 1.37(m), 2.21(m), 2.67(m), 6.80(m), 6.90(m), 7.30(m), 7.54(m), ^{13}C NMR ($CDCl_3$, 200 MHz) δ –0.19, 1.32,14.81,19.11, 29.17, 30.95, 123, 125.13, 126.93, 127.68, 128.46, 129.67, 131.72, 132.10, 133.90, 141.99, ^{29}Si NMR ($CDCl_3$, 600 MHz) δ –25.79, –21.72.

Poly[methyl (9-ethylanthracene)siloxane] (Entry 3, Table 3): ^{1}H NMR ($CDCl_3$, 200 MHz) δ 0.43, 1.37(m), 1.62(t), 3.80(q), 7.33(m), 7.55(m), 7.70(m), 8.13(m), 8.45(m), 8.55(m), ^{13}C NMR ($CDCl_3$, 200 MHz) δ 2.51, 15.94, 21.51, 123.21, 125.51, 125.78, 126.43, 126.85, 129, 71, 130.23, 131.89, 132.03, 133.99, 137.00, ^{29}Si NMR ($CDCl_3$, 600 MHz) δ –21.45 (broad).

Poly[methy(2-ethylferrocene)siloxane] (Scheme 6): ^{1}H NMR ($CDCl_3$, 200 MHz) δ –0.047, 0.041, 0.74(m), 1.14(m), 1.49(m), 2.23(m), 4.00 ^{13}C NMR ($CDCl_3$, 200 MHz) δ –0.16, 14.09,18.83, 22.69, 26.77, 67.12, 67.55, 68.39, 91.83, ^{29}Si NMR ($CDCl_3$, 600 MHz) δ –21.97 (broad).

Acknowledgements Contributions of all Chauhan group members whose names appear in the references are gratefully acknowledged. City University of New York at the College of Staten Island as well as polymer Ph. D. program of CUNY is also acknowledged for the facility, equipment, laboratory and student support.

References

1. McGrath M P, Sall E D, Tremont S. (1995) Functionalization of polymers by metal-mediated processes. Chem Rev 95:381–198
2. Whitesides G M, Mathias J P, Seto C T. (1991) Molecular self-assembly and nanochemistry: a chemical strategy for the synthesis of nanostructures. Science 254:1312–1319
3. Brinker C, Scherer G. (1990) Sol–gel science: the physics and chemistry of sol– gel processing. Academic Press, New York
4. Theng B K G. (1979) Developments in soil science. Formation and properties of clay–polymer complexes. Elsevier, Amsterdam
5. Lan T, Kaviratna P D, Pinnavaia, T J. (1995) Mechanism of clay tactoid exfoliation in epoxy-clay nanocomposites. Chem Mater 7:2144–2150
6. Giannelis E P, Krishnamoorti R, Manias E. (1999) Polymer-silicate nanocomposite: Model systems for confined polymers and polymer brushes, Adv Polym Sci 138:107–147.
7. Schwab J J, Lichtenhan J D. (1998) Polyhedral oligomeric silsesquioxane (POSS)-based polymers. Appl Organomet Chem 12:707–713.
8. Chauhan B P S, Rathore, J S, Chauhan M et al. (2003) Synthesis of polysiloxane stabilized palladium colloids and evidence of their participation in silaesterification reactions. J Am Chem Soc 125:2876–2877.
9. Chauhan B P S, Rathore, J S, Bandoo T. (2004) "Polysiloxane-Pd" Nanocomposites as recyclable chemoselective hydrogenation catalysts. J Am Chem Soc 126:8493–8500.
10. Chauhan B P S, Rathore J S. (2005) Regioselectively synthesis of multifunctional hybrid polysiloxanes achieved by Pt-nanoclusters catalysis. J Am Chem Soc 127:5790–5791.
11. Chauhan B P S, Rathore J S. (May 16–20, 2006) Synthesis of Isolable Nanocluster Catalysts and Their Applications in Macromolecular Hydrosilylation Catalysis. 39th Silicon Symposium, Frankenmuth, MI USA, Abs. No. 300
12. Chauhan B P S, Balagam, B Raghunath M. (May 16–20, 2006) Silyl Functionalization: An Efficient Route for Polyolefin Modification 39th Silicon Symposium, Frankenmuth, MI USA, Abs. No. 254.
13. Chauhan B P S, Balagam B. (2006) Silyl Functionalization of Polyolefines. Macromolecules 39:2010–2012.
14. Iraqi A, Seth S, Vincent C A et al. (1992) Catalytic hydrosilylation of polybutadienes as a route to functional polymers. J Mater Chem 2:1057–1064
15. Iraqi A, Cole-Hamilton D J. (1991) Polyketones via oxidation of polybutadienes catalysed by complexes of platinum (II). Polyhedron 10:993–995
16. Gahagan M, Iraqi A, Cupertino D C et al. (1989) A high activity molybdenum containing epoxidation catalyst and its use in regioselective epoxidation of polybutadiene J Chem Soc Chem Commun issue 21:1688–1690
17. Guo X, Farwaha R, Rempel G L. (1990) Catalytic hydrosilylation of diene-based polymers. 1. Hydrosilylation of Polybutadiene. Macromolecules 23:5047–5054
18. Narayanan P, Clubley B G, Cole-Hamilton D J et al. (1991) Polycarboxylic acids via catalytic hydrocarboxylation of polybutadienes. J Chem Soc Chem Commun issue 22:1628–1629
19. Lange J P, Schoon L, Villena A et al. (2004) Monolithic catalysts for the fixed-bed hydrogenation of polymers. Chem Commun issue 24:2864–2865
20. Guo X and Rempel G L. (1992) Catalytic hydrosilylation of diene-based polymers. 2. Hydrosilylation of styrene-butadiene copolymer and nitrile-butadiene copolymer. Macromolecules 25:883–886

21. Baum K, Baum J C, Ho T. (1998) Side-Loop Polymers Based on the Hydrosilylation of Polybutadiene. J Am Chem Soc 120:2993–2996
22. Hempenius M A, Michelberger W, Moller M. (1997) Arborescent Graft Polybutadienes. Macromolecules 30:5602–5605.
23. Kepczynski M, Lewandowska J, Romek M et al. (2007) Silicone nanocapsules templated inside the membranes of catanionic vesicles. Langmuir 23:7314–7320.
24. Pinho R O, Radovanovic E, Torriani I L et al. (2004) Hybrid materials derived from divinylbenzene and cyclic siloxane. Eur Poly J 40:615–622
25. Mu J, Liu Y, Zheng S. (2007) Inorganic–organic interpenetrating polymer networks involving polyhedral oligomeric silsesquioxane and poly(ethylene oxide). Polymer 48:1176–1184.
26. Chien L–C, Cada G. (1994) Photo-Cross-Linkable and Optically Active Side-Chain Liquid-Crystalline Copolymers. Macromolecules 27:3721–3726.
27. Coqueret X, Wegner G. (1991) Platinum-catalyzed hydrosilylation of allyl aryl ethers: kinetic investigation at moderately high dilution. Organometallics 10:3139–3145.
28. Zhu Z, Einset A G, Yang Ch-Y et al. (1994) Synthesis of Polysiloxanes Bearing Cyclic Carbonate Side Chains. Dielectric Properties and Ionic Conductivities of Lithium Triflate Complexes. Macromolecules 27:4076–4079.
29. Matisons J G, Provatas A. (1994) Characterization of Novel Cationic Aminohydroxysiloxanes. Macromolecules 27:3397–3405.
30. Coqueret X, Lablache-Combier A, Loucheux C. (1988) Functionalization of polysiloxanes by esterification of pendant glycidic groups: Catalyzed reaction in N,N-dimethylformamide. Eur Poly J 24:1137–1143.

Polysiloxane Based Interpenetrating Polymer Networks: synthesis and Properties

Odile Fichet, Frédéric Vidal, Vincent Darras, Sylvie Boileau, and Dominique Teyssié

Abstract This article summarizes a large amount of work carried out in our laboratory on polysiloxane based Interpenetrating Polymer Networks (IPNs). First, a polydimethylsiloxane (PDMS) network has been combined with a cellulose acetate butyrate (CAB) network in order to improve its mechanical properties. Second, a PDMS network was combined with a fluorinated polymer network. Thanks to a perfect control of the respective rates of formation of each network it has been possible to avoid polymer phase separation during the IPN synthesis. Physico-chemical analyses of these materials led to classify them as "true" IPNs according to Sperling's definition. In addition, synergy of the mechanical properties, on the one hand, and of the surface properties, on the other hand, was displayed.

1 Introduction

Interpenetrating polymer networks (IPNs) can be a smart solution in order to avoid some polymer defects. Indeed, these polymer associations generally lead to materials exhibiting better mechanical properties, increased resistance to degradation and a possibly improved combination of the properties of their components, i.e. a synergetic effect. All these characteristics come from the particular architecture of IPNs which combine two or more polymer networks synthesized in juxtaposition [1,2]. In addition, the entanglement of cross-linked polymers leads to forced "miscibility" compared to usual blends and the resulting materials exhibit a good dimensional stability over time. IPNs can be prepared through different synthetic pathways. In the *in situ* synthesis, all reactants are mixed together and the polymerization/cross-linking reactions leading to formations

O. Fichet
Laboratoire de Physicochimie des Polymères et des Interfaces (LPPI) – Université de Cergy-Pontoise – 5, mail Gay-Lussac, Neuville-sur-Oise – 95031 Cergy-Pontoise Cedex – France

F. Ganachaud et al. (eds.), *Silicon Based Polymers*,

of the two networks can be started simultaneously or sequentially one after the other. Hence in this process, the reaction mechanisms must be of different nature, otherwise a single copolymer network instead of a true IPN is formed through cross reactions.

Although the final material is stable, a phase segregation occurring during the synthesis cannot be avoided. Qualitative information about the extent of phase separation in the IPN can be derived straight from the relative transparency of the material. Indeed, the more transparent the material, the smaller the phase separation, since any light wavelength will be diffracted by any phase domain the size of which is of the same order of magnitude, making the material appear translucent to opaque [1]. Dynamic Mechanical Thermal Analysis (DMTA) is also an adequate tool in order to corroborate the interpenetrating extent of both networks inside the IPN architecture. Ideally, interpenetration should occur only through physical interactions and the polymer network mixture should be homogeneous on a molecular scale without covalent bonding between the different networks. In this case, the DMTA analysis only detects one phase, witnessed by just one mechanical relaxation, the corresponding domain size being estimated at about 5–50 nm depending on the width of the relaxation signal [3]. Thus, this "single" phase is characterized through the loss factor (tanδ)-temperature curve by only one peak located at a temperature between those of the mechanical relaxation of the single networks combined inside the IPN architecture.

Siloxanes have a set of extremely interesting properties, i.e. very low surface energy, excellent gas and moisture permeability, good heat stability, low temperature flexibility and biocompatibility [4] but, because of weak intermolecular interactions between polymer chains, they exhibit also poor mechanical properties remaining a drawback for certain applications. Thus, low failure stress and low tear resistance limit silicone rubber applications in biomedical domain [5]. Besides, fluorine-containing elastomers maintaining rubber-like elasticity in extremely severe environments including exposure to high temperatures and corrosive chemicals have been developed. Fluorinated polysiloxane single networks have been prepared by sol-gel reaction or by hydrosilylation reaction and they are mainly used as protecting coatings. However the synthesis pathways leading to the telechelic precursors of those networks do not allow a fine control of the molar mass, polymolecularity and functionality. In addition, fluorinated polysiloxane networks do not show convenient mechanical properties at room temperature which restrains also their development. Thus, it can be interesting to elaborate a material combining a fluorinated polymer and a non fluorinated silicone independently.

Combining silicones with any organic polymer into a new material brings about some unique challenges. Indeed, polydimethylsiloxanes (PDMS) are known for their incompatibility with other polymers due to their low solubility parameter and extraordinary flexibility which makes them difficult to entrap into a polymer combination. Some IPNs associating PDMS with a thermoplastic polymer [6] such as PMAA [7], polyurethane [8], PMMA [9,10], polystyrene [11] or polycarbonate [12], have led to biphasic materials exhibiting higher mechanical properties than those of a single PDMS network.

The present work shows how it is possible to combine a PDMS network with two different partner polymer networks into IPN architecture in order to overcome some of PDMS weaknesses. Although these associations are not thermodynamically favored, a homogeneous material can be obtained by adjusting the relative formation rates of the two partner networks. Thus, first, a PDMS network was combined to a cellulose acetate butyrate (CAB) network in order to improve the PDMS thermomechanical properties. Second, PDMS was associated with a fluorinated network in order to find an alternative to the elaboration of materials keeping only the advantageous characteristics of fluorinated polysiloxane single networks. For each series the weight proportions of the two partner networks vary from gross 25 to 75 wt%. All IPNs presented here have been synthesized according to an *in situ* synthesis strategy and characterized by different techniques. Thus, new properties or the exaltation of one property were searched in order to know what could arise from the particular IPN architecture.

2 Experimental Section

α,ω-(3-hydroxypropyl)polydimethylsiloxane (diOH-PDMS oligomer – Mn_{NMR} = 1100 g.mol^{-1}) was kindly provided by Rhodia and was dried at least 3 days under vacuum (0.1 mmHg) before use. α,ω-divinyl polydimethylsiloxane (divinyl-PDMS oligomer-Mn_{NMR} = 650 g.mol^{-1}) and cellulose acetate butyrate (CAB – Mn = 16 000 g.mol^{-1}– 4% acetate –50% butyrate –46% OH,) were purchased in ABCR and Acros Organics, respectively. The fluorinated acrylate (AcRf6) was synthesized as previously described [13].

Dibutyltindilaurate (DBTDL, Aldrich), ethylene glycol dimethacrylate (EGDMA, Aldrich), trimethylolpropane tris(3-mercaptopropionate) (Trithiol, Aldrich), and Desmodur® N3300 pluri-isocyanate cross-linker (Bayer) (NCO content: 21.8 ± 0.3 wt% according to the supplier) were used as received. This last compound is described as an isocyanurate mixture resulting from the condensation of three to several hexamethylene diisocyanate molecules and mainly composed of mono-, di- and tri-isocyanurates with a global functionality higher than 2 [14]. 2,2-azobis isobutyronitrile (AIBN, Acros Organics) was recristallized in methanol before use. Dicyclohexylperoxidicarbonate (DCPD, Groupe Arnaud) was dried under vacuum before use. Toluene (Carlo Erba, puro grade) and chloroform (Carlo Erba, puro grade) were distilled and dried before use.

PDMS-CAB IPNs were prepared by dissolving CAB in chloroform (5 mL per gram), then di-vinyl-PDMS, trithiol ([HS]/[C=C] = 1.1), Desmodur® N3300 ([NCO]/[OH] = 0.60), DBTDL ([DBTDL]/[OH] = 0.05), and AIBN ([AIBN]/[C=C] = 0.05) were added and the mixture was stirred under nitrogen atmosphere for 30 min. The mixture was then introduced in a mould made from two glass plates clamped together and sealed with a 1 mm thick Teflon® gasket. The mould was then kept at 50°C for 16 h. After 16 h, the sample was dried and post-cured for 2 h at 60°C.

PDMS/polyAcRf6 IPNs were synthesized by mixing together the appropriate weights of α,ω-(3-hydroxypropyl) polydimethylsiloxane (diOH-PDMS oligomer), Desmodur® N3300 pluri-isocyanate ([NCO]/[OH] = 1.2), DBTDL ([DBTDL]/[OH] = 0.014), AcRf6, EGDMA ([EGDMA]/[AcRf6] = 0.05) and DCPD

([DCPD]/[C=C]=0.05) in 2 mL toluene under argon. Then, the mixture was introduced into a mould. The mould was heated in an oven at 55°C for 15 h. Toluene was then taken off under vacuum.

For all the IPNs, the same proportions between monomer, cross-linker and catalyst or initiator in a same network were kept whatever the IPN composition.

The soluble amounts contained in the final materials were determined by Soxhlet extraction with an adequate solvent and the weight percentages were calculated as $EC(\%) = (W_0 - W_E)/W_0 \times 100$ where W_0 and W_E are the weights of samples before and after extraction respectively.

DMTA measurements were carried out on film samples with a Q800 apparatus (TA Instruments) operating in tension mode. Experiments were performed at a 1 Hz frequency and a heating rate of 3°C min^{-1} from –120 to 100°C.

The liquid-film contact angles (θ) were determined from image recordings by video microscopy after deposition of a 25 μL water drop at 20°C on the material surfaces. For these measurements, the samples, synthesized between two glass plates, were unmoulded and hung in air without contact with any material during drying. The values of equilibrium contact angles were computed after processing the images with a frame grabbing software (Drope Shape Analysis). For each composition, contact angles were measured on about six samples resulting from different syntheses; four drops per sample were analyzed. The reported contact angle values correspond to the average of all measurements with an error bar corresponding to twice the standard deviation.

3 Results and Discussion

All IPNs presented here have been synthesized according to an *in situ* synthesis strategy, i.e. all network precursors were mixed together in a suitable solvent in order to obtain a homogeneous starting solution. By choosing this IPN synthesis pathway, it is important to check that no side reaction occurs during the synthesis. For both series and throughout the studied weight composition range, it was checked that soluble fraction contained in IPNs was lower than 10 wt% which is characteristic of a satisfactory cross-linking degree of the networks inside the IPNs.

3.1 PDMS/CAB IPN

First, PDMS network was combined to a cellulose acetate butyrate (CAB) network into an IPN architecture in order to improve the thermomechanical properties of PDMS network (Scheme 1). The linear CAB can be cross-linked through its free OH groups with a Desmodur® N3300 pluri-isocyanate. The alcohol/isocyanate reaction is catalyzed by DBTDL leading to urethane cross-links. Simultaneously, PDMS oligomers must be cross-linked independently in order to form the PDMS network. In order to carry out independent cross-linking reactions, divinyl-PDMS has been chosen as the PDMS precursor. This oligomer does not react according to a free-radical thermal initiated polymerization. Thus this possibility was discarded and a thiol-ene addition cross-linking was considered instead. The silicone network was formed through a thiol-ene addition initiated with AIBN between α,ω-divinyl-polydimethylsiloxane and trimethylolpropane tris(3-mercaptopropionate) (trithiol) as a cross-linking agent.

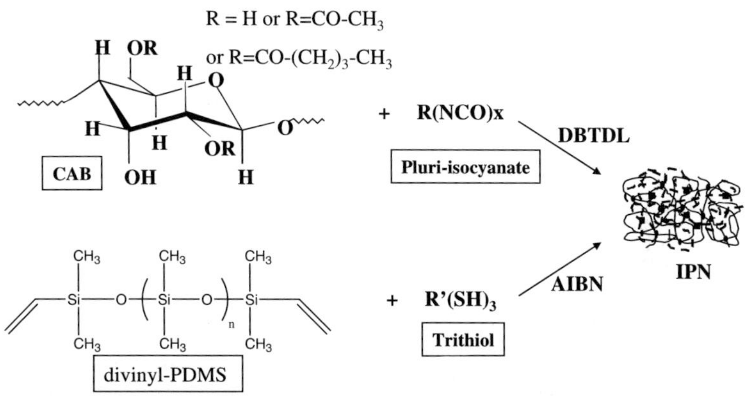

Scheme 1 Synthesis pathway of PDMS/CAB IPNs

Previous kinetic studies [15] have shown that thiol functions from the divinyl-PDMS cross-linker do not react with isocyanate groups from the CAB cross-linker and that the obtained IPNs are not grafted. However, thiol groups form stable species with the DBTDL catalyst which in turn results in decreasing the actual available quantity of catalyst in the medium thus slowing down the CAB network formation. Consequently, in order to obtain a transparent material withstanding the presence of thiol functions the DBTDL proportion must be increased. Indeed, it was shown that the NCO conversion level must be at least 40% before the complete PDMS network formation in order to avoid phase segregation. Thus, the exact DBTDL amount has been determined for each PDMS/CAB weight proportion in order to obtain a transparent IPN which is characteristic of the absence of phase separation during the synthesis.

The absence of phase segregation has been confirmed by DMTA measurements. Indeed, PDMS/CAB IPNs show only one mechanical relaxation lying between those of the single CAB and PDMS networks (Fig. 1).

These results show that the networks are correctly interpenetrated and no phase separation is observed at the DMTA level. Thus, these IPNs exhibit many characteristics, which would allow considering them as close to "true" IPNs.

In order to check whether the mechanical properties of the PDMS network are really improved by the CAB network introduction in a final IPN material, the stress-strain curves of PDMS/CAB IPNs with different compositions were recorded (Fig. 2). The PDMS and CAB single networks break when less than 10 and 5% strains are applied respectively. On the other hand, the elongation at break is over 150% for the PDMS/CAB (50/50) IPN. Thus, the poor mechanical properties of the PDMS

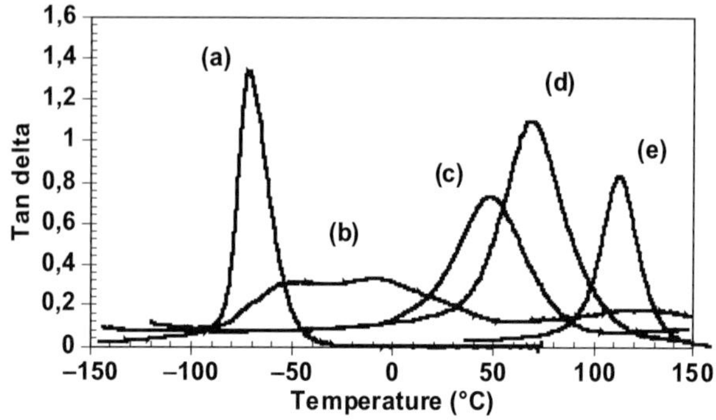

Fig. 1 tanδ versus temperature of PDMS/CAB IPNs with different weight compositions (b) 75/25, (c) 50/50 and (d) 25/75. PDMS (a) and CAB (e) single network curves are reported for comparison.[16]

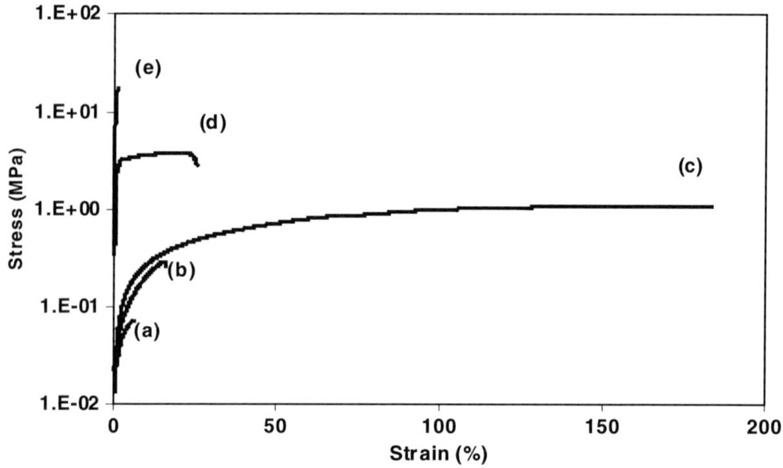

Fig. 2 Stress-strain curves of PDMS (a) and CAB (e) single networks and of PDMS/CAB IPNs with different compositions (b) 75/25, (c) 50/50 and (d) 25/75. Deformation rate: 1 mm.min^{-1} – T=35°C.[16]

network are clearly and significantly reinforced by combining it into an IPN with 50 wt% CAB, for example. This stress-strain behaviour is indicative of a synergistic effect involving the elongation at break and arising from the IPN architecture. For other IPN compositions, the extent of this synergetic effect is limited but still exists.

3.2 *PDMS/polyAcRf6 IPN*

The fluorinated network has been elaborated starting from 3,3,4,4,5,5,6,6,7,7,8,8,8-tridecafluorooctyl acrylate (AcRf6) by free-radical co-polymerization with ethylene glycol dimethacrylate in the presence of DCPD as the initiator. In this case, the PDMS network could not be synthesized by thiol-ene addition otherwise a co-network would have been obtained. Thus PDMS network was formed starting from α,ω-(3-hydroxypropyl)polydimethylsiloxane (diOH-PDMS oligomer) by DBTDL catalyzed addition between the hydroxy end groups and an pluri-isocyanate cross-linker (Scheme 2). No side reaction has been detected between these precursors.

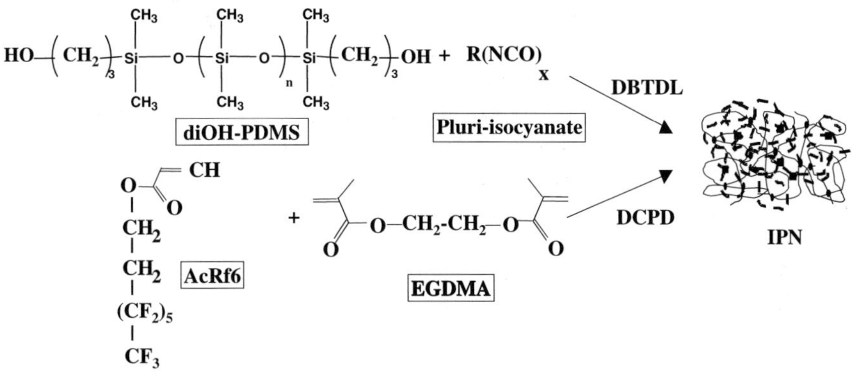

Scheme 2 Synthesis pathway of PDMS/polyAcRf6 IPNs

The extent of network interpenetration in these IPNs has been shown by different techniques and studied as a function of PDMS/polyAcRf6 composition [13]. As previously shown, DSC and DMTA are suitable techniques to detect the occurrence of phase separation in IPNs. Here, DSC and DMTA measurements show only one glass transition and only one broad mechanical relaxation, respectively, the positions of which depend on the IPN composition. However, the fact that the Tg values of the two single networks are close to each other (Tg(PDMS) = −25°C and Tg(polyAcRf6) = +3°C) can explain why only one broad transition is observed in the different IPNs. For this reason, other techniques have been used in order to obtain more information about the interpenetration degree of the two partner networks. Thus, for all IPN compositions the specific volumes are higher than the values of the linear combination of those of the single networks and this behavior cannot be matched with that of a blend in which polymers are not miscible. In other words, PDMS/polyAcRf6 IPNs possess densities lower than those predicted by additive volume law: that suggests the presence of repulsion forces between partner networks. This is in agreement with the known incompatibility of fluorinated compounds with non-fluorinated ones. However, those repulsions have been overcome by the counteracting cross-linking effect since no macroscopic phase separation is observed (IPNs are transparent). These results have been also confirmed by measurements of IPN refractive indices and their comparison with those calculated from Lorentz-Lorentz equation.

Finally, the hydrophobic character of a polymeric material can be estimated from contact angle measurements of a water drop deposited on its surface, the values of the contact angle depending on the chemical composition of the surface. Fluorinated polymers having good hydrophobic and oleophobic properties and silicones being hydrophobic as well, the contact angles measured on PDMS/polyAcRf6 IPNs should be larger than 90°. The association of two networks in IPN architecture can lead to a material the properties of which correspond to a value in between those of the two single partner networks. The contact angle of water drop deposited on PDMS/polyAcRf6 IPN surface has been measured as a function of polyAcRf6 weight proportion (Fig. 3). Remarkably, the combination of the fluorinated and PDMS networks into an IPN architecture leads to a linear increase of the contact angle from 114 to 123° for single PDMS network to PDMS/polyAcRf6 (25–75) IPN respectively. For polyAcRf6 contents higher than 75 wt%, the contact angle sharply decreases down to 117°, which corresponds to the angle measured on the polyAcRf6 single network surface. Thus the contact angle is maximum for the PDMS/polyAcRf6 (25/75) IPN and contrary to what is generally observed on copolymer surfaces [17], the contact angle does not vary monotonously with the polyAcRf6 weight proportion.

In order to check that this phenomenon is intrinsic to the material and does not result for example from the migration of unreacted starting materials towards the surface, contact angles have been measured on the surface of the same IPNs before and after extraction in a Soxhlet (Fig. 3). Contact angles remain unchanged after this treatment i.e. the observed phenomenon is specific to those IPNs. The high contact angle values of fluorinated compounds are due to the surface concentration of CF_3 groups. The above IPN density measurements have shown that repulsive interactions are established in IPNs between PDMS and poly-AcRf6 networks, resulting in a volume expansion of the material. During the IPN synthesis, the fluorinated grafts probably reorganize so as to minimize those repulsions and can in particular concentrate at the surface. Thus the surface concentration of the fluorinated grafts could as a result be higher in the IPNs than in the single network leading to an interesting synergy of the surface properties.

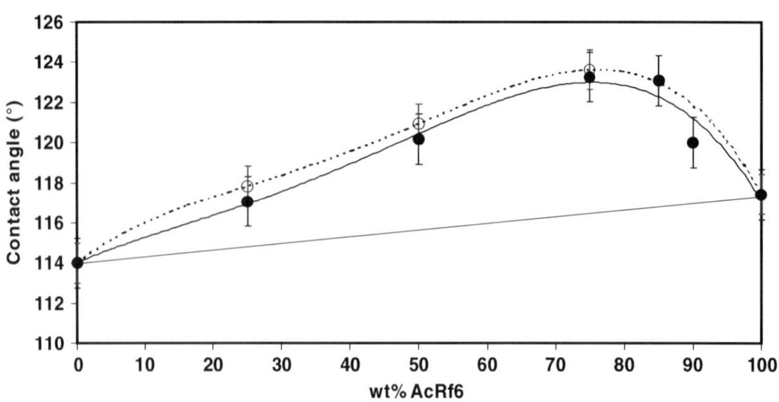

Fig. 3 Water contact angles of the surface of PDMS/polyAcRf6 IPNs vs polyAcRf6 weight proportion measured on IPNs before (●) and after (○) Soxhlet extraction with F113 for 48 hours.[13]

4 Conclusion

These results summarize a large amount of work carried out in our laboratory on polysiloxane based Interpenetrating Polymer Networks (IPNs). First, a polydimethylsiloxane (PDMS) network has been combined with a cellulose acetate butyrate (CAB) network in order to improve its mechanical properties. Thanks to a perfect control of the respective formation rates of networks it has been possible to avoid polymer phase separation during the IPN synthesis. Indeed, PDMS/CAB IPNs are transparent and only one mechanical relaxation was detected by DMTA measurements which are characteristic of "true" IPNs. In addition, a synergy effect is observed on the stress-deformation curves. Second, a PDMS network was combined with a fluorinated polymer network and the resulting IPNs can also be considered as "true" IPNs. In this case, a synergy of the surface properties was displayed.

References

1. Sperling LH, Mishra V. (1997) The current status of interpenetrating polymer networks. In: Kim SC, Sperling LH (ed). IPNs around the world: science and engineering. Wiley, New York, p. 1
2. Sperling LH. (1994) Interpenetrating Polymer Networks. In: Sperling LH, Klempner D, Utracki LA, (ed) Advances in chemistry series, vol. 239. American Chemical Society, Washington DC, p. 3
3. Sophiea D, Klempner D, Sendijarevic V et al. Interpenetrating polymer networks. In: Sperling LH, Klempner D, Utracki LA, (ed) Advances in chemistry series, vol. 239. American Chemical Society, Washington DC, p. 39
4. Noll W, Chemistry and Technology of Silicones (1968) Academic Press New-York
5. Skalsky M, Vaughan JD, Chapman RE. (1987) Progress in Biomedical Engineering. In: Planck H, Syre I, Davner M, Egbers G (ed) Polyurethanes in Biomedical Engineering II, Elsevier, New-York, p. 75
6. Mazurek M. (2000) In: Jones RG, Ando W, Chojnowski J (ed), Silicon-containing Polymers, Kluwer Academic.Publishers p. 113
7. Turner JS, Cheng YL. (2003) Morphology of PDMS-PMAA IPN Membranes. Macromolecules 36:1962–6
8. Vlad S, Vlad A, Oprea S. (2002) Interpenetrating polymer networks based on polyurethane and polysiloxane. Eur. Polym. J. 38:829–35
9. Huang GS, Li Q, Jiang LX. (2002) Structure and damping properties of polydimethylsiloxane and polymethacrylate sequential interpenetrating polymer networks. J. Appl. Polym. Sci. 85:545–51
10. He XW, Widmaier JM, Herz JE, Meyer GC. (1994) In: Klempner D, Frisch KC (ed), Advances in Interpenetrating Polymer Networks, Technomic Publishing Co., Lancaster, p. 321
11. Hamurcu EE, Baysal BM. (1995) Interpenetrating polymer networks of poly(dimethylsiloxane) with polystyrene, polybutadiene and poly(glycerylpropoxytriacrylate) Macromol. Chem. Phys. 196:1261–76
12. Boileau S, Bouteiller L, Ben Khalifa R et al. (2000),In: Clarson SJ, Fitzgerald JJ, Owen MJ, Smith SD (ed) Silicones and Silicone-Modified Materials. ACS Symp. Ser.,729. ACS, Washington DC p 383
13. Darras V, Fichet O, Perrot F et al. (2007) Polysiloxane – Poly(fluorinated acrylate) Interpenetrating Polymer Networks: Synthesis and Characterization. Polymer, 48:687–95

14. Ni H, Aaserud DJ, Simonsick WJ Jr., Soucek MD. (2000) Preparation and characterization of alkoxysilane functionalized isocyanurates Polymer: 41:57–71.
15. Fichet O, Vidal F, Laskar J, Teyssié D. (2005) Polysiloxane – Cellulose Acetate Butyrate Interpenetrating Polymer Networks - Synthesis and kinetic formation study. Part I. Polymer: 46:37–47
16. Vidal F, Fichet O, Laskar J, Teyssié D. (2006) Polysiloxane – acetate butyrate cellulose Interpenetrating Polymers Networks close to true IPNs on a large composition range. Part II. Polymer 47:3747–53
17. Van de Grampel R, Van Geldrop J, Laven J, Van der Linde R. (2001) P[CF3(CF2)(5)CH(2)MA-co-MMA] and P[CF3(CF2)(5)CH(2)MA-co-BA] copolymers: Reactivity ratios and surface properties. J. Appl. Polym. Sci. 79:159–65

Simple Strategies to Manipulate Hydrophilic Domains in Silicones

David B. Thompson, Amanda S. Fawcett, and Michael A. Brook

Abstract Hydrophilic silicone polymers offer advantageous properties in a variety of applications. However, it is not always straightforward to control the placement of hydrophilic domains in a hydrophobic silicone elastomer. A facile method for the preparation of poly(ethylene oxide)(PEO)-modified PDMS elastomers is described. Hydrosilane rich elastomers are fabricated by adding poly(hydromethylsiloxane) to a standard addition cure elastomer formula. After curing the elastomer, the silicones can be modified using hydrosilylation with mono- and di-allyl pol(ethylene oxide) of varying molecular weights to give PEO-rich silicone surfaces. The efficiency of the grafting process, as measured by PEO on the surface, depends both on molecular weight and functionality of the PEO. By contrast, when the allyl functional PEO is added directly to the elastomer preparation (co-cure), silicone elastomers with internal PEO domains are formed: SiH rich polymers present preferentially at the external interface.

1 Introduction

The high hydrophobicity of silicones can complicate their use in some applications. For example, proteins can undergo denaturation in contact with silicones [1]. In such cases, the siloxane can be modified to include a hydrophilic domain. This is typically accomplished by functionalizing the silicone with a hydrophilic polymer such as poly(ethylene oxide)(PEO). Silicone surfactants of this type have found widespread use as stabilizers for polyurethane foams, and have been investigated as a structurant to prepare siloxane elastomers for biomaterials

M.A. Brook
Department of Chemistry, McMaster University, 1280 Main Street West,
Hamilton, ON, Canada, L8S 4M1

F. Ganachaud et al. (eds.), *Silicon Based Polymers*,

applications [2]. This can be done externally, allowing for the presentation of a hydrophilic domain at bio-interfaces, or internally, where PEO domains and channels within the elastomer can allow the incorporation of hydrophilic molecules for drug delivery [3].

Although several routes to PEO-modified silicone elastomers have been described [2,4,5], we were interested in developing general methodologies that will allow for the facile functionalization of siloxane elastomers internally or externally using the same feedstocks. Such materials could have application as biomaterials, optionally with a facility to deliver drugs. Several strategies are available to prepare such materials, including RTV condensation of alkoxysilane modified PEO. However, given the commercial availability of mono- and diallyl-ethers of low molecular weight PEO (<2000 MW, also known as poly(ethylene glycol)(PEG)) a direct hydrosilylation route was chosen.

Several features of this route to grafted PEO polymers are attractive. PEO is not very soluble in silicone, a property that can be exploited to direct reaction to interfaces. The ability to control degree of functionality at one or both ends of the PEO polymer permits independent control of crosslink density and total PEO content. Finally, any residual OH or allyl groups can be used for subsequent functionalization. We describe below strategies to 'graft in' (co-cure) and 'graft to' (post-cure) silicone elastomers with PEO.

2 Results and Discussion

2.1 *Synthesis of Hydride Functional PDMS Elastomer 1*

Sylgard 184 is a commercially availably platinum-cure kit for the preparation of non-functional polydimethylsiloxane elastomers. The kit consists of two parts, a 'base', containing vinyl functional siloxanes and platinum catalyst, and a 'curing agent' containing hydride functional siloxanes. The pliable, transparent elastomers produced by this kit are a commonly used starting material for investigations into the modification of silicone rubbers [5,6], particularly those associated with biomaterials [7–10]. This elastomer is crosslinked via hydrosilylation of vinyl silicones with hydrosilicones. It is possible to directly modify the properties of the silicone during cure by adding compounds with functional groups capable of hydrosilylation. For example, the addition of vinyl- or hydrosilyl- functional materials to the uncured Sylgard silicones will ultimately lead to their incorporation into the resulting elastomer via covalent bonds.

We chose to utilize this approach to introduce poly(ethylene oxide)(PEO) into or onto silicone elastomers. As allyl-modified PEO is commercially available or readily synthesized, it was necessary to ensure the starting silicone materials had sufficient SiH functionality to permit covalent tethering. Therefore, addition of poly(hydromethyl)siloxane (PHMS; DC 1107 fluid) to the conventional

Sylgard kit was examined in order to prepare hydride functional PDMS elastomer **1**.

It was found that a ratio of 10 parts (by weight) elastomer base to 1 part cure to 1 part PHMS gave high levels of Si-H functionalization both on the surface and interior of the elastomer (by ATR-FTIR – interior measurements were made after sectioning the cured elastomer). As expected, addition of PHMS affected both the cure-time and physical properties of the resultant elastomers; adding higher proportions of PHMS (10:1:2, 10:1:1.5, etc.) gave elastomers which were unacceptably brittle. Functionalization could be maximized by storing the resulting elastomers for several days under vacuum. This prevented undesired hydrolysis of SiH groups by adventitious water while the Pt catalyst remained active.

2.2 *Post-Cure Process for PEO Functional PDMS Elastomers*

In a procedure related to those previously described by our group [2, 11], SiH rich elastomers **1** were modified post-cure with mono and di-allyl functional PEO of different molecular weights. The procedure involves a biphasic system whereby excess PEO in THF solution was reacted at the interface with SiH groups on the non-dissolved, functional elastomer. As in any 'grafting-to' approach, it was necessary to consider the factors that affect the graft density of PEO at the surface. The steric bulk of grafted PEO can hinder the approach of additional PEO chains, resulting in lower PEO reaction at the surface. These effects can be ameliorated somewhat by shifting to a less bulky (lower MW) polymer, although this runs the risk of reducing the total mass of hydrophilic polymer that will be bound at the interface. In order to better understand the interplay between these factors, three different molecular weights of PEO (250, 550 and 1100 g/mol) were used in these experiments. Both the mono- and di-allyl versions of each size polymer were reacted, from which conclusions could be drawn about the presence of loops and chains (di-allyl, **L**, mono-allyl, **C**, Fig. 1) of PEO at the silicone interface.

ATR-FTIR surface characterization was utilized to quantify the results of these experiments. Typically, this method involved normalization of the signal of interest against the C-H methyl stretch from PDMS; where necessary, signals from a 'blank' were also background-subtracted. It was found that most (80–95%) of the surface SiH groups were consumed between the hydrosilylation and alcoholysis/hydrolysis during workup. At the same time, significant increases in the CH_2 stretch for PEO were observed for all reacted disks. The efficiency of grafting (the number of surface grafted chains per surface area) was affected by both the molecular weight of the polymer, and whether mono- and di-functional PEO was used. The results were analyzed by determining total amount of PEO on the surface and the relative grafting efficiencies for each polymer length could be determined by dividing the normalized signal strength by the number of ethylene oxide units per polymer (MW 250 = 4, MW 550 = 11, MW1100 = 23; all surface areas were ~1 cm^2).

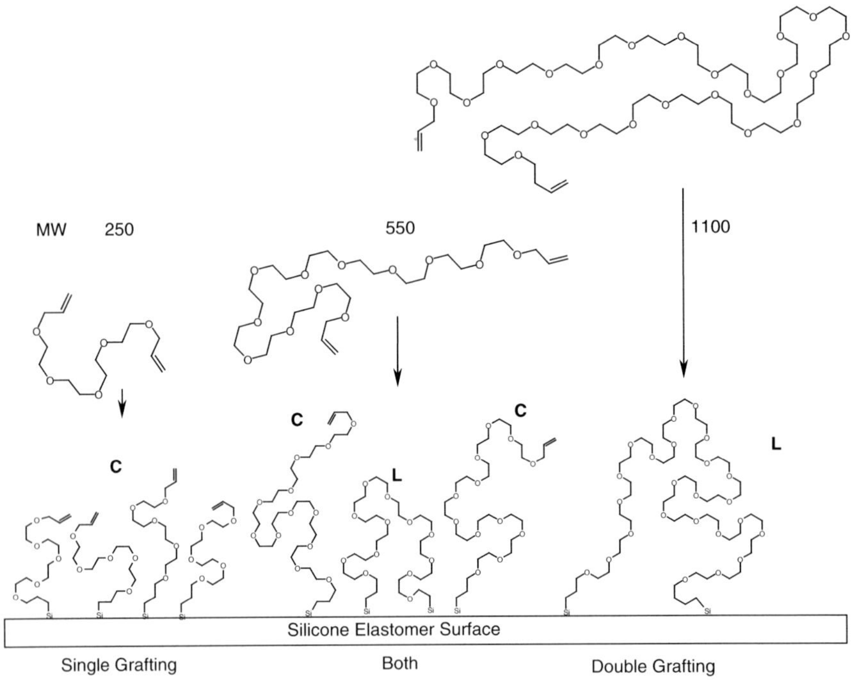

Fig. 1 Schematic showing preference for double grafting with increasing PEO molecular weight (C=chain, L=loop)

For the post-cure grafting of mono-allyl PEO, the results followed expected trends. The total amount of PEO at the interface increased with polymer size. This increase was not linear: 250 MW and 550 MW polymers showed a total surface PEO value of 60% and 93%, respectively, of that of the 1100 MW polymer. Examination of grafting efficiency (number of chains grafted to the surface, Table 1) also showed a strong dependence on molecular weight. The 250 MW polymer had the highest grafting efficiency, whereas the 550 grafted at only 57% of this value. The 1100 MW polymer showed half the grafting efficiency of the 550 monoallyl PEO. Based on these results, the 550 MW polymer shows the best compromise between grafting efficiency (number of chains grafted) and total mass of PEO polymer at the interface, whereas 1100 MW polymer shows slightly superior grafted mass of polymer at the interface.

Results for the di-allyl PEO grafted disks also illustrate that the efficiency of grafting is dependent on the molecular weight of the polymer. In this case, however, the 550 MW polymer gave the greatest yield of polymer at the interface, followed by the 250 MW. As with the mono-allyl polymers, grafting efficiency decreased with increasing molecular weight of PEO.

A comparison of grafting results for mono- and di-allyl polymers provides information about the degree to which ‘chains’ and ‘loops’ contributed to the observed surface properties. Interestingly, the 250 MW di-allyl polymer grafted slightly more efficiently

Table 1 Results of elastomers prepared via post-cure methodology

Property	1	Control[f]	Mono Allyl PEO			Di Allyl PEO		
MW (g/mol)	NA	NA	250	550	1100	250	550	1100
Contact Angle (°)[a]	97.5	105	95.5	100.3	99.4	102.3	101.9	102.7
Roughness (nm)[b]	19.906	15.08	17.89	20.91	11.19	10.57	20.86	11.51
SiH (by ATR-IR)[c]	0.25	0.263	0.019	0.017	0.053	0.062	0.016	0.034
PEO[d]	na	0	0.112	0.172	0.185	0.122	0.145	0.113
Graft Efficiency[e]	na	0	0.028	0.016	0.008	0.031	0.013	0.005

[a] Sessile water drop.
[b] Veeco profilometer rms roughness (Rq)
[c] ATR-FTIR signal height for SiH stretch normalized against CH_3 stretch
[d] ATR-FTIR signal for CH_2 stretch normalized against CH_3 stretch with background subtraction
[e] The number of grafted PEO chains/surface area was determined by taking the total PEO observed by IR and diving by the number of PEO units per chain (4 for 250, 11 for 550, 23 for 1100); surface area for all samples was the same.
[f] control sample was exposed to identical reaction conditions and washing protocols as the reaction with 550 mono-allyl polymer, but in the absence of Pt catalyst

than its mono-allyl analogue. This small difference is attributed to a component of 'unstable' grafting of PEO with the mono-allyl polymer through the free hydroxyl group at the non-allyl terminus. In the presence of platinum catalyst, it is possible for hydroxyl groups to dehydrogenatively couple to hydrosilane functionalities. While reaction with alkenes is highly favored, it is likely that a small percentage of the mono-allyl polymer underwent this process, grafting 'upside down'. As the resulting alkoxysilane groups are not stable to solvolysis/hydrolysis, they would be cleaved during the subsequent washing protocols (see experimental section). At any rate, the very high graft efficiency of the di-allyl polymer suggests that di-reaction (formation of 'loops' at surface) did not take place to a significant degree during the grafting process (Fig. 1).

For the post-cure modification with 550 and 1100 MW di-allyl PEO, fewer chains grafted to the surface than the mono-allyl versions of the same molecular weight (Table 1). This result suggests that, to differing degrees, the polymers were able to doubly react at the surface, forming loops (**L** Fig. 1). The grafting efficiency of the 550 di-allyl polymer was approximately 20% lower than that of its mono-allyl analogue, suggesting that this surface is comprised of a mix of chains and loops at the interface (Fig. 1). The grafting efficiency of the longest di-allyl polymer, meanwhile, was only 60% of the mono-allyl PEO of the same size. Reaction of both allyl groups on the diallyl PEO would give an efficiency of 50%: the value of 60% suggests that the surface also contains a few straight chains. However, di-reacted loops seem likely to comprise the bulk of the hydrophilic polymer at the interface. These results suggest that there is a size requirement for double reaction of PEO polymer at the silicone surface; the larger PEO polymers cast a larger shadow leading to a significant enhancement in the rate of intramolecular grafting (second allyl group on a polymer) over that for intermolecular coupling (first of two allyl groups on PEO). The strategy using low MW PEO is particularly interesting, as it provides the possibility of secondary grafting to the residual allyl groups (**C** Fig. 1).

2.3 Co-Cure Process for PEO Functional PDMS Elastomers

Grafting of PEO to silicone elastomers was also attempted in a co-cure fashion; SiH functionalization, PEO functionalization and platinum catalyzed hydrosilylative curing of the PDMS elastomer were undertaken concurrently (Table 2). In this case, it was anticipated that the PEO would be distributed in homogeneous domains throughout the final rubber. To permit comparison with the compounds described above, mono- and di-allyl PEO of varying molecular weights were independently added to Sylgard 184 base containing excess PHMS, in the ratio of 10:1:1 used previously.

It was discovered that the addition of 0.05 parts PEO to the 10 parts base, 1 part curing agent, 1 part PHMS mixture allowed for the best compromise of functionalization, incorporation of PEO, cure times, and physical properties. Additionally, due to the extreme differences in polarity between PEO and PDMS, it was found that addition of 0.5 parts dichloromethane was necessary to facilitate incorporation of PEO into the mixture.

This protocol led to significantly rougher surfaces when compared with the post-cure materials described above. This difference is attributed to the altered kinetics of the cure and the resulting effect on degassing/solvent evaporation: control experiments demonstrated that roughness did not arise from the presence of the solvent alone. In essence, bubbles of gas escaping under vacuum are 'captured by crosslinking' at the surface. The differences in roughness suggest that cure is inhomogeneous, with a greater degree of crosslinking occurring at the external surface. While notable, this roughness is not believed to be a contributing factor to the distribution of PEO within the silicone matrix. Additionally, while there is variation in the properties of the co-cure silicone-PEO elastomers between the various hydrophilic polymers used, no correlation between polymer size/functionality and resulting properties was discernable.

Unlike the post-cure modified samples, these co-cure elastomers were not optically transparent, but rather slightly cloudy and translucent, which is attributed

Table 2 Results of elastomers prepared via co-cure methodology

	1	Control[e]	Mono Allyl PEO			Di Allyl PEO		
MW (g/mol)	NA	NA	250	550	1100	250	550	1100
Contact Angle (°)a	97.5	103.6	107.6	103.8	106.5	105.8	106.5	108.6
Roughness (nm)b	19.906	44.3	224.24	348.39	197.21	532.67	778.47	687.23
SiH (by ATR-IR)c	0.25	0.291	0.484	0.424	0.396	0.494	0.367	0.37
PEO[d]	na	0	0.012	–0.001	0.015	0.005	0.001	–0.008

[a] Sessile water drop.
[b] Veeco profilometer rms roughness (Rq)
[c] ATR-FTIR signal height for SiH stretch normalized against CH_3 stretch
[d] ATR-FTIR signal for CH_2 stretch normalized against CH_3 stretch with background subtraction
[e] The control sample was an elastomer prepared under identical conditions (base, curing agent, PHMS, solvent) but without the addition of PEO

to the presence of discrete PEO domains within the siloxane. ATR-FTIR, however, showed no evidence for PEO on the surface. Additionally, unlike the post-cure results, these reactions were not accompanied by a decrease in SiH functionality at the surface. Instead, in all cases the relative concentration of SiH groups at the surface was significantly larger than in the control experiments. These observations suggest that the PEO is completely sequestered within the interior of the elastomers during the cure. This conclusion was further confirmed by cutting one of these disks (550 mono-allyl PEO co-cure) in half and measuring the ATR-FTIR spectrum of the interior of the disk. The spectrum showed a corrected PEO value of 0.163, with a SiH value of 0.191. This PEO value for the inside is comparable to the surface value for the post-cure surfaces, while the SiH value is lower than that found at the surface of the SiH elastomer from part 1.

Silicones have very low surface energy [12,13]. As a consequence, and facilitated by their high flexibility (Tg ∼ −123°C), silicones generally migrate to air interfaces. Because of this, even hydrophilically modified surfaces, as in the post-cure samples described above, have very high contact angles: small amounts of silicone bleed to the air interface. We note that contact angles are affected by surface roughess [2], by the chemical constituents of the surface, and orientation at the interface. For example, contact angles of PEG block co-polymers were highly dependent on the solvents to which the polymer was exposed and the hydrophobicity of the surface [14]. Thus, while it is possible to synthetically direct PEO only to the interface with our post-cure methodology, the resulting hydrophilic character will manifest itself only when the object is submerged in polar, typically aqueous, media, but not in air.

A similar effect is observed with silicone elastomers prepared with the co-cure method: the surfaces are hydrophobic and deficient in PEO, because PDMS and PHMS constituents are directed to the air interface. Interestingly, in this case, however, the silicones partition differently at the surface. ATR-FTIR demonstrated a relative increase in SiH functionality over PDMS when compared to the control. These results can only be explained by preferential migration of SiH polymer to the surface when sequestering PEO in the interior, perhaps as a result of the reduced steric bulk of each monomer unit. The resulting ‘inside out’ elastomers with a hydrophilic interior and a SiH rich exterior may offer a potential route to asymmetrically structured siloxanes by subsequent reactions with other olefinic groups.

3 Conclusion

We have developed strategies to allow the preparation of structured siloxane-poly (ethylene oxide) elastomers. The properties of the elastomers are controlled by exploiting the interfacial properties of silicones, the locus of grafting and the extent of reaction. The single step synthesis of hydride rich PDMS elastomers using commonly available starting materials can be simultaneously (co-cure) or sequentially (post-cure) used to incorporate mono- and di-allyl PEO. With the co-cure methodology, PEO segregates to the inner part of the elastomer. Simultaneously, the PHMS preferentially migrates to the outer surface, when compared to PDMS. For elastomers prepared via post-cure modification, efficiency of grafting is inversely proportional to PEO MW, although more EO units end up on the surface with increasing molecular weight. With di-allyl PEO, the degree to which 'loops' of di-reacted polymer are present at the interface is also heavily dependent on polymer size: of the three polymers examined, the highest molecular weight 1100 MW di-allyl PEO seems to mostly form polymer loops, though at the expense of lower graft densities: the 250 MW polymer, appears to exclusively form singly reacted chains. Reactions with the remaining allyl groups may offer new routes to layered polymer surfaces.

4 Experimental Section

Mono-allyl PEO (250 MW, 550 MW, and 1100 MW) were a gift from Clariant. Sodium hydride, allyl bromide, diethylene glycol dimethyl ether, and Karstedt's catalyst were purchased from Aldrich. The Sylgard 184 kit and poly(methylhydrosiloxane)(PHMS)(DC1107, 30 Cs MW ~ 2000 g.mol^{-1}) were purchased from Dow Corning. Sylgard is a platinum cured silicone elastomer filled with between 10–30% Me_3Si-modified, hydrophobic silica. Hexanes, dichloromethane, and tetrahydrofuran were purchased from Caledon and dried using pressurized alumina columns. Absolute ethanol was purchased from Industrial Alcohols and was used without further purification.

NMR spectra were recorded using Bruker Biospin AV200 spectrometer (at 200 MHz for protons). Infrared spectra were recorded on a Bio-Rad FTS-40 attenuated total reflection Fourier transform IR (ATR-IR) using a horizontal ATR apparatus with a cadmium selenide crystal. Surface roughness was measured by an interferometerometer, which generates a 3D profile (Veeco WYKO NT1100 optical profiler).

4.1 Synthesis of 250, 550, and 1100MW DiallylPEO

In a typical synthesis (shown for mono allyl PEO MW 250): mono allyl PEO (2.05 g, 8.22 mmol) and THF (40 mL) were combined in a 100 mL round-bottomed flask, and stirred with a magnetic stir bar. Sodium hydride (0.63 g, 26.2 mmol, excess) was slowly added over a period of 15 minutes and the resulting mixture was left to stir for an additional 15 min. Excess allyl bromide (2.77 mL, 32.0 mmol) was added dropwise and the reaction was stirred for 30 min. The resulting mixture was gravity filtered and excess solvent removed using a rotary evaporator and residual solvent was removed under high vacuum overnight. Yield 250 [63%, 1.28 g, 5.15 mmol], 550 [52%, 1.03 g, 1.88 mmol], 1100 [69%, 1.39 g, 1.26 mmol].

250MW: ^{1}H NMR ($CDCl_3$) δ = 5.87 (m, 2H); 5.19 (m, 4H), 4.01 (d, 4H, J = 5.6 Hz), 3.63 – 3.58 (m, 19H).

550MW: ^{1}H NMR ($CDCl_3$) δ = 5.87 (m, 2H); 5.19 (m, 4H), 4.01 (d, 4H, J = 6.0 Hz), 3.63 – 3.58 (m, 48H)

1100MW: ^{1}H NMR ($CDCl_3$) δ = 5.87 (m, 2H); 5.19 (m, 4H), 4.01 (d, 4H, *J* = 6.2 Hz), 3.63 – 3.58 (m, 96H)

4.2 Preparation of SiH Functionalized PDMS

In a beaker, Sylgard 184 base (11.30 g), Sylgard 184 curing agent (1.13 g), and DC1107 (1.13 g) were sequentially combined. The solution was stirred vigorously and 11.60 g were transferred to a 100 mm × 20 mm Petri dish. The dish was placed under vacuum for three days to cure at room temperature. After initial foaming, the silicone cured to an optically transparent, bubble free elastomer. Approximately 45 disks (0.64 cm diameter, ~0.22 cm thick) were punched out of the surface and washed with THF (5 × 5 mL × 30 sec) to remove any excess DC1107. The disks were then dried under high vacuum for 24 h.

ATR-FT-IR: 2961; 2163 (Si-H); 1050 (Si-O-Si); 1070 cm^{-1}

4.3 Post-Cure Synthesis of PDMS-PEO Elastomers with 250, 550, and 1100MW Mono- and Di-Allyl PEO

In a typical synthesis PEO (0.10 g), THF (3.0 mL), and 5 × SiH functionalized PDMS elastomer disks (*ca.* 63.0 mg/disk) were combined in a sample vial and stirred for one hour to allow the system to reach equilibrium. Karstedt's catalyst (10 μL) was added and the vials were left to stir for 16 h. The disks were washed with H_2CCl_2 (3 × 5 mL), EtOH (3 × 5 mL), water (3 × 5 mL, after which PEO was not detected in the wash water), EtOH (3 × 5 mL), and H_2CCl_2 (3 × 5 mL) and dried under vacuum for 24 h.

ATR-FT-IR: 2961; 2873; 2163 (Si-H); 1050 (Si-O-Si)

4.4 Co-Cure Synthesis of PDMS-PEO Elastomers with 250, 550, and 1100MW Mono- and Di-Allyl PEO

In a typical synthesis (shown for mono allyl PEO MW 250), Sylgard 184 base (2.01 g), Sylgard 184 curing agent (0.20 g), DC1107 (0.20 g), and a solution of PEO (0.01 g) in DCM (0.10 g) were combined sequentially in a beaker. Simultaneously, a control experiment was prepared wherein the PEO was omitted. The solution was stirred vigorously and ~1 g was transferred to a 35 mm × 10 mm Petri dish. The dish was then placed under vacuum for four days to cure. Approximately 7 disks of 0.65 cm diameter × 0.13 cm thickness, (*ca.* 42 mg/disk) were punched out of each surface and washed with H_2CCl_2 (3 × 5 mL), EtOH (3 × 5 mL), water (3 × 5 mL, after which PEO was not detected in the wash water), EtOH (3 × 5 mL), and H_2CCl_2 (3 × 5 mL) and dried under vacuum for 24 h.

ATR-FT-IR: 2961; 2873; 2163 (Si-H); 1050 (Si-O-Si)

Acknowledgements We thank Clariant for the providing polyethers, and Dow Corning for providing PMHS. We gratefully acknowledge the financial support of the Natural Sciences and Engineering Council of Canada.

References

1. Zelisko, P., Brook, M. A. *Stabilization of α-Chymotrypsin and Lysozyme Entrapped in Water-In-Silicone Oil Emulsions, Langmuir* **2002**, *18*, 8982–8987.
2. Chen, H., Zhang, Z., Chen, Y., Brook, M. A., Sheardown, H. *Protein Repellant Silicone Surfaces by Covalent Immobilization of Poly(Ethylene Oxide), Biomaterials* **2005**, *26*, 2391.
3. Ratner, B., Kwok, C., Walline, K., Johnston, E., Miller, R. J. *Silicone blends and composites for drug delivery* WO/2004/000382 (to Genzyme Corporation), **2003**.
4. Chen, H., Brook, M. A., Chen, Y., Sheardown, H. *Surface properties of PEO-silicone composites: reducing protein adsorption, J. Biomater. Sci.-Polym. Ed.*, **2005**, *16*, 531–48.
5. Yu, K., Han, Y. *A stable PEO-tethered PDMS surface having controllable wetting property by a swelling–deswelling process, Soft Matter* **2006**, *2*, 705.
6. Efimenko, K., Wallace, W. E., Genzer, J. *Surface modification of Sylgard-184 poly(dimethyl siloxane) networks by ultraviolet and ultraviolet/ozone treatment, J. Colloid Interface Sci.* **2002**, *254*, 306.
7. Stevens, M. M., Mayer, M., Anderson, D. G., Weibel, D. B., Whitesides, G. M., Langer, R. *Direct patterning of mammalian cells onto porous tissue engineering substrates using agarose stamps Biomaterials*, **2005**, *26*, 7636–7641
8. Chen, H., Chen, Y., Sheardown, H., Brook, M. A. *Immobilization of heparin on a silicone surface through a heterobifunctional PEG spacer, Biomaterials* **2005**, *26*, 7418–7424.
9. Brown, X. Q., Ookawa, K., Wong, J. Y. *Evaluation of polydimethylsiloxane scaffolds with physiologically-relevant elastic moduli: interplay of substrate mechanics and surface chemistry effects on vascular smooth muscle cell response, Biomaterials* **2005**, *26*, 3123–3129.
10. Park, S. A., Kim, I. A., Lee, Y. J., Shin, J. W., Kim, C.-R. *Biological Responses of Ligament Fibroblasts and Gene Expression Profiling on Micropatterned Silicone Substrates Subjected to Mechanical Stimuli, J. Biosci. Bioeng.* **2006**, *102*, 402–412
11. Chen, H., Brook, M. A., Sheardown, H. D., Chen, Y., Klenkler, B. *Generic Bioaffinity Surfaces, Bioconjugate Chem.* **2006**, *17*, 21.
12. Owen, M. J. "*Siloxane Surface Activity*", in: *Silicon-based Polymer Science: A Comprehensive Resource* (Eds.: J. M. Zeigler, F. W. G. Fearon), American Chemical Society (ACS Adv. Chem. Ser. 224), Washington, DC, **1990**, Chap. 40, p. 705.
13. Owen, M. J. "*Surface Chemistry and Applications*", in: *Siloxane Polymers* (Eds.: S. J. Clarson, J. A. Semlyen), Prentice Hall, Englewood Cliffs, NJ, **1993**, Chap. 7, p. 309.
14. Popescu, D. C., Rossi, N. A. A., Yeoh, C. T., Durand, G. G., Wouters, D., Leclere, P. E. L. G., Thune, P., Holder, S. J. Sommerdijk, N. A. J. M. *Surface-Induced Selective Delamination of Amphiphilic ABA Block Copolymer Thin Films, Macromolecules*, 2004, **37**, 3431–3437.

Aldehyde and Carboxy Functional Polysiloxanes

Elke Fritz-Langhals

Abstract Aldehyde functional polysiloxanes (**1**) (Fig. 1) can be selectively prepared in high yield and purity by oxidation of the corresponding carbinols with technical bleach and catalytic amounts of TEMPO. This simple oxidation method is also successfully used for the preparation of the corresponding carboxylic acids **2** (Fig. 1). The stability, reactivity and chemical properties of new compounds **1** and **2** are investigated.

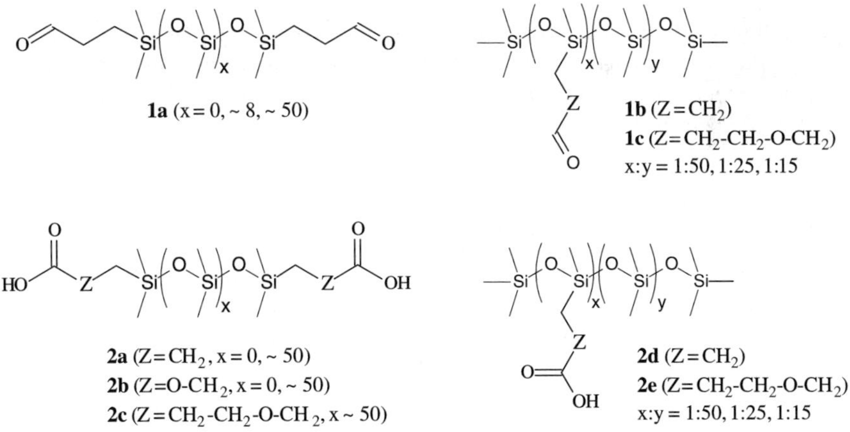

1a (x = 0, ~ 8, ~ 50)

1b (Z = CH_2)
1c (Z = CH_2-CH_2-O-CH_2)
x:y = 1:50, 1:25, 1:15

2a (Z = CH_2, x = 0, ~ 50)
2b (Z = O-CH_2, x = 0, ~ 50)
2c (Z = CH_2-CH_2-O-CH_2, x ~ 50)

2d (Z = CH_2)
2e (Z = CH_2-CH_2-O-CH_2)
x:y = 1:50, 1:25, 1:15

Fig. 1 Aldehyde and carboxy functional polysiloxanes

E. Fritz-Langhals
Wacker Chemie AG, Consortium für elektrochemische Industrie,
Zielstattstraße 20-22, D-81379 München, Germany
e-mail: elke.fritz-langhals@wacker.com

F. Ganachaud et al. (eds.), *Silicon Based Polymers*,

1 Introduction

Polysiloxanes (silicones) are highly hydrophobic, inert polymers. As a result, many silicone-based materials display good water repellency and weather resistance. Furthermore, the polysiloxane chain's flexibility gives rise to the macroscopic properties of smoothness and lubricity. This makes silicones attractive candidates for the modification of surfaces, such as textiles, paper and leather. Polysiloxanes, however, have poor adhesive properties, due to their limited interaction with other materials. Functional groups have to be introduced for this purpose. They are inserted at the end of the silicone polymer chain or into the side chains.

Aldehyde functionalities are very reactive toward amino and hydroxy groups. As a result, polymers **1** can bind to cellulose or protein surfaces. Their reactivity toward nucleophiles enables them to take part in various crosslinking reactions, e.g. with amines. Carboxy functional polysiloxanes **2** are also able to bind to surfaces because of the carboxy group's high polarity. They can undergo crosslinking reactions, e.g. with epoxides. Their salts are good surfactants.

2 Aldehyde Functional Polysiloxanes

2.1 Synthesis

To date, three methods have been described for the synthesis of aldehyde functional polysiloxanes (**1**) in patent literature, but these are not practiced. The hydroformylation of olefinic siloxanes with carbon monoxide and hydrogen under high pressure and temperature conditions is possible (Scheme 1, eq. (1)), but produces a mixture of

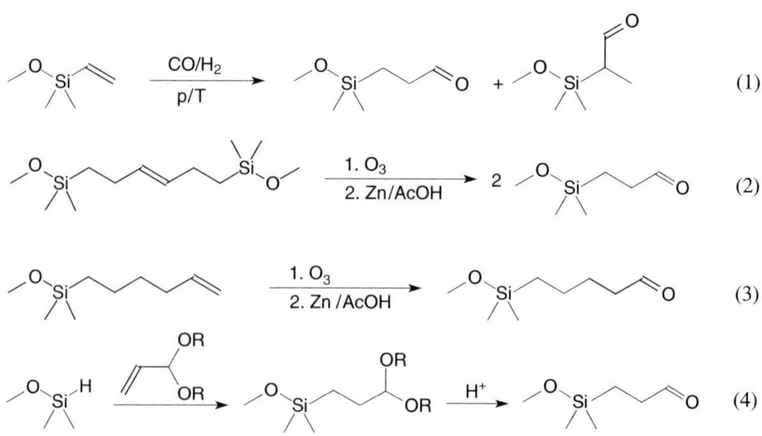

Scheme 1 Methods for the synthesis of aldehyde functional polysiloxanes described in the literature [1–5]: (1) hydroformylation, (2) and (3) ozonolysis, (4) hydrosilylation of acrolein acetals and acid hydrolysis

structural isomers [1]. Moreover, a general problem with the hydroformylation reaction is that high conversion is difficult to achieve. Another method is the ozonolysis of vinylic polysiloxanes, followed by the reductive cleavage of the ozonides with zinc (Scheme 1, eq. (2) and (3)). However, the purities and yields of the products are not reported [2, 3].

A platinum catalyzed hydrosilylation reaction is not possible for olefins with free aldehyde groups, because the silane preferentially adds to the C-O double bond [4]. The corresponding acetals, however, can be used [5]. After the hydrosilylation, the acetal protecting group has to be removed under acid conditions (Scheme 1, eq. (4)). This gives rise to side reactions such as aldol condensations [6] and the degradation of the polysiloxane chain.

As an alternative, we considered the selective oxidation of the corresponding carbinols, e.g. **3a** (Scheme 2). These are readily available on an industrial scale, for example by hydrosilylation [7] or by termination of α,ω-dihydroxypolysiloxanes with the 2,2-dimethyl[1,2]oxasilolane **4a** [7, 8]. The oxidation method of choice was the reaction with bleach and catalytic amounts of "TEMPO" (2,2,6,6-tetramethyl-piperidine-1-oxyl), described by Anelli et al. [9] for simple aliphatic or benzylic alcohols [10, 11]. It proceeds in biphasic mixtures at ambient temperature under almost neutral or weakly alkaline conditions and is a very fast reaction with a reaction half life ($\tau_{1/2}$) of about 4 s [12].

The oxidation of the di- (x = 0), oligo- (x ~ 8) and poly- (x ~ 50) siloxane diols **3a** with 1.2 equivalents of technical bleach and 2.5 mol % 4-hydroxy TEMPO or 4-acetoxy TEMPO gives the dialdehydes **1a** in 85–90% yield with almost

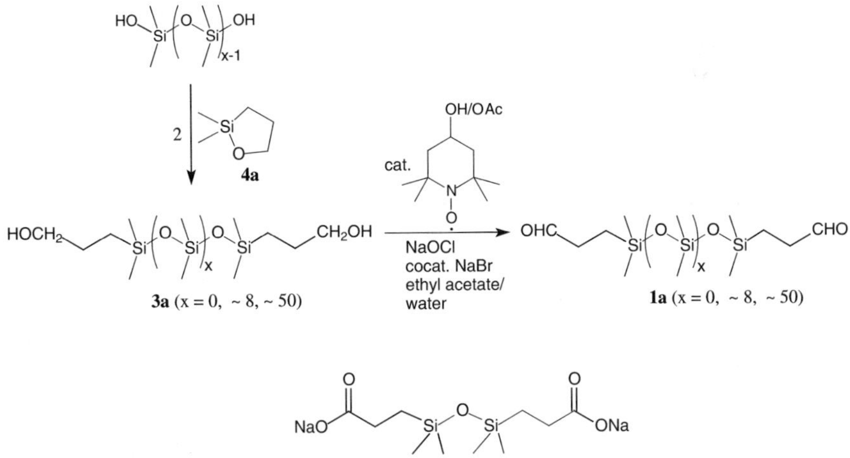

2a

Scheme 2 Synthesis of aldehyde functional polysiloxanes **1a** by oxidation of the polysiloxane carbinols **3a** with bleach (~2 M aqueous NaOCl) and 2.5 mol % 4-hydroxy TEMPO or 4-acetoxy. TEMPO [13] in ethyl acetate/water; sodium bromide (2.5 mol %) is used as a cocatalyst. Dichloromethane is used as solvent for the dimeric compound **3a** (x = 0). The precursors **3a** are made by reaction of α, ω-dihydroxy polysiloxanes with the 1,1-dimethyloxasilolane **4a** or its polymerized form $-(SiMe_2-CH_2-CH_2-O)_n-$

complete conversion [14,15]. The only detected side product is the *disiloxane* carboxylic acid **2a** [16]. It is completely separated from the product as a disodium salt in the aqueous phase. Thus, purities of about 95% are achieved [17]. Analogously, polysiloxanes **1b** and **1c** (Fig. 1), which contain side chain aldehyde groups, can be obtained from the corresponding carbinols. They can be prepared by hydrosilylation of allyl alcohol or 2-(allyloxy)ethanol, respectively. The aldehyde functional silicone resin **1d** can also be synthesised using the TEMPO/bleach method (Scheme 3). The precursor is prepared by reaction of the commercially available polysiloxane resin SILRES® REN 168 with 2,2-dimethyl[1,2]oxasilolane **4a**. The oxidation is slow in dichloromethane and ethyl acetate. It is considerably faster with toluene as the solvent – the reaction is complete after three minutes. This pronounced solvent effect is probably due to improved swelling of the resin, which contains phenyl groups, in toluene.

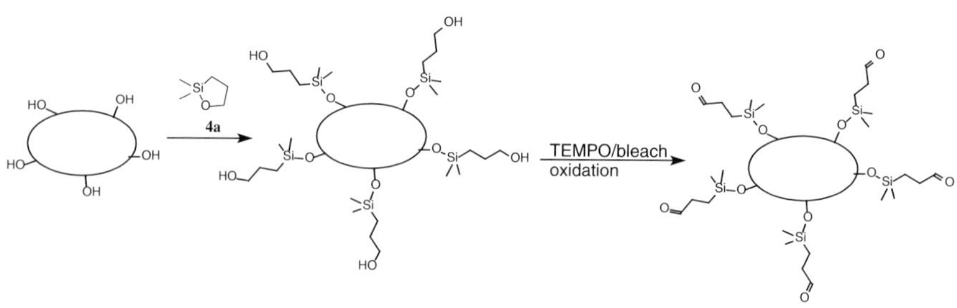

1d

Scheme 3 Synthesis of the aldehyde functional polysiloxane resin **1d**

2.2 Stability and Reactivity

The stability of aldehydes **1a** and **1b** depends on the concentration of aldehyde groups in the material: the disiloxane dialdehyde **1a** (x=0) polymerizes slowly at room temperature [18], whereas the α,ω–dialdehyde **1a** (x ~ 50) is completely stable for several months at ambient temperature. No oxidation is observed on exposure to air for 34 h. Similarly, the side chain aldehydes **1b** with x:y=1:50 and x:y=1:25 are stable oils with constant viscosity (Fig. 2). **1b** with x:y=1:15, however, displays spontaneous crosslinking, as can be seen from the material viscosity increasing with time (Fig. 2). The material becomes completely insoluble in common solvents after 6 days.

It is assumed that the crosslinking is caused by the reaction of two aldehyde groups. They form a hemiacetal hydrate in the presence of water (Scheme 4). The formation of trimeric structures known in low molecular chemistry is less probable for polymers, but cannot be excluded.

Fig. 2 Viscosity (in pa) of **1b** with x:y = 1:50 (x + y = 145), x:y = 1:25 (x + y = 70) and x:y = 1:15 (x + y = 75) over time t (h)

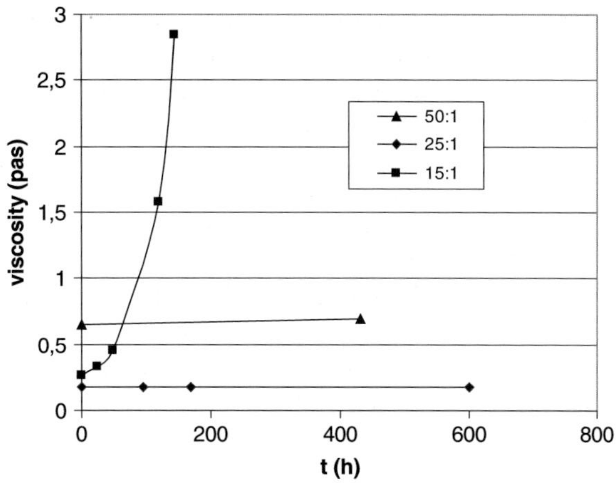

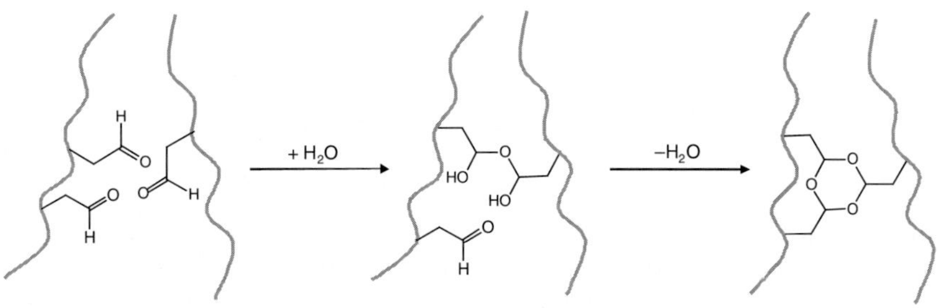

Scheme 4 Proposed mechanism for crosslinking of polysiloxane aldehydes **1b,** x:y = 1:15 and **1c**

The glycolic aldehydes **1c** (Fig. 1), with Z = CH_2-CH_2-O-CH_2, x:y = 1:50, 1:25, and 1:15, show very fast and spontaneous crosslinking after solvent removal. The solvent free products are completely insoluble in common solvents. The presence of an α-oxygen enhances the reactivity of the aldehyde group [11] and thus the tendency to crosslink.

3 Carboxy Functional Polysiloxanes

3.1 Synthesis

The classical methods, e.g. the malonic ester condensation reaction with subsequent hydrolysis [19], the Grignard reaction with carbon dioxide [20], and the hydrosilylation of acrylonitrile combined with the acid hydrolysis of the resulting propane nitrile [21], are restricted to *di*siloxanes (Scheme 5). Another method is the

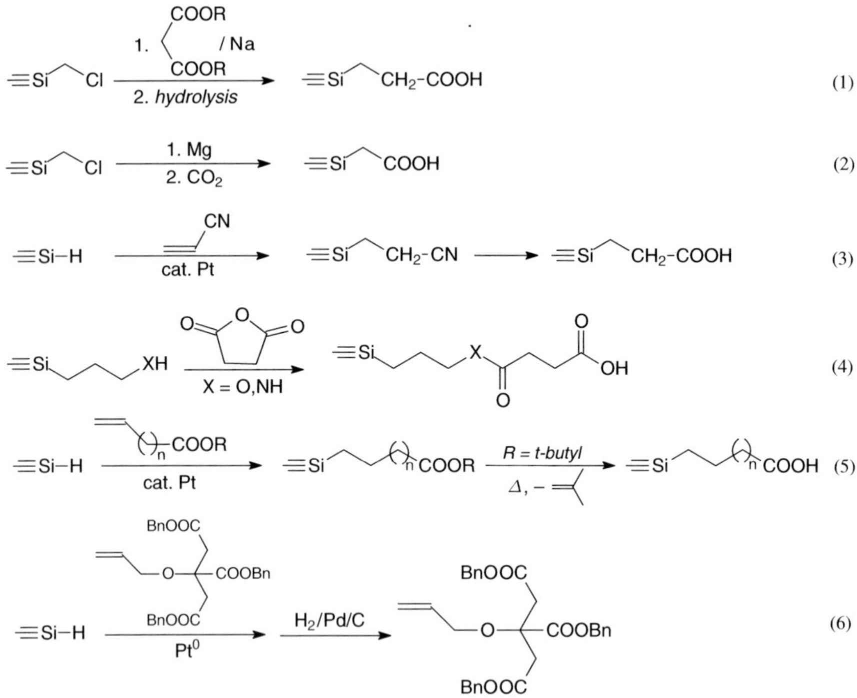

Scheme 5 Methods for the synthesis of carboxy functional polysiloxanes described in the literature [19–25]: (1) malonic ester condensation reaction [19], (2) Grignard reaction [20], (3) hydrosilylation of acrylonitrile [21], (4) reaction of carbinols with succinic anhydride [22], (5) hydrosilylation of 10-undecenylic acid [23, 24], (6) hydrosilylation of allyl citrate [25], bn = benzyl

reaction of carbinols with anhydrides, e.g. succinic anhydride [22]. Hydrosilylation of ω-unsaturated carboxylic acid derivatives is a good alternative for the synthesis of polymeric α,ω- and side chain carboxy functional polysiloxanes [23–25]. However, 10-undecenylic acid is the only ω-unsaturated carboxylic acid available on an industrial scale [26].

The disiloxane dicarboxylic acids **2a** ($Z=CH_2$, $x=0$) and **2b** ($Z=OCH_2$, $x=0$) can be prepared quantitatively in high purity from the corresponding diols with 2.4 equivalents of bleach and TEMPO (Scheme 6). During the reaction the pH is kept neutral by adding sodium hydroxide solution. The free carboxylic acids are isolated from the organic phase after acidification. The polysiloxane α,ω-carboxylic acids **2a** and **2b** were synthesized by redistribution with an α,ω-dihydroxypolysiloxane and catalytic amounts of a PNC-catalyst according to [27].

The polymers **2c**, **2d** and **2e** are synthesized directly from the polymeric carbinols by the TEMPO/bleach reaction. Under the weakly alkaline conditions a perfect emulsion is formed due to the presence of the sodium salts, which act as surfactants (Fig. 3). The free carboxylic acids can be easily isolated as oils by lowering the pH to 4–5.

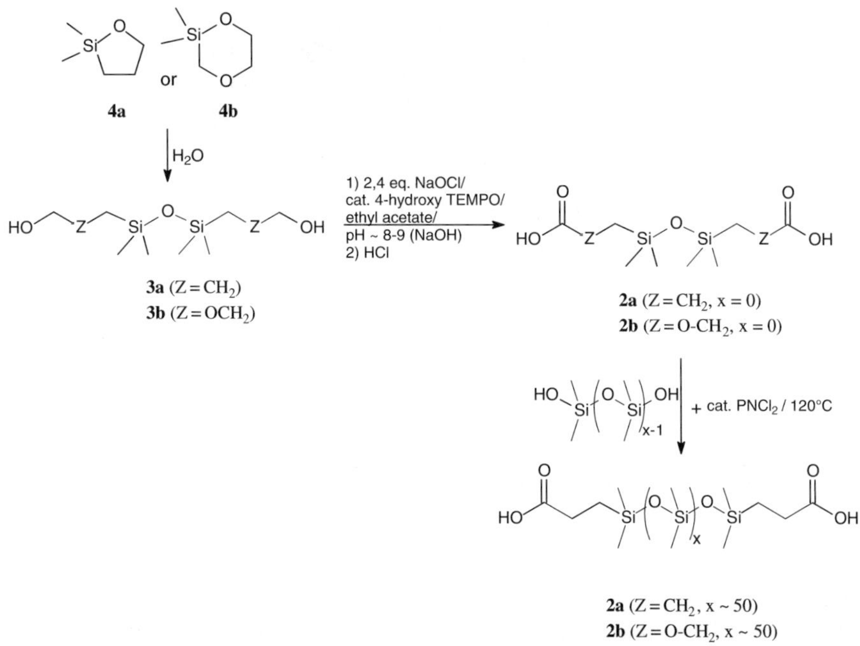

Scheme 6 Synthesis of the disiloxane dicarboxylic acids **2a**, **2b** by oxidation of the carbinols **3a**, **3b**, and synthesis of the polysiloxane carboxylic acids **2a**, **2b** (x ~ 50) by redistribution. **3a**, **3b** are made by hydrolysis of the cyclic compounds **4a** and **4b** [28]

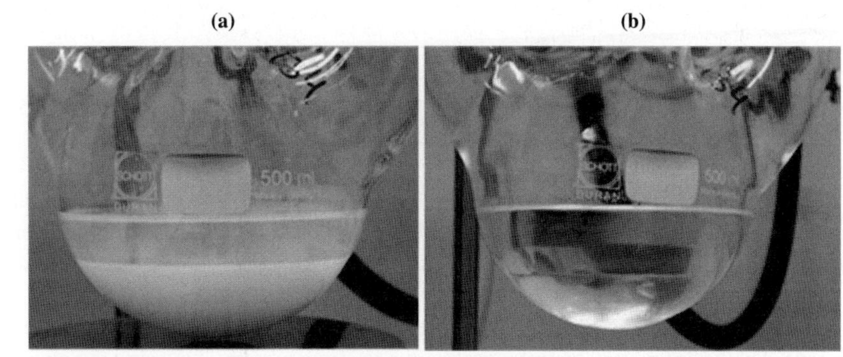

Fig. 3 Synthesis of **2c** in ethyl acetate/water, (a) reaction mixture at pH 7–8, (b) after acidification at pH 4–5 (See also Plate 2 in the Colour Plate Section)

3.2 Stability

Neutralization of the *polysiloxane* carboxylic acids **2a** (Z=CH_2, x ~ 50) and **2b** (Z=O-CH_2, x ~ 50) in dichloromethane with a buffered aqueous solution (pH 7.0) produces an emulsion, because the salts of **2a** and **2b** act as surfactants. From these salts, however, dimethylsilanols **5a** and **5b** are spontaneously produced and can be found in the aqueous phase together with the dimers **2a** and **2b** (Scheme 7). Only the α,ω-dihydroxy polysiloxane remained in the organic phase. Evidently, the carboxylic functional group is detached from the silicone backbone by nucleophilic attack of the carboxylate at the silicon centre. This backbiting reaction is favored by the formation of an intermediate five- or six-membered ring. Kinetic measurements indicate that the decomposition is almost complete after two hours.

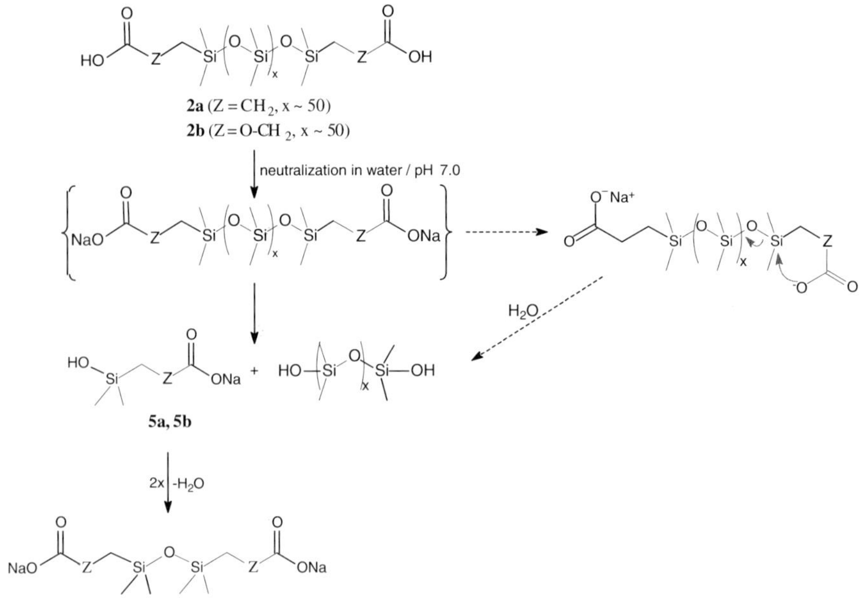

Scheme 7 Decomposition of the sodium salts of **2a** (Z=CH2, x ~ 50) and **2b** (Z=O-CH2, x ~ 50) in the presence of a neutral aqueous solution (phosphate buffer pH 7.0)

The same experiment was carried out with the polymeric side-chain carboxylate **2d**. Here, the degradation leading to the release of the methylsilanediol **5c** was very slow. After about 8 days, 50% of units bearing a carboxy group were detached (see diagram, Fig. 4).

The degradation is assumed to be the result of two consecutive backbiting steps (Fig. 4). The first step may be unfavorable due to steric reasons, lowering the overall rate of the degradation. As expected, the molecular weight decreases in parallel with the degradation reaction. At the beginning, the molecular weight of the polymer was ~ 10,000 and after 8 days, when about 54% **5c** was detected, the remaining polymer had a molecular weight of ~ 5,000. We also found about 8 % of dimethylsilanediol **5d**, which is formed during the reaction from the hydroxylic ends of the polysiloxane chain.

In contrast, the polysiloxane carboxylates **2c** and **2e** (Fig. 1), containing elongated spacers, are completely stable even under weakly alkaline conditions (pH 8). Here, no favorable ring can be formed by an intramolecular backbiting reaction.

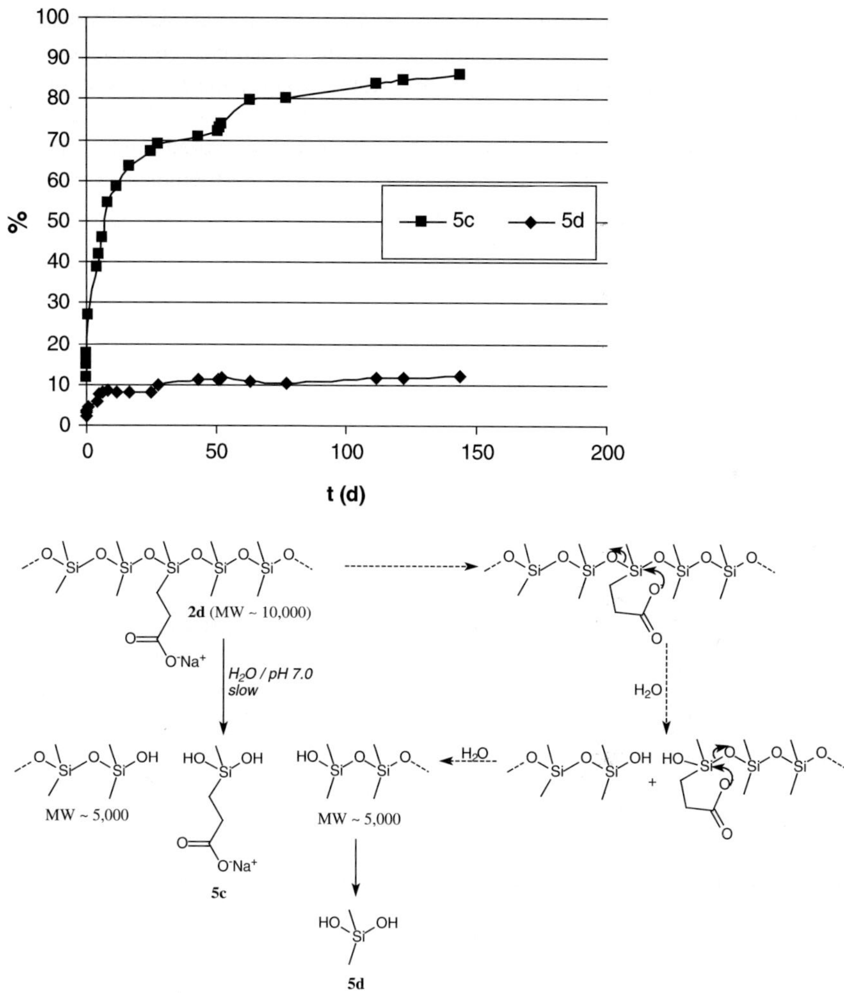

Fig. 4 Degradation of the carboxy functional polysiloxane **2d** in the presence of water (phosphate buffer, pH 7.0)

Therefore, the length of the spacer determines the stability of the carboxy functional polysiloxanes.

4 Conclusion

Both aldehyde and carboxy functional polysiloxanes (**1** and **2**) can be selectively prepared in high yield by TEMPO-bleach oxidation of the corresponding carbinols.

The stability of aldehyde functional polysiloxanes (**1**) depends on the concentration of the aldehyde group and the nature of the spacer. Spontaneous crosslinking of the polymers is favored by decreasing equivalent weight and by the presence of an activating β-oxygen.

Rules of stability have also been developed for carboxy functional polysiloxanes (**2**). The stability of this class of functional polymers depends on the length of the spacer. Carboxy functional polysiloxanes with short spacers, e.g. CH_2-CH_2 or CH_2OCH_2, degrade by a backbiting reaction, whereas compounds with longer spacers are completely stable.

5 Experimental

Starting materials The polymeric carbinols **3a** and **3b** (Fig. 5) are commercially available (WACKER IM 11, IM 15; FLUID OH 15 D, 40 D). **3c** and the side chain carbinols **3d** and **3e** (Fig. 5) can be prepared by hydrosilylation of allyl alcohol and allyloxyethanol, using commercial hydride functional polysiloxanes (commercial source e.g. ABCR – Gelest).

Synthesis of aldehyde functional polysiloxanes (1) – typical procedure 478 g **3d** (x:y = 1:25; x + y = 70, 0.2 mol), 400 ml ethyl acetate and 0.46 g (2.7 mmol) 4-hydroxy TEMPO are mixed in a three-necked round-bottom flask equipped with a mechanical stirrer, a thermometer and a dropping funnel. Saturated $NaHCO_3$ (100 ml) and a solution of 0.6 g sodium bromide (cocatalyst) in 6 ml water are added. The mixture is cooled to ~10°C, and 110 ml of technical bleach (NaOCl, c = 2.2 M in water, pH adjusted to ~9.5 by addition of 5 N sulfuric acid, 0.24 mol) are added slowly within half an hour with vigorous stirring. The temperature is held

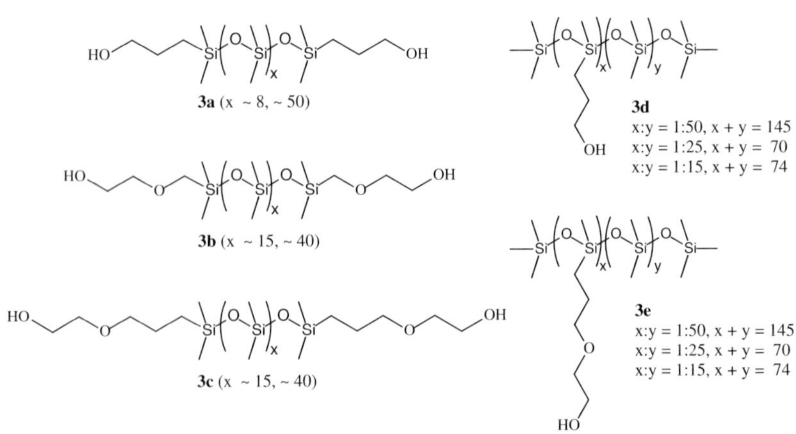

Fig. 5 Polysiloxane carbinols

below 15°C by cooling. Once the reaction is over, the phases are separated. If necessary, some ethyl acetate and water are added to complete the phase separation. The organic layer is evaporated in vacuo. Yield: 480 g **1b**, content of aldehyde groups: 90–95% (quantitative ^{1}H-NMR analysis).

Synthesis of carboxy functional polysiloxanes (2) – typical procedure 121 g **3e** (x:y = 1:50; x + y = 145, 30 mmol), 500 ml ethyl acetate and 0.37 g (1.6 mmol) 4-acetoxy TEMPO are mixed in a four-necked round-bottom flask equipped with a mechanical stirrer, a thermometer, a pH meter and a dropping funnel. Saturated $NaHCO_3$ (500 ml) and a solution of 80 mg sodium bromide (cocatalyst) in 0.5 ml water are added. The mixture is cooled to ~ 2°C, and 35.2 ml of technical bleach (NaOCl, c = 2.1 M in water, 0.072 mole) are added slowly over twenty minutes with vigorous stirring. The temperature is held below 10°C by cooling. The pH is held between 6 and 8 by addition of aqueous NaOH. After addition, stirring is continued for 2 h, and the reaction mixture is then allowed to stand overnight. 20% phosphoric acid is added until a pH of 4.5–5 is reached. The phases are separated and the organic layer is evaporated in vacuo. Yield: 119 g **2e**, content of carboxy groups: 29 mmol, 97% (quantitative ^{1}H-NMR analysis).

Acknowledgements I want to thank Dr. J. Stohrer for valuable discussions, and Dr. C. Briehn and Dr. O. Minge for providing some of the starting materials.

References

1. J.M. Frances (Rhone Poulenc Chimie) EP 392 948 (13.4.1989); J.M. Frances, J. Cavezzan, B. Bouvy (Rhone Poulenc Chimie) EP 402 274 (2.6.1989), F. Metz, J.M. Frances (Rhone Poulenc Chimie) US 5,021,601 (5.9.1988).
2. D.L. Bailey, Snider, W.T. Black (Union Carbide) US 2,947,770 (13.5.1955); D. Graiver, A.Q. Khieu, B.T. Nguyen (Dow Corning) US 5,739,246 (6.3.1997).
3. D. Graiver, B. Nguyen, F.J. Hamilton, J. Harwood, *ACS Symp. Series* **2000**, Chapter 30, 445–459.
4. *Comprehesive Handbook on Hydrosilylation*, B. Marciniec, A. Mickiewicz (Ed.), Elsevier 1992. A hydrosilylation of an olefin with a free aldehyde group, however, is described in: L.J. Sutton, J.J. Kennan (Dow Corning) WO 07011465 (25.1.2007).
5. J.L.Speier (Dow Corning) US 2,803,637 (18.1.1954).
6. W.E. Dennis and J. W. Ryan, *J. Org. Chem.* **1970**, 35, 4180–4183.
7. T. Kammel, K. Stowischek, B. Pachaly; O. Schäfer (Wacker Chemie AG) DE 10206121 (26.6.2003).
8. O. Schäfer; T. Kammel, B. Pachaly (Wacker Chemie AG) EP 1370601 (1.3.2001); G.T. Bruno, L.D. Kennan, A.K. Roy, A.J. Tselepsis (Dow Corning) US 5595593 (12.10.1995).
9. P.L. Anelli, C. Biffi, F. Montanari, S. Quici, *J. Org. Chem.* **1987**, *52*, 2559–2562.
10. A.E.J. De Nooy, A.C. Besemer, H. van Bekkum, *Synthesis* **1996**, 1153–1174.
11. W. Adam, C.R. Saha-Möller, P.A. Ganeshpure, *Chem. Rev.* **2001**, *101*, 3499–3548.
12. E. Fritz-Langhals, *Org. Proc. Res. Dev.* **2005**, *9*, 577–582.
13. C. Ochs, E. Fritz-Langhals (Wacker Chemie AG) EP1737872 (10.11.2005).
14. Instead of the popular TEMPO, 4-hydroxy TEMPO (which is cheaper than TEMPO by a factor of ten) or 4-acetoxy TEMPO can be used.
15. Due to the high reaction rate, the reaction is controlled by the dosage of bleach.
16. No *polysiloxane* dicarboxylic acid was found, because it decomposes under the reaction conditions.

17. The purity was determined by quantitative ^{1}H-NMR analysis.
18. See also [5].
19. L.H. Sommer, J.M. Masterson, O.W. Steward, R.H. Leitheiser, *J. Am. Chem. Soc.* **1956**, *78,* 2010–2015.
20. L.H. Sommer, J.R. Gold, G.M. Goldberg, N.S. Marans, *J. Am. Chem. Soc.* **1949**, *71,* 1509.
21. H. Suzuki (Toa Gosei Chemical Industry Co.) JP1999–341101/19991130.
22. T. Matsuo, Y. Kizaki (Chisso Corp.) US 6576734 (16.12.1999); H. Friedrich, P. Hoessel (BASF AG) EP 842656 (19.11.1996); J.J. Coo-Ranger, P.M. Zelisko, M.A. Brook, *Polym. Preprints* **2004**, *45*, 674–675.
23. J. Bindl, H. Petersen, K. Bachhuber, M. Ott (Wacker Chemie AG) EP 675150 (31.3.1994).
24. H. Furukawa, Y. Morita, T. Okawa, H. Ueki, S. Okawa (Dow Corning Toray Silicone Co) US 6133394 (29.1.1998); M. Takarada, Y. Yoshikawa (Shin Etsu Chemical) EP 569189 (1992).
25. M.A. Brook, S. Rodica, WO 07014471 (2.8.2006).
26. H. Baumann, M. Bühler, H. Fochem, F. Hirsinger, H. Zoebelein and J. Falbe, *Angew. Chem. Int. Ed.* **1988**, *27*, 41–62.
27. R. Hager, O. Schneider, J. Schuster (Wacker Chemie AG) EP 0626 415 (26.5.1994).
28. O. Schäfer, A. Kneissl (Wacker Chemie AG) WO 2005123812 (17.6.2004).

Molecular Devices. Chiral, Bichromophoric Silicones: Ordering Principles in Complex Molecules

Heinz Langhals

Abstract The interaction of chromophores is proposed as a basis for the construction of devices with high complexity. Silicones are promising elements of structure to control the function of such devices.

1 Introduction

There is a demand in modern technology for IT devices with high speeds of operation and high integration. Moore's law [1] predicts a doubling of complexity for storage every two years and is one popular expression for the tendencies to increase the performance of devices concerning size and speed. One can ask if there is a natural limit for the speed of operation, what factors are limiting this speed and what type of devices are expected for such a technology.

2 Optical Technology

The speed of operation is interlinked with the properties of electromagnetic radiation at the corresponding frequencies. The frequencies of operation in modern IT devices have reached some GHz in the region of microwave radiation. A further increase would step to infrared radiation, visible light and so on; see Fig. 1.

Ionization is coming about as a novel problem with such an increase of operating frequencies; see Fig. 1. Only very special structures allow ionization in

H. Langhals
Department of Chemistry, LMU University of Munich, Butenandtstr. 13, D-81377 Munich, Germany

F. Ganachaud et al. (eds.), *Silicon Based Polymers*,

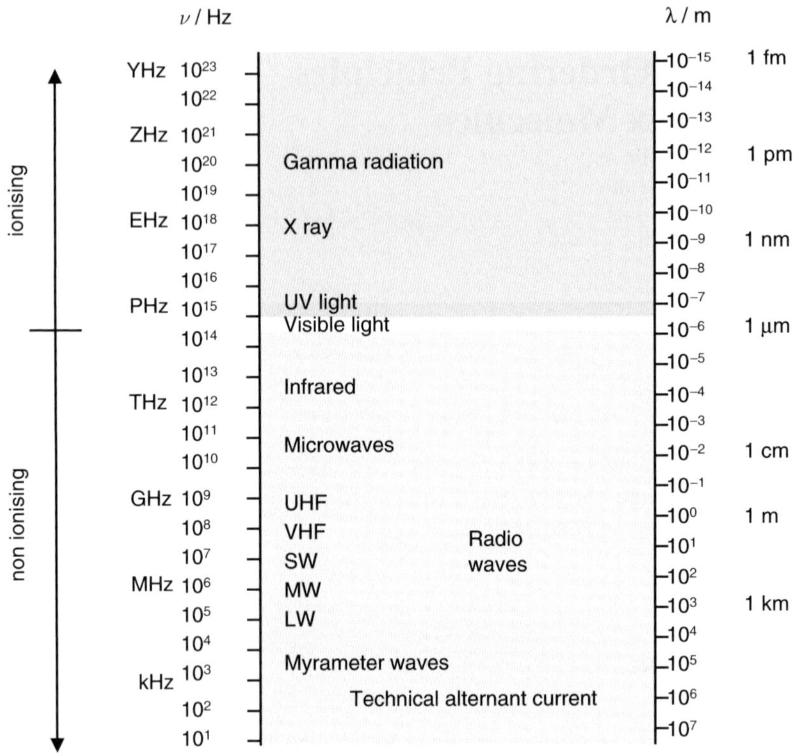

Fig. 1 Electromagnetic radiation (See also Plate 3 in the Colour Plate Section)

the near infrared and the visible so that this problem can be efficiently controlled. More and more chemical structures are becoming suitable for ionization with further increasing frequencies so that this process will finally become ubiquitary. Ionization will cause damage of materials. Such defects are becoming more and more important if increasing integration forces a diminishing of the size of operating structures. Thus, even small defects may cause a complete breakdown of the entire function. As a consequence, PHz operating devices (Petahertz, 10^{15} Hz) are supposed to be a good compromise for high frequency of operation where problems concerning damage by ionization can be still controlled by suitable chemical structures and thus, may be the limiting technology for IT devices and should be targeted for future developments. Petahertz technology means the application of light; see Fig. 1.

The entire devices should be appreciably smaller than the wavelengths of radiation in order to concentrate the operation to one single device. The wavelength of light is in the order of half a micron and this corresponds to 500 nm and 5000 Å, respectively. Dimensions some magnitudes smaller can hardly be reached with conventional concepts in electronics. Thus, molecular devices are required for the construction of complex functional structures.

3 Molecular Devices for Optical Technology

The components of electrical IT devices find their equivalents with elements of molecular structures for molecular devices; see Table 1. Macroscopic metallic conductors may be replaced by conjugated π-systems where electrons can be shifted because of delocalization. Many of the macroscopic concepts can still be retained because the wavelengths of light of the order of 0.5 μm correspond to macroscopic dimensions. However, the rules of quantum mechanics have to be applied to chromophores as molecular resonators; this will be the central item of this chapter. Molecular resonators, chromophores, are of special importance for molecular devices because of their ability to store energy and transfer it to other devices. A non fluorescent chromophore corresponds to a strongly damped resonator and a fluorescent chromophore to a slightly damped one being able to conserve the energy of excitation during the fluorescence lifetime and transfer it to other structures. Transistors are to be replaced by molecular switches for the transfer of energy [2,3]. It is important to avoid quenching of the energy transfer because this would terminate further processing of the energy of excitation. A coupling of chromophores will be of special importance for such processes and may be a promising concept for molecular processing and is the center of further treatment.

Table 1 Macroscopic and molecular IT devices compared

IT Devices	Macroscopic devices	Molecular devices

Conductors	Metallic wires	π-systems
Resonators	Resonating circuits	Chromophores
Switching devices	Transistors	Molecular switches

4 Coupled Chromophores: Bichromophores

The general interaction of two or more chromophores was theoretically treated by Kuhn [4] and Davydov [5,6], experimentally investigated by Scheibe [7] and Jelley [8] and clearly expressed by Förster [9]. A simplest example of two interacting π-systems in given for two molecules of ethylene in Fig. 2. Starting for each molecule with two π-orbitals, the HOMO electronically occupied and the LUMO not, leads to remaining four orbitals when interacting. However, the energetic levels are split so that two electronic transitions will result, one with lower energy (α) and the other with higher (β) than the initial transitions. The energy of transition (E) is interrelated with the wavelengths of absorption (λ) by Einstein's equation (1).

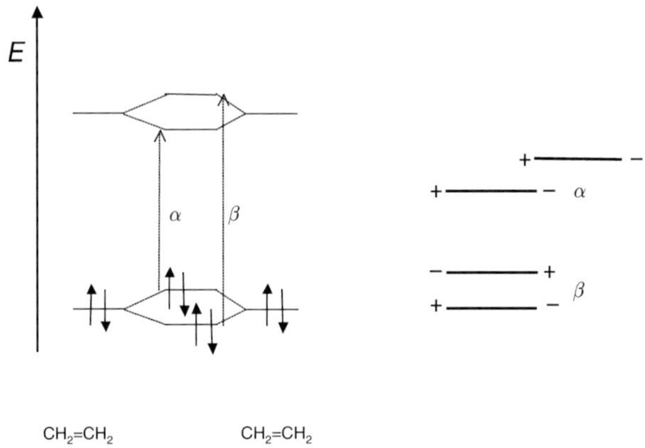

Fig. 2 FMO of two interacting ethylene molecules (*left*) and orientation of transition moments (*right*) for two arrangements

$$E = h \cdot \nu = h \cdot c/\lambda \tag{1}$$

The orientation of the transition moments discriminates which of the transition is allowed and which one not. Only one transition is preferred for parallel arrangements of the moments according to Fig. 2, right, because of electrostatic interaction. The more hypsochromic β transition is allowed if the two moments are opposed, whereas the staggered arrangement will prefer the more bathochromic α transition; see Fig. 3.

The characteristic splittings of the absorption by the exciton interaction shown in Fig. 3 are of interest both to control UV/Vis-spectra by the orientation of interacting chomophores and to monitor the orientation of molecular segments.

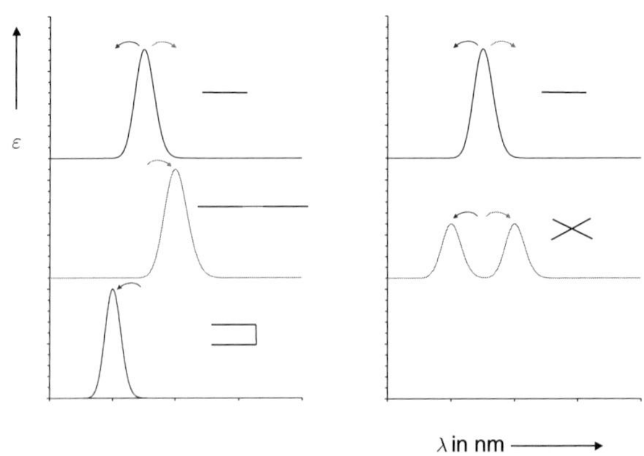

Fig. 3 Splitting of UV/Vis absorption (*black line*) caused by various orientation of the transition moments in bichromophores. *Left side*: α orientation (*light gray line*) and β orientation (*dark gray line*). Magenta line on the *right side*: Skew arrangement (See also Plate 4 in the Colour Plate Section)

A suitable building block is required for the construction of such complex structures of multichromophores.

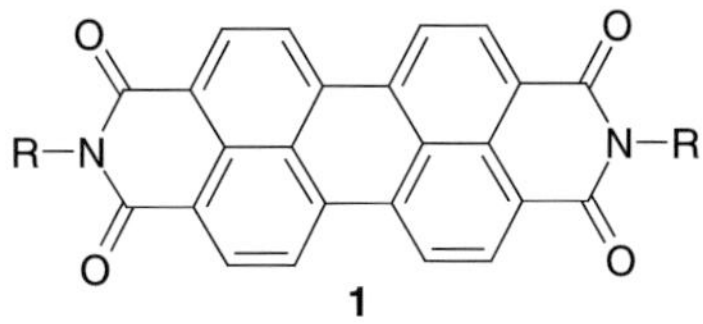

1

5 Perylene Tetracarboxylic Bisimides as Building Blocks for Molecular Electronics

The perylene tetracarboxylic bisimides **1** [10, 11] exhibit unique properties such as high chemical and photochemical persistency as a prerequisite for complicated chemical synthesis and sufficiently long-term operating of functional structures. Moreover, there is only one electronic transition in the visible along the *N-N* connection line [12] that facilitates planning molecular functions. There are orbital nodes at the nitrogen atoms of **1** in HOMO and LUMO [13] making these positions suitable for linkage without interfering with operating functions. On the other hand, the energy of the electronic transition can be controlled with substituents at the aromatic core. This makes **1** an ideal component of complex functional structures.

6 Linearly Arranged Multichromophores

Linearly arranged perylene bisimides [14] can be synthesized according to the reaction sequence of Scheme 1 where the starting material **1** was prepared from the technical perylene bisanhydride **5** and primary amines.

The low solubility of perylene carboxylic bisimides is a general obstacle for their application in homogeneous solution. This can be solved by the introduction of long-chain secondary alkyl groups ("swallow-tail-substituents") [15, 16,17] to the nitrogen atoms such as R = 1-hexylheptyl.

The carboxylic bisimide **1** is partially hydrolyzed [18] to form the anhydride carboxylic imide **2** according to Scheme 1. The bichromophore **4** is directly formed from **2** and stochiometric amounts of hydrazine, whereas an excess of hydrazine allows the preparation of the amino derivative **3**. A condensation of **2** with **3** is an alternative to prepare **4**. The trichromophore **6** is obtained from **3** and the bisanhydride **5**.

The UV/Vis absorption spectra of **1**, **4** and **6** are reported in Fig. 4 and show clearly the effect of coupling chromophores, A bathochromic shift is observed because of the preference of the α transition according to Fig. 2. Moreover, there is an amplification of the absorption because the absorptivity of **6** exceeds the expected threefold absorptivity of **1** by far; see Fig. 4. The linear alignment of chromophores

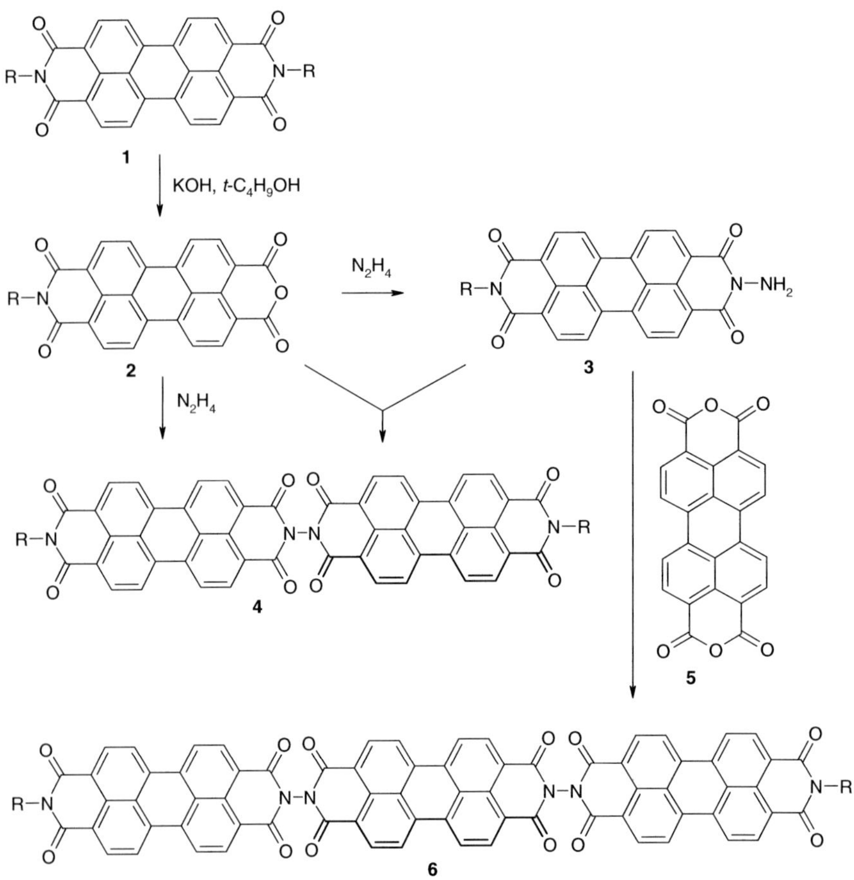

Scheme 1 Syntheses of the bi- and trichromophoric perylene dyes **4** and **6**, respectively

does not interfere with high fluorescence quantum yields because close to 100% is found for **6**. This experimental result is important for the construction of complex arrangement of chromophores where the preservation of the energy of excitation is a prerequisite for the processing of information.

7 Molecular Dynamics

Novel properties can be obtained with the combination of exciton effects with dynamic molecular processes such as for the coupling of the perylene derivative **7** to form the bichromophore **8**. **7** forms yellow fluorescent solutions [19] as is shown in Fig. 5. The coupling of two chromophores such as in **8** induces an appreciable increase of the absorptivity by exciton interaction; both

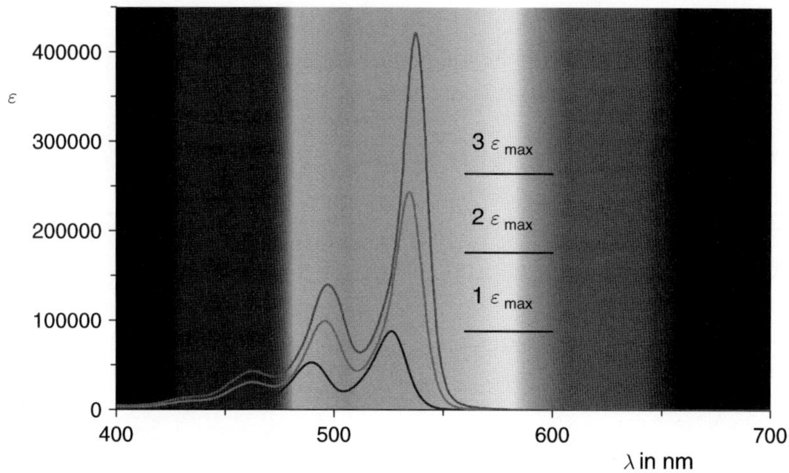

Fig. 4 UV/Vis absorption spectra of **1** (*blue*), **4** (*magenta*), and **6** (*red*) in chloroform (See also Plate 5 in the Colour Plate Section)

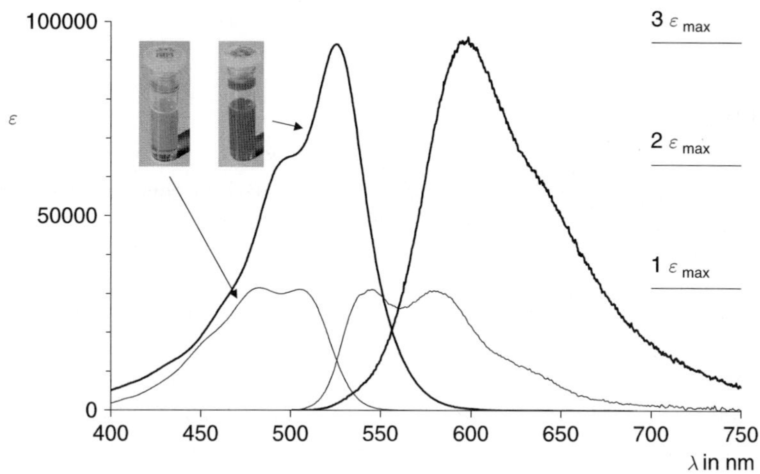

Fig. 5 UV/Vis absorption and fluorescence spectra of **7** (*thin lines*) and **8** (*thick lines*) in chloroform (See also Plate 6 in the Colour Plate Section)

chromophores arrange with orthogonal π-systems because of steric interactions; see Scheme 2. The absorption of light is a vertical transition so that the geometry remains unchanged in **8*** that relaxes subsequently to the more stable arrangement of **8a***. This causes a batochromic shift concerning the fluorescence to the unfavorable ground state **8a** that relaxes to the initial structure **8**. This cyclic process causes an increase of the Stokes' shift and a brilliant, intensely red fluorescence; see Fig. 5.

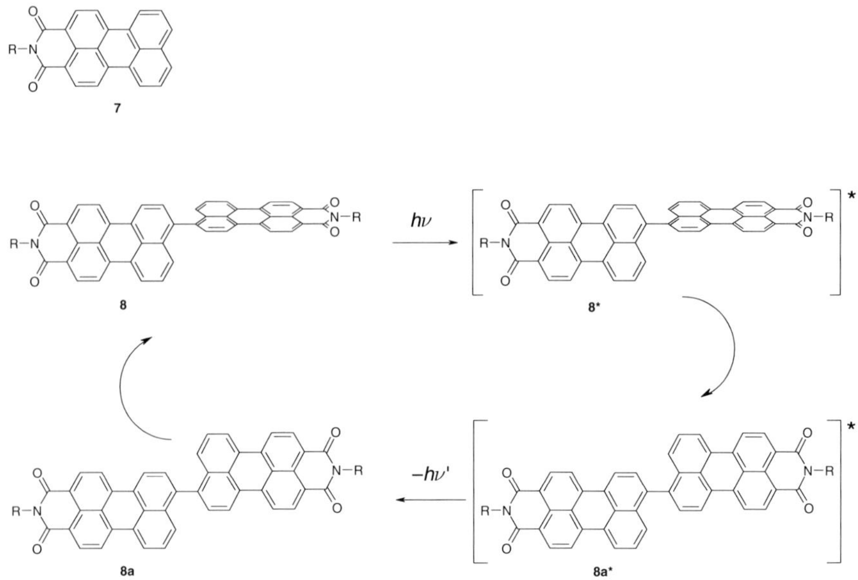

Scheme 2 Perylene-3,4-dicarboxylic imides **7** and molecular dynamics of the bichromophore **8**

8 Chirality

Even more complex functionalities can be established with chiral functional units. There is a very pure π-π^* transition of **1** in the visible. As a consequence, no CD effects are observed in the visible for chiral aliphatic substituents R. This allows the construction of complex chiral structures without interference with the optical properties of **1**.

However, very strong CD effects are introduced by exciton interactions for interacting chromophores in chiral arrangements such as for the axial chiral [20] dyes [21] **9**-*(P)* and **9**-*(M)*; see Fig. 6. Both enantiomers of **9** exhibit high fluorescence quantum yield and ellipticity of even circularly polarized light is expected in such arrangements of chromophores. Their orientation in space can be generally controlled by attached molecular segments such as the axially chiral binaphthyl unit in **9**. Silicones are therefore of special interest because of their unique type of interactions with other structures.

9 Silicones as Element of Structure for Controlling Intramolecular Interactions

The linkage of two chromophores with a silicone spacer is an attractive concept for establishing a complex arrangement because the shape of such a linker will determine the orientation of the terminal chromophores in space and thus,

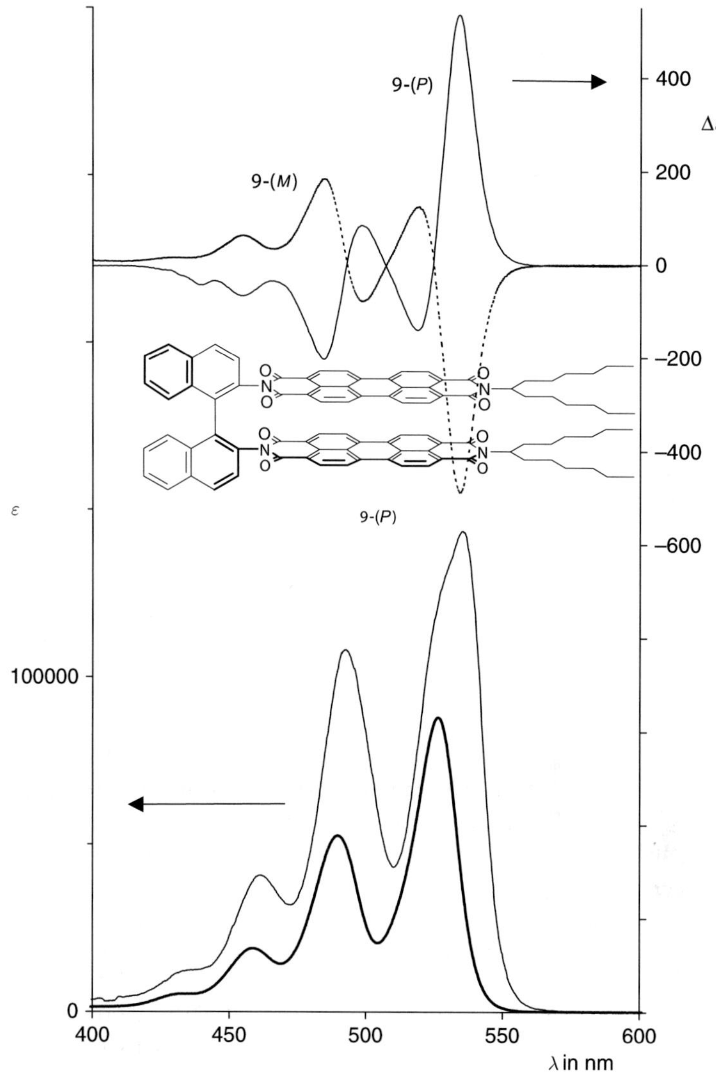

Fig. 6 UV/Vis absorption spectrum of **9** (*thin, lower line*) compared with 1 (*thick line*) in chloroform. Upper lines: CD spectra of the enantiomers of **9**

control exciton interactions. Therefore, the perylene anhydride carboximide **2a** was condensed with the chiral 2,3-dihydroxy propyl amine to form the chiral dye **10** with a terminal diol group; see Scheme 3. This can be further condensed with the silicone dialdehyde **11** to form the bichromophoric dye **12** [22, 23]. The control of the stereochemistry of **12** proceeds very efficiently because the *meso* diastereomer **12a** is formed exclusively from the racemic diol (less than 0.4% racem form). Identical UV/Vis absorption and fluorescence spectra of **12a**

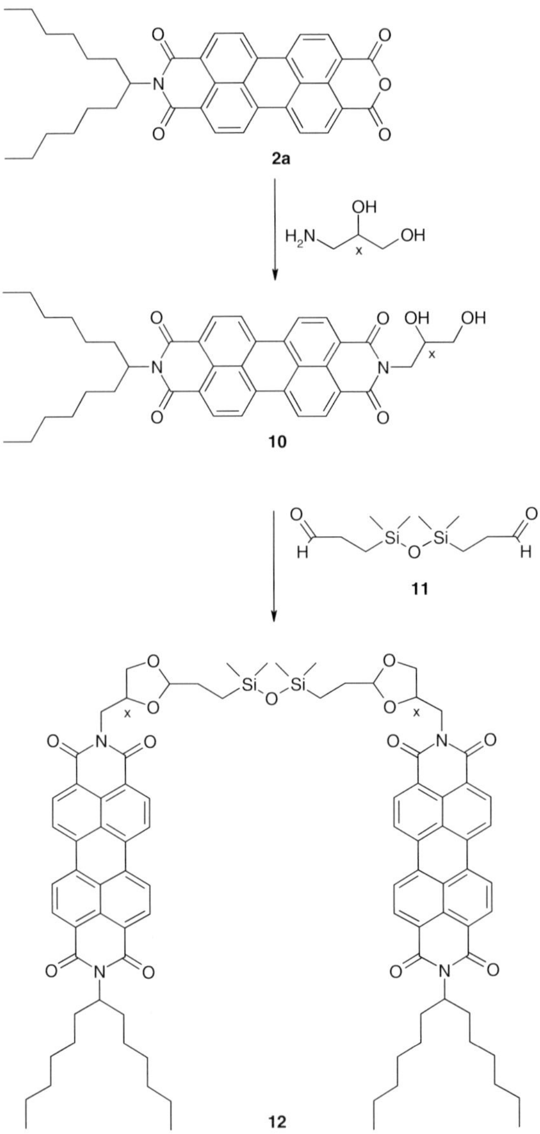

Scheme 3 Synthesis of the chiral silicone perylene bichromophores **12**

and **1** exclude exciton interaction; see Fig. 7. A completely different behavior is exhibited by the pure enantiomers **12b** (*RR*) and **12c** (*SS*), respectively, where exciton interactions not only alter the vibronic structure of the absorption spectra but also the fluorescence spectra from yellow to red so that even a visual discrimination becomes possible to decide if the starting material **10** was a pure enantiomer or a racemate. Moreover, even an artificial racemate formed from equimolar amounts of **12b** and **12c** exhibit the absorption spectra of the

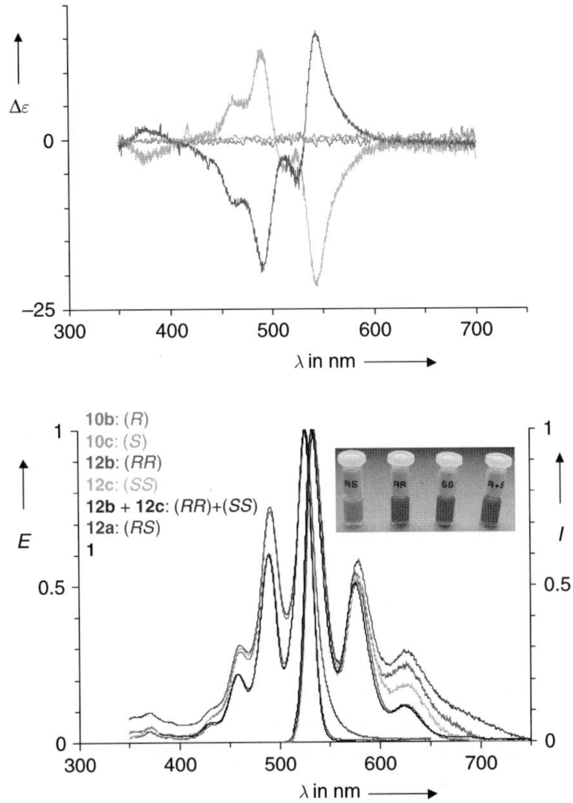

Fig. 7 Bottom: UV/Vis absorption (*left*) and fluorescence spectra (*right*) in chloroform. Blue: 1:1-Mixture of **12b** and **12c**, *red:* **12b**, *turquoise*: **12c**, *brown*: **12a**, *black*: reference of standard perylene dyes (**1**). *Top*: CD spectra of **12b** (*red*) and **12c** (*turquoise*) in chloroform and neglible CD effect of the enantiomers **10b** (*orange*) and **10c** (*green*) (See also Plate 7 in the Colour Plate Section)

pure enantiomers (**12b** + **12c**). However, a complete transformation proceeds to the *meso* form **12a** if a catalytic amount of acid is added to this racemate (**12b** + **12c**); see Fig. 8. This reaction proceeds precisely with first order and the isosbestic point at 541 nm indicates the exact stochiometry of this reaction. Thus, the molecular shape seems to be dominated by the silicone backbone, whereas the interactions of chromophores seem to be of minor importance.

10 Conclusion and Outlook

The spectral properties of **12** indicate a controlling of the shape of silicones across a rather large molecular distance. Remarkably, there are no rigid rings in this type of backbone such as commonly applied in carbon chemistry. This makes silicones

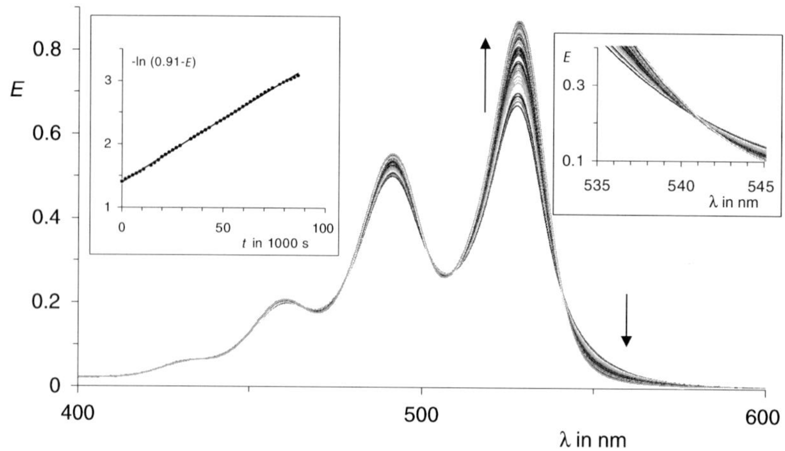

Fig. 8 Change of the UV/Vis spectra by the acid-catalysed transformation of a 1:1 mixture of **12b** and **12c** into **12a**. Right insert: amplification of the region of the isosbestic point at 541 nm. Left insert: kinetic analysis of the transformation according to pseudo first order at 527.2 nm (See also Plate 8 in the Colour Plate Section)

interesting structures for the construction of the backbone for complex functional materials. On the other hand, silicones can be efficiently prepared by the condensation of silanol units. Furthermore, a broad variation is possible to construct complex structures by the combination of such sub units. This is expected to be a good prerequisite for future developments.

References

1. G. Moore, *Electronics*, April 19, **1965**.
2. H. Langhals, S. Saulich, *Z. Naturforsch*. **2003**, *58b*, 695–697.
3. H. Langhals, S. Saulich, *Chem. Eur. J*. **2002**, *8*, 5630–5643.
4. W. Kuhn, *Trans. Faraday Soc*. **1930**, *26*, 293.
5. A. S. Davydov, *Zhur. Ekspt*l. *i Teoret. Fiz*. **1948**, *18*, 210–218; *Chem. Abstr*. **1949**, *43*, 4575f.
6. A. S. Davydow, *Theory of Molecular Excitations*, trans. H. Kasha and M. Oppenheimer, Jr., McGraw-Hill, New York **1962**.
7. G. Scheibe, *Angew Chem*. **1936**, *49*, 563.
8. E. Jelley, *Nature* **1936**, *138*, 1009.
9. Th. Förster, *Naturwissenschaften* **1946**, *33*, 166.
10. H. Langhals, *Helv. Chim. Acta*. **2005**, *88*, 1309–1343.
11. H. Langhals, *Heterocycles* **1995**, *40*, 477–500.
12. L. B.-Å. Johansson, H. Langhals, *Spectrochim. Acta* **1991**, *47A*, 857–861.
13. H. Langhals, S. Demmig, H. Huber, *Spectrochim. Acta* **1988**, *44A*, 1189–1193.
14. H. Langhals, W. Jona, *Angew. Chem*. **1998**, *110*, 998–1001; *Angew. Chem. Int. Ed. Engl*. **1998**, *37*, 952–955.
15. S. Demmig, H. Langhals, *Chem. Ber*. **1988**, *121*, 225–230.
16. H. Langhals, *Ger. Offen*. DE 3703495 (5. Februar **1987**); *Chem. Abstr*. **1989**, *110*, P59524s.

17. H. Langhals, S. Demmig, T. Potrawa, *J. Prakt. Chem.* **1991**, *333*, 733–748.
18. H. Kaiser, J. Lindner, H. Langhals, *Chem. Ber.* **1991**, *124*, 529–535.
19 H. Langhals, F. Süßmeier, *J. Prakt. Chem.* **1999**, *341*, 309–311.
20 V. Prelog, G. Helmchen, *Angew. Chem.* **1982**, *94*, 614–631; *Angew. Chem. Int. Ed. Engl.* **1982**, *21*, 567.
21. H. Langhals, J. Gold, *Liebigs Ann./Recueil* **1997**, 1151–1153.
22. H. Langhals, O. Krotz, *Angew. Chem.* **2006**, *118*, 4555–4561; *Angew. Chem. Int. Ed. Engl.* **2006**, *45*, 4444–4447.
23. H. Langhals, K. Fuchs, *Coll. Czech. Chem. Commun.* **2006**, *71*, 625–634.

Modified Azo-Polysiloxanes for Complex Photo-Sensible Supramolecular Systems

Nicolae Hurduc, Ramona Enea, Ana Maria Resmerita, Ioana Moleavin, Mihaela Cristea, and Dan Scutaru

Abstract Here we show the possibility to obtain different types of azo-polysiloxanes capable to respond to light stimuli. The azo-polymers were prepared starting from a polysiloxane containing chlorobenzyl groups in the side-chain, using a two-step substitution reaction. In the first step, the polysiloxanes were modified with azo-benzene groups and, in the second one, different systems, as functions of the envisaged application (nucleobases, donor/acceptor or ammonium quaternary groups) were connected to the side-chain. The photochromic behavior in the presence of UV irradiation or natural visible light was investigated, in solution or in the solid state. Even the maximum conversion degree from *trans*- to *cis*- configuration of the azo groups is slightly lower in the solid state as compared with the solution, the response rates are similar on the time-scale. The *cis-trans* relaxation behavior is different for the systems containing nucleobases, as compared with the donor/acceptor ones. In the case of the azo-polysiloxanes containing quaternary ammonium groups, the polymer aggregation capacity was investigated. The critical aggregation concentration is situated at lower values that can be explained by the azobenzenic group aggregation capacity to generate a hydrophobic micelle core.

1 Introduction

The azo-group capabilities to change their configuration under UV-VIS irradiation induce the possibility of the polymer conformational photo-control, with potential application in optoelectronics or biology fields. One of the consequences of the

---

N. Hurduc
Technical University "Gh. Asachi" Iasi, Department of Natural and Synthetic Polymers, Bd. Mangeron 71, 700050-Iasi, Romania
e-mail: nhurduc@ch.tuiasi.ro

F. Ganachaud et al. (eds.), *Silicon Based Polymers*,

photo-induced conformational change is the possibility to obtain a surface relief grating (SRG) [1–4] that presents a potential interest for increasing the efficiency of the photo-luminescent diodes or in the field of organic micro-lasers. In the field of the biomedical materials, the film surface relief can influence the cells adhesion mechanism, the cell multiplication rate etc. [5–7]. Although several models have been proposed to explain the photoinduced mass transport, the mechanism responsible for the material structuring at the surface is not completely elucidated [8–10]. Some parameters such as the isomerization pressure, the gradient of the electric force, the asymmetric diffusion, the mean-field forces or the permittivity gradients are assumed to play an important role in the different proposed mechanisms [9–13]. Our recent studies demonstrated the existence of a special state, named *conformational instability*, that appears as a consequence of a continuously *trans-cis-trans* isomerization during the azo-material UV irradiation, that can explain the appearance of a fluid state, even if the temperature is situated 30–40°C below T_g value [14]. The possibility to generate a directional flow (if a polarized UV laser source is used), demonstrated by Karageorgiev et al. [14], open a new interesting research direction related to the molecules nanomanipulation, especially biomolecules. But to realize this purpose the biomolecule immobilization to the azo-polymeric surface is necessary. This is the reason why we decided to connect to the polysiloxane not only azobenzenic, but nucleobases or donor/acceptor group systems capable to generate strong physical interactions.

Another possibility for azo-polysiloxane applications is to obtain photo-sensible micelles [15–18]. The interest for this application is explained by the possibility to use polymeric micellar aggregates for controlled release of substances such as drugs [19, 20]. There are few reports on the use of light as an external stimulus for small amphiphilic molecules by incorporating the azobenzene chromophore into surfactant [16] or for light-responsive micellar aggregates formed by amphiphilic block copolymers [17,18].

We describe in this article the possibility to generate different azo-polysiloxanic structures with potential applications in biomolecules immobilisation and nanomanipulation, in surface relief grating techniques or in the field of photo-sensible micelles. The azo-polysiloxanes molecular weights are situated between 6100 and 7150 $g.mol^{-1}$, having polydispersity index values of 1.80–2.05.

2 Experimental

The azo-polysiloxanes were obtained in a two-step reaction, starting from a polysiloxane containing chlorobenzyl groups in the side-chain. In the first step, the polysiloxane was modified [21] with 4-hydroxyazobenzene (50–75% substitution degree) and, in the second one, the unreacted chlorobenzyl groups were substituted with nucleobases or donor/acceptor groups.

The polysiloxanes containing chlorobenzyl groups were obtained by the hydrolytic polycondensation of the 73:27 mol/mol mixture of [2-(4-chloromethyl phenyl)ethyl] methyldichlorosilane and [1-(4-chloromethylphenyl)ethyl]-methyldichlorosilane followed by the cationic equilibration in the presence of triflic acid (divinyltetramethyldisiloxane was used as the chain end blocker to obtain a vinyl–Si ended chlorobenzyl-substituted polysiloxane). Details concerning the polymers characterization were previously reported [21].

To obtain the azo-polymers (Scheme 1), the polysiloxanes containing chlorobenzyl groups in the side-chain were modified with 4-(phenylazo)phenol. In a typical reaction, 1.4 mmol of polysiloxanes was dissolved in 6 mL DMSO and mixed with the corresponding amount (as a function of the imposed substitution degree) of sodium salt of 4-(phenylazo)phenol and 0.5–0.7 g tetra-butyl-ammonium hydrogen-sulphate. The mixture was introduced in a flask and heated 5 h at 90°C. The polymer was precipitated in methanol and washed 5–6 times with methanol to eliminate the unreacted products. The polymer was dried under vacuum (10 mm Hg and 35°C).

In the second step (Scheme 1), 0.5 g azo-polysiloxane was dissolved in 7 ml DMSO under stirring; the necessary quantity of nucleobase (as a function of the imposed substitution degree) and ~0.1 g K_2CO_3 were added and then the reaction mixture was heated 9 h at 55°C (under nitrogen atmosphere). The polymer was precipitated in methanol and washed 3–4 times with methanol to eliminate the unreacted products and dried under vacuum. In the case of donor/acceptor groups, 0.5 g azo-polysiloxane was dissolved in 7 ml DMSO under stirring; the necessary quantity of sodium phenoxide and 0.1 g Bu_4NHSO_4 were added and then the reaction mixture was heated 4–5 h at 80°C. The polymer purification was similar.

The amphiphilic polymers were obtained, too, in a two step reaction, in agreement with Scheme 1. The azo-polysiloxane obtained in a first step (as previously described) and containing unreacted chorobenzyl groups (40%) was quaternized with triethyl and tributylamines in solution (CH_2Cl_2) at 30°C, using reaction times situated between 5 and 7 h.

All the chemical products (analytical quality) were purchased from Aldrich and were used without supplementary purification. The polymers were characterised by ^{1}H-NMR, SEC, DSC, TGA and UV-VIS spectroscopy. The photochromic behaviour was investigated on the solid state, using thin films deposed on the surface of a quartz cell. The azo-polymeric films were irradiated using a UV lamp (50 W) equipped with 350 nm filter, irradiance of 7 mW/cm^2 at 30 cm from the filter (the distance between the UV light source and film surface was 30 cm). In all the UV irradiation experiments, the film surface was cooled with compressed air, the temperature being maintained at 20°C (±1°C). The ^{1}H-NMR spectra were recorded on a Brucker 300 MHz apparatus. SEC experiments were carried out in THF solution at 30°C, at a flow rate 1 ml/min using a Spectra Physics 8800 gel permeation chromatograph. DSC thermograms were recorded on a Mettler DSC 1 calorimeter with 10 K/min heating rate. For the photo-isomerization kinetic curves, a BOECO S1 UV spectrophotometer (at 350 nm wave length) was used. In order to obtain information concerning the polymeric chains geometry, corresponding to the *cis*- or *trans*-azobenzenic groups, molecular simulations were made using Accelrys [22] software (Materials Studio 4.0). The single polymeric chain conformation was obtained using a Molecular Mechanic procedure, Forcite module (Dreiding and

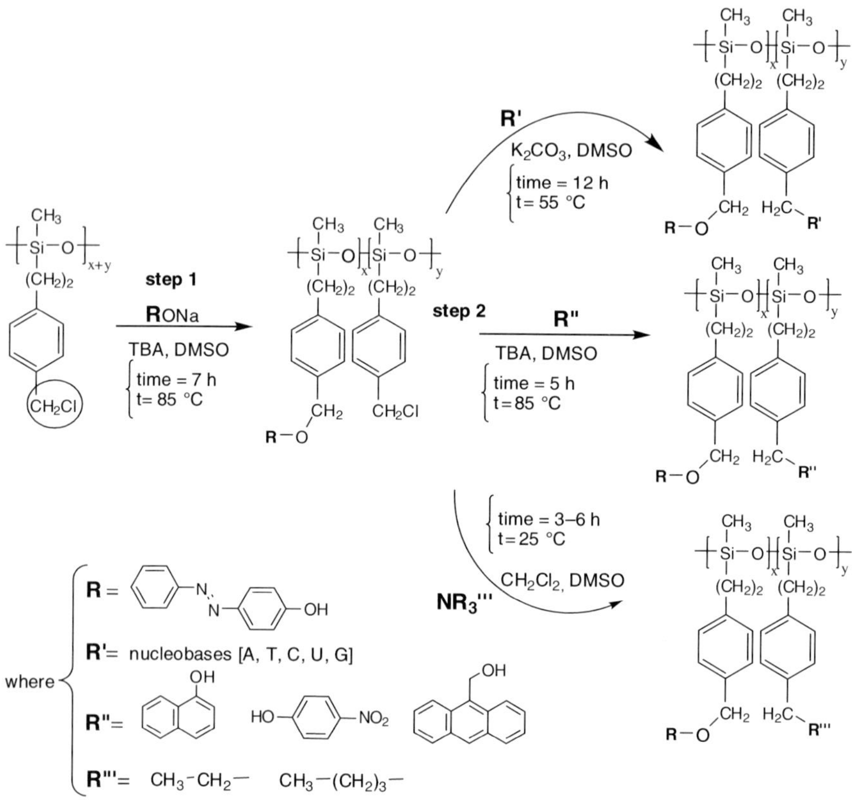

Scheme 1 Azo-polymers synthesis reaction scheme

PCFF force fields, alternatively with molecular dynamics, in order to identify the global minimum of the energy value). The investigated single chain has a polymerization degree DP=25, similar to the results obtained by synthesis.

3 Results and Discussion

All the investigated polymers were obtained starting from a polysiloxane containing chlorobenzyl groups in the side-chain. Details concerning the synthesis and characterization were previously reported [21]. As a function of the final polysiloxane chemical structure, different reaction conditions were used.

The first investigated azo-polysiloxane contains nucleobases in the side chain. This azo-polymer class presents a potential interest in the fields of optoelectronic or biomolecule immobilization and nanomanipulation. The optoelectronic applications are the consequence of the azo-materials capacity to generate surface relief grating (SRG) under UV irradiation. In spite of the fact that the SRG problem is studied by many research groups, until now, the mechanism of the surface structuration is not completely elucidated. It is well known that essentially for the surface structuration is the *trans-cis* isomerization process of the azo-groups, under UV-VIS irradiation. As a consequence, a very careful study regarding the photochromic behavior of the azo-polysiloxanes was effectuated. The photochromic response at the light stimuli was evaluated in different conditions: *trans-cis* isomerization under UV irradiation; *cis-trans* relaxation in the presence of the natural visible light; *cis-trans* relaxation only thermally activated (in the dark). It must be underlined that there are no reported studies in the literature concerning the *cis-trans* relaxation process in the presence of natural visible light, a very important aspect in our opinion regarding the elucidation of SRG formation mechanism.

In our view, one of the difficulties to understand the surface structuring mechanism is due to the fact that, up to this moment, there is no well established standard methodology concerning the UV irradiation procedure. For example, usually in the papers it is not specified if the polymeric film UV irradiation is done in the presence or in the absence of natural visible light, in the presence or in the absence of neon tube light etc. All these additional visible light sources influence in a strong manner the *cis-trans* relaxation process and implicitly the isomerisation equilibrium.

Table 1 presents some characteristics of the modified polysiloxanes. As one can see, the molecular weights are situated between 6100 and 7150 $g.mol^{-1}$ having polydispersity index values of 1.80–2.05. The modified polysiloxanes chemical structures were confirmed by ^{1}H-NMR spectroscopy. In the case of azo-polysiloxanes modified with nucleobases, the polymer substitution degree was monitored using the signals corresponding to the $-CH_2Cl$ groups (4.5 ppm), that shifts to 5.1 ppm after the connection of the azobenzene and to 4.7–4.8 ppm after the connection of the thymine, adenine or uracil. Moreover, there are additional characteristic signals: for thymine at 1.7 ppm ($-CH_3$) and 11.3 ppm (-NH-); for adenine at 8.15 ppm (=CH-); for uracil 11.4 ppm (-NH-) and 5.6–5.8 ppm (-CH=CH-). A typical spectrum corresponding to an azo-polysiloxane modified with adenine is presented in Fig. 1.

The relatively low T_g values situated slight above the room temperature (Table 1) imposed supplementary precautions during UV irradiation in the solid state. Taking into consideration the heating effect of the UV light, we have used a cooling system with compressed air that succeeded to maintain the temperature film surface at 20–21°C. We are strongly interested to study the photo-isomerization process in the solid state, below the T_g value, because our intention was to obtain new information concerning the photo-fluidization mechanism. In a previous study [14] we introduced a new concept, named *conformational instability* that can explain the apparition of a fluid phase during the UV irradiation. This special state is the result of a continuously *trans-cis-trans* isomerization process during irradiation, the configurational change of the azobenzenic groups being accompanied by conformational modification at the entire polymeric chain level. These continuous conformation changes hinder the phase stabilization in a solid state. Moreover, the phase stabilisation in a solid state will be hard to achieve too, due to the fluctuation of the azo-groups dipole-moment values during isomerization process (i.e. in the case of azobenzene, the dipole-moment is 0.1 D for *trans* configuration and 3.5 D for *cis*) [23, 24]. This combination between the conformational instability and the dipole-moment

Table 1 Some characteristics of the synthesized azo-polysiloxanes

Spl. no.	Substituent type	Substitution degree (%)	M_n*	PI	Tg (°C)
1	azobenzene	91	7350	1.80	35
2	azobenzene	72	7600	1.91	45
	adenine	15			
3	azobenzene	70	7350	1.88	35
	thymine	16			
4	azobenzene	74	7250	1.95	27
	uracil	14			
5	azobenzene	73	7150	1.95	37
	cytosine	16			
6	azobenzene	68	7400	1.90	30
	p-nitro-phenol	21			
7	azobenzene	69	7350	2.05	35
	naphthalene	22			
8	azobenzene	68	7500	2.03	31
	9-methoxy-anthracene	19			
9	azobenzene	40	6900	–	–
	triethylamine	58			
10	azobenzene	40	7050	–	–
	tributylamine	57			

PI – Polydispersity index
T_g – glass transition
* – obtained by SEC

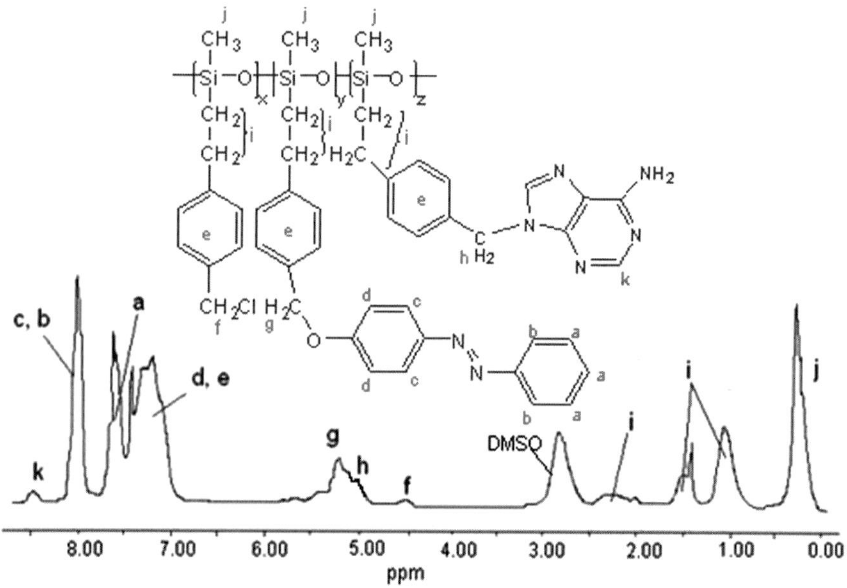

Fig. 1 ^{1}H-NMR spectrum corresponding to an azo-polysiloxane modified with adenine

fluctuation can explain the material fluidization as a consequence of the UV irradiation process.

The molecular simulation evidenced a disordered chain geometry corresponding to the azo-polysiloxanes, which can favour the presence of the amorphous phases. The disordered geometry is imposed by the presence of two structural units (SU) types, named α and β (Fig. 2). The ratio between the two SU in the polysiloxane is 75% β and 25% α, their distribution being random [21]. The homopolymers corresponding to each SU present ordered chains, but having different geometry. For the polysiloxane based on β-SU, a helicoidally chain geometry was evidenced, having a high value of the helix diameter (Fig. 3). For the β-SU type, a linear chain growth is present (Fig. 4). Both simulation were effectuated for a DP=50 (higher as compared with the experimental values), in order to evidence the helical chain conformation for the homopolymer based on β-SU. The combination between these two different tendencies of chain growth has as a result disordered structures, as one can see in Fig. 5.

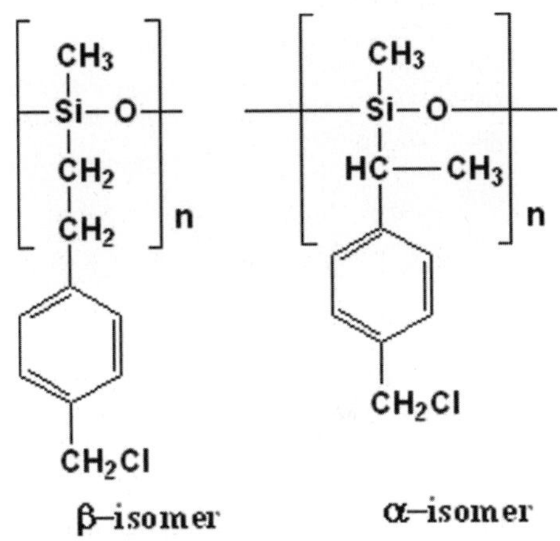

Fig. 2 α- and β-type polysiloxane structural units

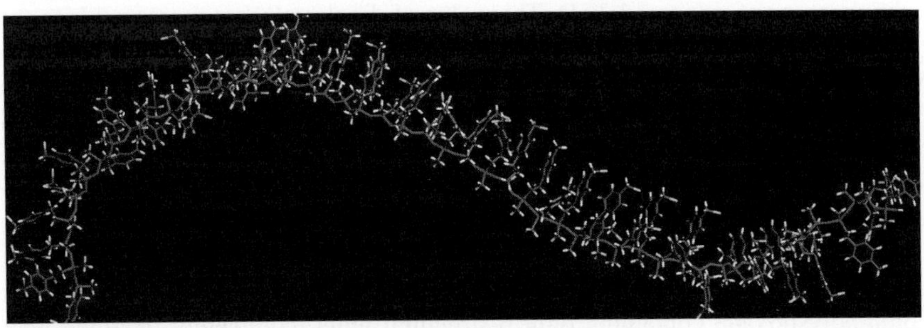

Fig. 3 Minimum energy conformation corresponding to the homopolymer containing only β-structural units (See also Plate 9 in the Colour Plate Section)

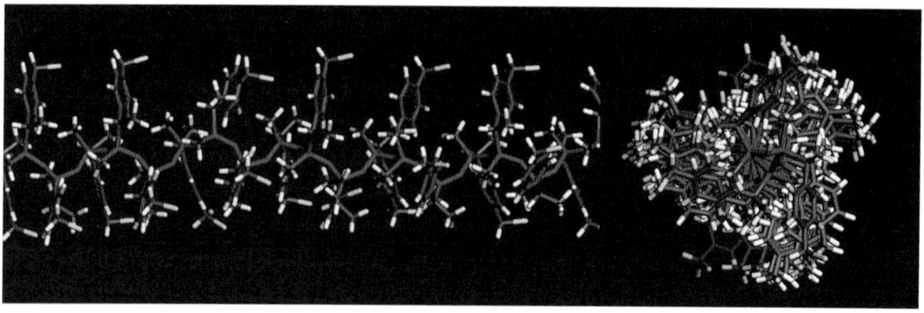

A B

Fig. 4 Minimum energy conformation corresponding to the homopolymer containing only α-structural unit: (**A**) longitudinal view; (**B**) axial view (See also Plate 10 in the Colour Plate Section)

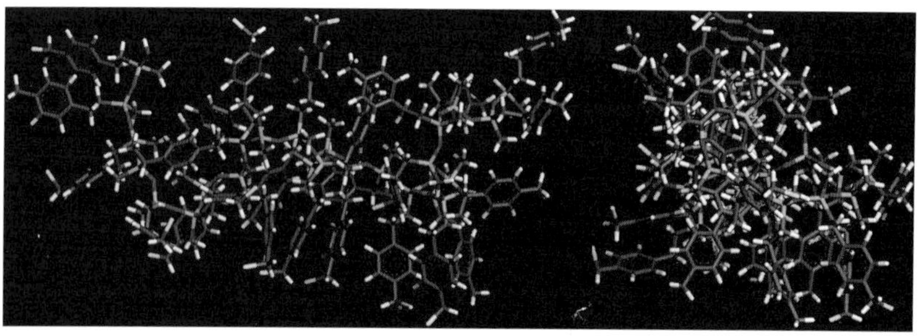

A B

Fig. 5 Minimum energy conformation corresponding to the polysiloxane containing α (25%) and β (75%) structural units: (**A**) longitudinal; (**B**) axial view (See also Plate 11 in the Colour Plate Section)

The presence of the azobenzenic units in the polysiloxanic side-chain is expected to induce an ordering tendency, due to their dipole-moment values. If the azobenzenic structure has not a dipole-moment (0.1 D) the azo-phenolic groups has a dipole moment value of 1.15 D. This value of dipole-moment corresponds to the *trans*-configuration and was estimated by molecular modelling simulation, using Vamp module (AM1 field-force) [22]. In Fig. 6 is presented a polysiloxane with DP=25, modified with azobenzene (>90%) in agreement with the experimental characteristics of the sample 1 (Table 1). As one can see, the displacement of the azobenzenic groups reported the polysiloxanic chain is not regular, due to the presence of different structural units (α and β) types.

The ordering chain tendency induced by the presence of the azo-phenolic groups in the polysiloxanic chain is evidenced by the minimum energy conformation of a homo-polysiloxane containing only β-structural units (completely modified with azobenzene). A very nice helix structure is obtained (Fig. 7) if compared with the unmodified polysiloxane presented in Fig. 3.

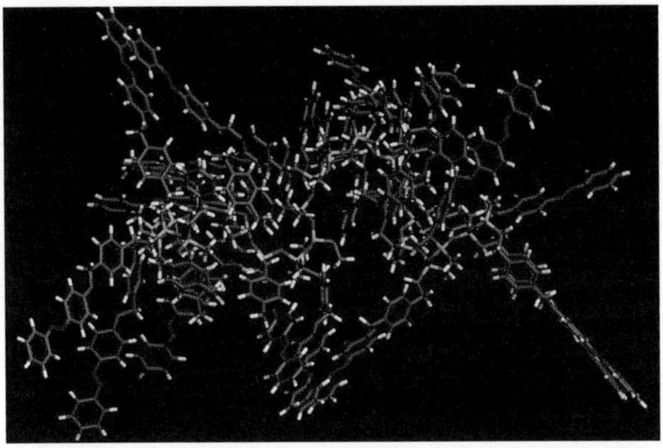

Fig. 6 Minimum energy conformation corresponding to a polysiloxane modified with azobenzene (DP=25) (See also Plate 12 in the Colour Plate Section)

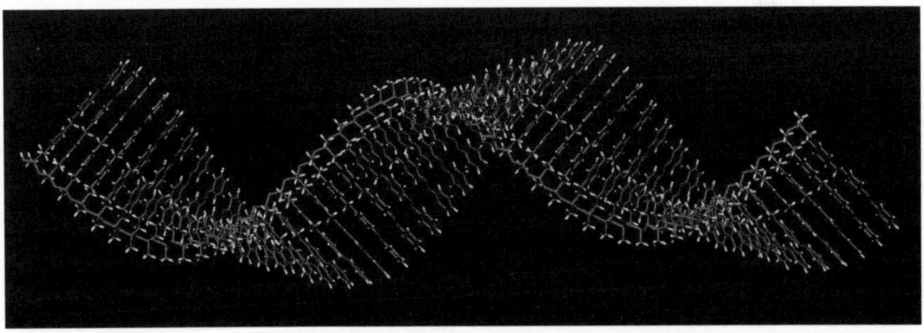

Fig. 7 Chain geometry of an azo-polysiloxane based on β – structural unit type (See also Plate 13 in the Colour Plate Section)

The molecular simulations results are confirmed by the DSC analysis that reflects in the case of Sample 1, the presence of amorphous structures, with a T_g value situated at 35°C. Supplementary comments are necessary concerning the DSC curves profile. As one can see in Fig. 8, a slow exothermal signal appears immediately after the glass transition that can be attributed to an ordering process generated by the azobenzenic group interactions. But the enthalpy corresponding to this exothermal signal is very small (1.086 J/g) reflecting only primary ordering phenomena. This exothermal signal is succeeded by an endothermic one corresponding to a melting process of the ordering phase that appears at 54°C. Similar aggregation and crystallization processes above 50°C, due to azobenzene association dipoles, were reported recently by Yager et al. [25]. But, taking into consideration the signal intensity, only a small quantity of material is engaged in the crystalline phase. This aspect is important for azo-materials applications, because the photoisomerization process can be disturbed by a too small value of the free volume (characteristic of the crystalline phases).

To verify this hypothesis, a careful study concerning the azo-polymers behaviour in the presence of UV/VIS irradiation was carried out. The *trans*- and *cis*-configuration

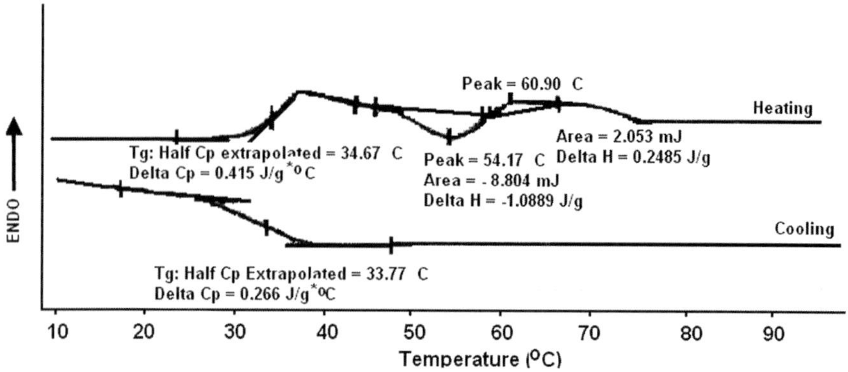

Fig. 8 Second DSC heating/cooling cycle, corresponding to the Sample 1 (rate 10 K/min)

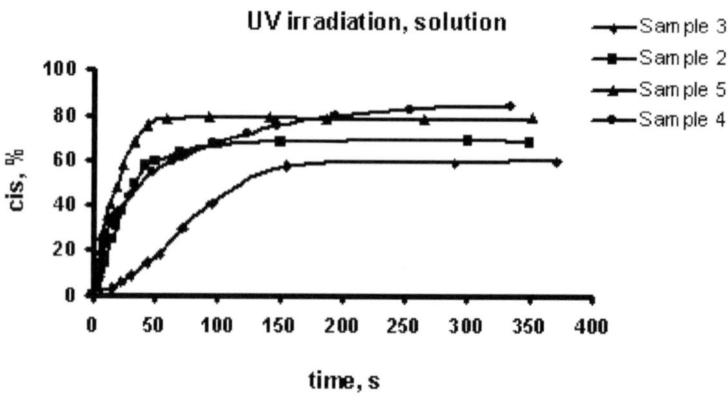

Fig. 9 The kinetic curves corresponding to the *trans-cis* isomerization process in solution (samples 2–5, Table 1)

content during UV irradiation was calculated based on the absorption values measured at 360 nm, characteristic of the π-π* transition of the *trans* isomer. Exposed to daylight or to common neon tubes, the samples contain always a small proportion of remaining *cis* isomer. In order to study *trans/cis* photo-isomerization starting from the 'pure' *trans* state, the samples were maintained in the dark overnight. Upon UV irradiation, the band at 360 nm decreases strongly, the intensity being proportionally with the *trans*-isomer content. In Fig. 9 are presented the kinetic curves corresponding to the *trans-cis* isomerization process under UV irradiation (in solution) of the azo-polysiloxanes modified with nucleobases (Table 1). The best results concerning the response rate under UV irradiation were obtained for the azo-polysiloxane modified with cytosine. One can appreciate that, from the view point of the maximum *trans-cis* conversion degree, the azo-polysiloxanes modified with adenine, cytosine and uracil have similar behaviours. A lower value corresponding both to the maximum conversion degree and to the rate response was obtained in the case of thymine. Taking into consideration that there are no major differences

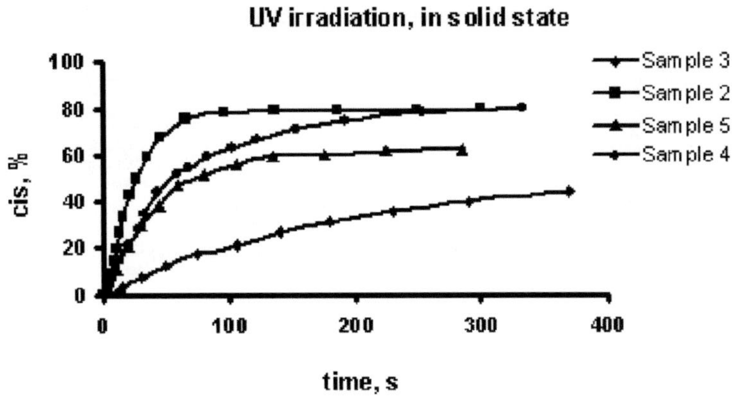

Fig. 10 The kinetic curves corresponding to the *trans-cis* isomerization process in the solid state (samples 2–5, Table 1)

concerning the modification degree between the samples, the only explanation can be the presence of some specific interactions between thymine and azobenzenic groups that disturb the *trans-cis* isomerization process.

Figure 10 presents the kinetic *trans-cis* photoisomerization process, under UV irradiation in the solid state. In this case, significant differences appear between samples behaviour, as a function of the nucleobase chemical structures. It is interesting to note that, in the case of azo-polysiloxane substituted with adenine (sample 2 -Table 1), the behaviours in the solid state and in solution are similar. That means that the polysiloxane chain flexibility, combined with the amorphous polymer ordering assure enough free volume for the *trans-cis* isomerization process.

The decrease of the maximum conversion degree concerning the *cis*-azobenzenic configuration, obtained as a result of UV irradiation, can be explained for the other samples by the decrease of the free volume value, induced by the possibility of the nucleobases to generate H-bonding by themselves. The possibility to form such H-bonds is a known process, especially in the case of thymine [26, 27].

The next step of the photochromic behaviour study was the comparison of the *cis-trans* relaxation process between natural visible light induced relaxation and thermal induced one (in the dark). It must be underlined that no studies concerning the influence of the natural visible light upon the relaxation process are mentioned in the literature. Usually, the relaxation process is studied by using VIS-monochromatic laser sources or in the dark. Figure 11 presents the relaxation kinetic curves, obtained as a result of the natural visible light action.

A very interesting behaviour is obtained for the azo-polysiloxane modified with adenine (Sample 2 – Table 1). In spite of the fact that the *trans-cis* isomerization process, as a result of the UV irradiation, is very fast in the solid state (similar to the behaviour in solution), the relaxation takes place in two steps, the recuperation rate in the first step being only 15 %. This behaviour can be explained by some associations that can take place between the *cis*-azobenzenic groups and adenine. The same curve profile, but less evident, is obtained for the azo-polysiloxane modified with cytosine

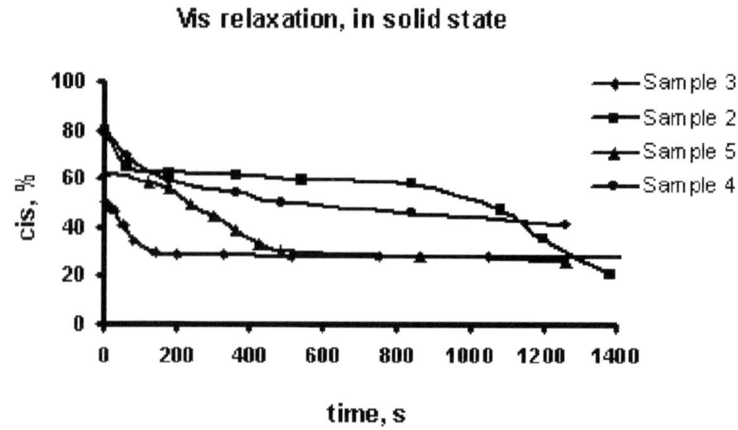

Fig. 11 The kinetic curves corresponding to the *cis-trans* relaxation process, induced by natural visible light, in the solid state (samples 2–5, Table 1)

(Sample 5 – Table 1) which has a chemical structure similar to that of adenine. A faster relaxation process is obtained in the case of the azo-polysiloxane modified with thymine, this behaviour being in agreement with the idea of the H-bonding formation that can create a 3D-physical network which forces the azobenzenic groups to relax faster.

Figure 12 presents the *cis-trans* relaxation process in the dark (only thermal activation) which is much slower that in the presence of natural visible light. For all the systems, an important percentage of relaxation takes place in the first 24–30 h; after this period, the relaxation rate decreases dramatically. As in the case of the natural visible light activated relaxation process, the dark relaxation processes of the azo-polysiloxanes containing adenine and cytosine are very similar.

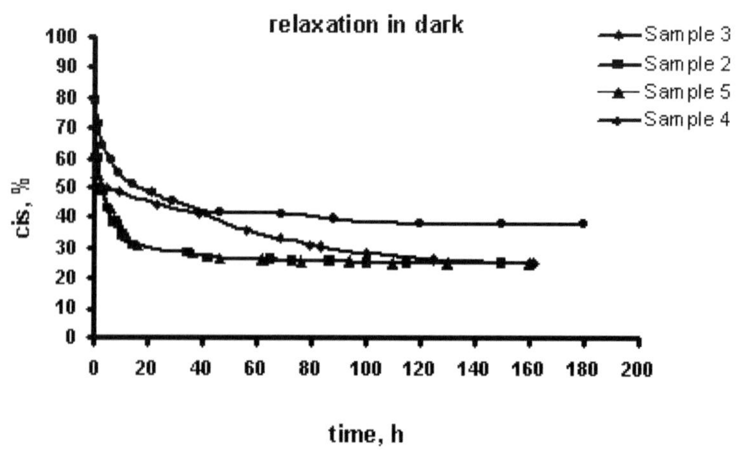

Fig. 12 The kinetic curves corresponding to the *cis-trans* relaxation (in the dark) process on the solid state (samples 2–5, Table 1)

A special comment is necessary regarding the content of *cis*-form of the azobenzenic groups after 150h of relaxation. Our previous studies developed on similar polysiloxanes chains (but fully substituted with azophenol) evidenced a complete *cis-trans* relaxation process after 160h [14]. For the azo-polysiloxanes containing nucleobases, one can observe a high residual value corresponding to the *cis*-form, situated above 20%, even if the relaxation process is done in the presence of the natural visible light. Taking into consideration that this residual percentage is close to the value corresponding to the substitution degree with nucleobases, it is possible to assume some stable associations between nucleobases and the *cis*-form of the azobenzenic groups. Supplementary studies are necessary to elucidate this aspect.

A second azo-polysiloxane polymer group investigated is based on donor/acceptor systems (Samples 5, 6 and 7 – Table 1). The donor/acceptor systems were connected to the polymeric chain using the reaction with the chlorobenzyl groups (Scheme 1). The chemical structures were confirmed by ^{1}H-NMR spectroscopy, the substitution degree being calculated using the signals attributed to the -CH_2Cl groups (4.5 ppm), that shift to 5.0–5.1 ppm after the connection of the azobenzene or donor/acceptor groups. A typical ^{1}H-NMR spectrum is presented in Fig. 13.

Concerning the photochromic behaviour, the situation is different if we compare this polymer group with the azo-polysiloxanes modified with nucleobases, especially concerning the relaxation process. For the photochromic behaviour in solution and in the solid state, there are some differences concerning the maximum conversion degree of the azo-groups in the *cis*-form (about 15–20%). Figs. 14 and 15 present the photochromic response under UV irradiation of the Samples 6–8 in solution and in the solid state, respectively.

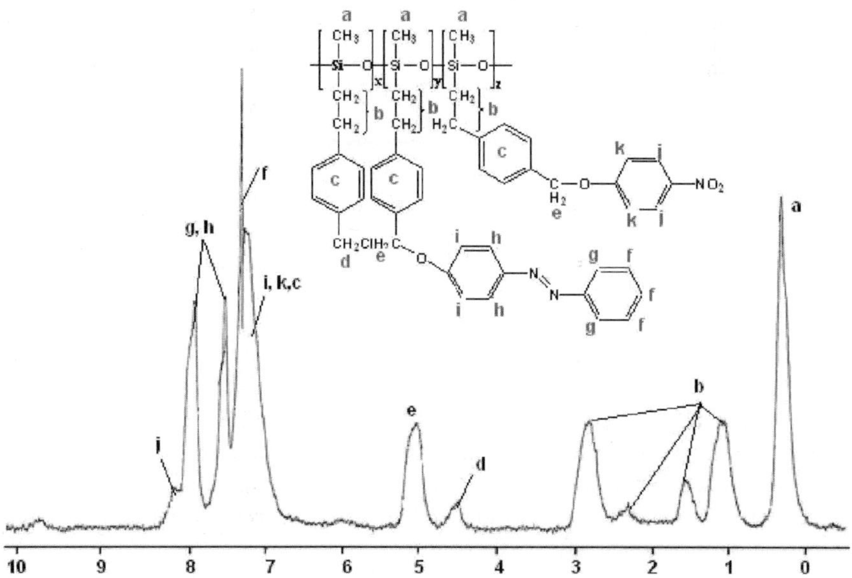

Fig. 13. ^{1}H-NMR spectrum corresponding to an azo-polysiloxane modified with p-nitro-phenol (Sample 6 – Table 1)

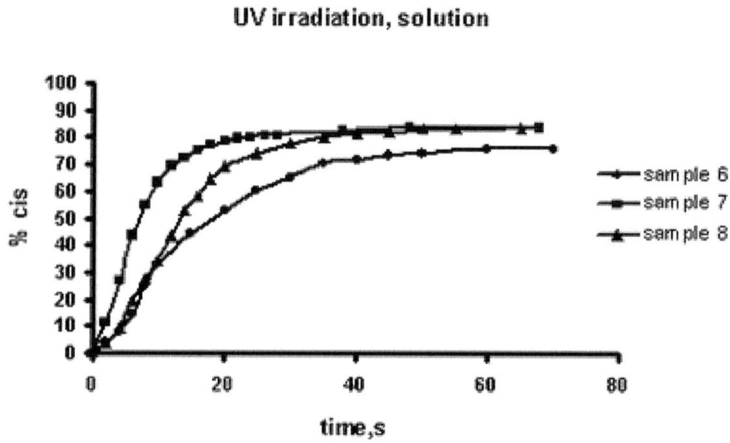

Fig. 14 The kinetic curves corresponding to the *trans-cis* isomerization process in solution corresponding to Samples 6–8 (Table 1)

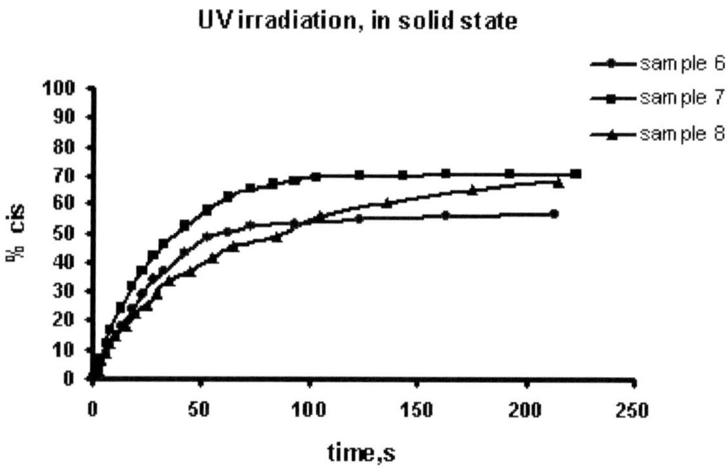

Fig. 15 The kinetic curves corresponding to the *trans-cis* isomerization process in the solid state corresponding to Samples 6–8 (Table 1)

One can observe that the response rate in solution to the UV irradiation is faster for the systems containing naphthalene and anthracene units, comparatively with the *p*-nitrophenol containing azo-polysiloxane. In the solid state, the differences concerning the response rate diminish, but the maximum *cis*-group conversion degree is situated only at about 52%.

If the relaxation process is stimulated by natural visible light, the *cis-trans* isomerization is complete after 500 s (Fig. 16).

One can underline that, in the case of the azo-polysiloxanes containing donor/acceptor groups, the relaxation process is total, no associations between *cis*-azobenzene and donor/acceptor groups (that can disturb the relaxation process) being evidenced.

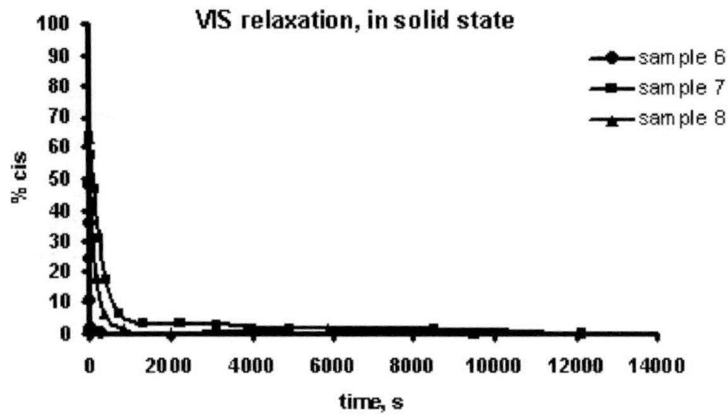

Fig. 16 The kinetic curves corresponding to the *cis-trans* relaxation process, induced by natural visible light, in the solid state, for Samples 6–8 (Table 1)

The next discussed azo-polysiloxane type compounds contain quaternary ammonium groups, the idea being to generate photosensitive micelles, capable to aggregate/disaggregate under UV irradiation [28–30]. Here we report only the possibility to obtain amphiphilic azo-polysiloxanes and their capacity to aggregate in aqueous solution, their photochromic behaviour being under investigation. Both amphiphilic azo-polysiloxanes contain 40% azo and 57–58% quaternary ammonium groups.

To evaluate the aggregation polysiloxane capacity, the classical method using pyrene fluorescence spectroscopy was used [31]. To calculate the critical aggregation concentration (*cac*) value, the first (named I_1) and the third (named I_3) absorption peaks corresponding to the fluorescence pyrene spectrum were used. In aqueous solution, the I_1/I_3 ratio value is situated around 1.70–1.75 [32]. For the two amphiphilic samples investigated (9 and 10 – Table 1), for concentrations lower than 10^{-3} g/L, no aggregation process was evidenced, the I_1/I_3 ratio being ~1.7. By increasing the azo-polysiloxane concentration, the pyrene was progressively incorporated in the hydrophobic core of the aggregates that begin to form in the aqueous solution, the result being reflected by the diminution of the I_1/I_3 ratio value. This diminution is not sudden, as in the case of low-molecular amphiphiles (Fig. 17).

The *cac* values were estimated as the first inflexion point from the curves and they are equal to: 2×10^{-3} g/L for sample 9 and 10^{-2} g/L for sample 10, respectively. These relatively low values can be explained by the presence on the polysiloxane side-chain of the azobenzenic hydrophobic groups with a high aggregation tendency (H-type or J-type – Fig. 18) [33, 34].

Taking into consideration the high flexibility of the polysiloxanic main-chain and the hydrophobic/hydrophilic separation tendency of the side groups (of azobenzenic and quaternary ammonium type) an azobenzenic core formation of the micelles, with J-type aggregation arrangement is expected. This assumption concerning the J-type aggregation process is asserted by the UV-VIS spectral studies that show a slight slow red-shift displacement of the maximum absorption peak from 336 to 342 nm, as a function of the azo-polysiloxane concentration in water (Fig. 19).

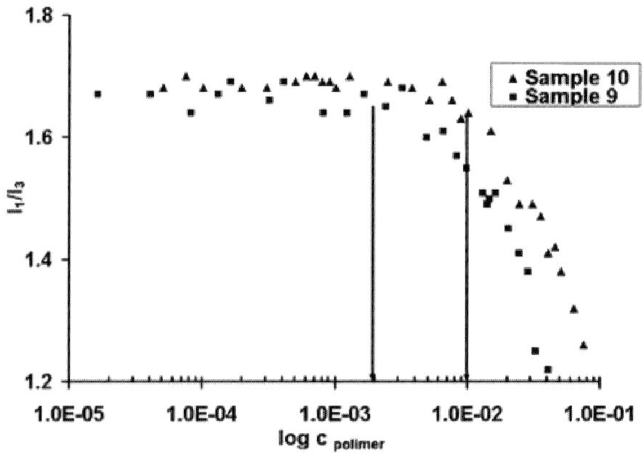

Fig. 17 Plot of the I_1/I_3 ratio as a function of the azo-polysiloxane concentration corresponding to Samples 9 and 10

Fig. 18 H-type and J-type aggregation possibility of the azobenzenic groups

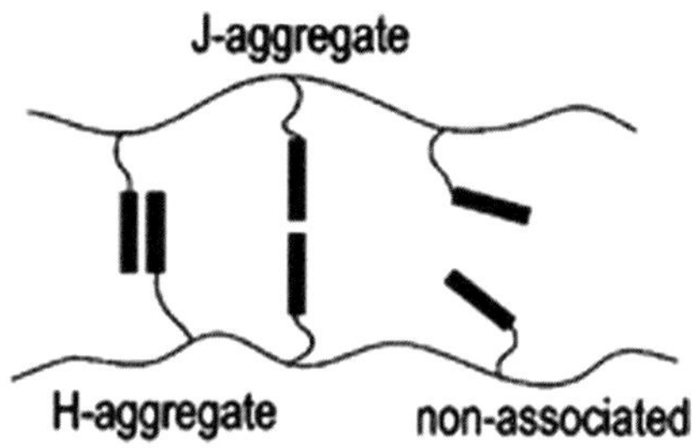

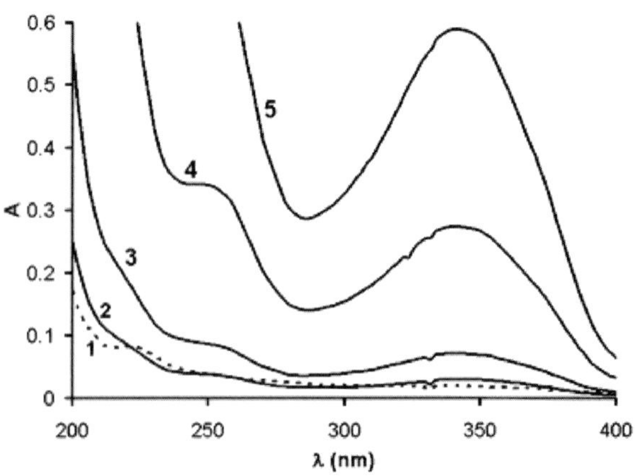

Fig. 19 UV-VIS spectra corresponding to the aqueous solution of the Sample 9: (1) 2.32×10^{-4} g/L; (2) 2.32×10^{-3} g/L; (3) 2.32×10^{-2} g/L; (4) 4.64×10^{-2} g/L; (5) 2.32×10^{-1} g/L

4 Conclusions

The synthesis and characterization of different azo-polysiloxanic compounds containing nucleobases, donor/acceptor or quaternary ammonium groups is reported. These polymeric systems present a potential interest in the field of molecules and biomolecules immobilization and nanomanipulation, taking into consideration the fact that the presence of the azobenzenic groups confers to them the possibility to generate a photo-fluid phase, under UV irradiation. The photochromic behaviour of the azo-polysiloxanes in solution and in the solid state was investigated. Even if the maximum conversion degree from *trans*- to *cis*-groups is lower in the solid state as compared with the solution, the response rates are situated in a similar scale-time, as a consequence of the high flexibility of the polysiloxanic main-chain. The relaxation *cis-trans* process is different for the systems containing nucleobases, as compared with the donor/acceptor groups. In the case of the azo-polysiloxanes containing nucleobases, the relaxation process is not complete neither under natural visible light irradiation, nor in the dark. This phenomenon can be explained by some association processes that take place between *cis*-azobenzenic groups and nucleobases. For the donor/acceptor systems, the relaxation process is complete. In the case of the azo-polysiloxanes containing quaternary ammonium groups, the polymer aggregation capacity was investigated. The critical aggregation concentration is situated at lower values, between 2×10^{-3} and 10^{-2} g/L. These lower values can be explained by the azo-benzenic group aggregation capacity that probably forms a hydrophobic micelle core. The red shift of the 366 nm maximum absorption peak, induced by the micelle formation process, suggests a J-type ordering of the azobenzenic groups inside the core.

Acknowledgements The authors would like to thank the Ministry of Education for the financial support. The research activity was realised in the frame of MATNANTECH Project – CEEX 107/2006 and CNCSIS – GR 100/2007 (code CNCSIS 273). The molecular simulations were effectuated on the frame of Multifunctional, High Performance Polymers – Interdisciplinary Platform for Education and Research (contract no. 69/2006 CNCSIS)

References

1. Natanson A, Rochon P. (2002) Photoinduced motions in azo-containing polymers. Chem Rev 102:4139–4175
2. Hubert C, Malcor E, Maurin I, Nunzi J-M et al. (2002) Microstructuring of polymers using a light-controlled molecular migration processes. Appl Surf Sci 186:29–33
3. Fukuda T, Matsuda H, Shiraga T et al. (2000) Photofabrication of surface relief grating on films of azobenzene polymer with different dye functionalization. Macromolecules 33:4220–4225
4. Yager K G, Barrett C. (2001) All-optical patterning of azo polymer films. Current Opinion in Solid State & Materials Science 5:487–494
5. Calvacanti E A, Shapiro I M, Composto R J, et al. (2002) RGD Peptides immobilized on a mechanically deformable surface promote osteoblast differentiation. J Bone Min Res 17:2130–2140

6. Yoshinari M, Matsuzaka K, Inoue T, Oda Y, Shimono, M. (2004) Effects of multigrooved surfaces on osteoblast-like cells in vitro: Scanning electron microscopic observation and mRNA expression of osteopontin and osteocalcin. J Biomed Mat Res Part A 68 A:227–234
7. Craighead H G, James C D, Turner A M P (2001) Chemical and topographical patterning for directed cell attachment. Current Opinion in Solid State & Materials Science 5: 177–184
8. Cojocariu C, Rochon P. (2004) Light-induced motions in azobenzene-containing polymers. Pure Appl Chem 76:1479–1497
9. Saphiannikova M, Geue T M, Hennenberg O, Morawetz K, Pietsch U. (2004) Linear viscoelastic analysis of formation and relaxation of azobenzene polymer gratings. J Chem Phys 120:4039–4045
10. Yager K G, Barrett C. (2004) Temperature modelling of laser-irradiated azo-polymer thin films. J Chem Phys 120:1089–1096
11. Kim W H, Bihari B, Moody R, Kodali N B, Kumar J, Tripathy S K. (1995) Self-assembled spin-coated and bulk films of a novel poly(diacetylene) as second-order nonlinear optical polymers. Macromolecules 28:642–647
12. Delaire J A, Nakatani K. (2000) Linear and nonlinear optical properties of photochromic molecules and materials. Chem Rev 100:1817–1846
13. Pedersen T G, Johansen P M. (1997) Mean-field theory of photoinduced molecular reorientation in azobenzene liquid crystalline side-chain polymers. Phys Rev Lett 79:2470–2473
14. Hurduc N, Enea R, Scutaru D et al. (2007) Nucleobases modified azo-polysiloxanes, materials with potential application in biomolecules nanomanipulation. J Polym Sci: Part A: Polym Chem 45:4240–4248
15. Karageorgiev P, Neher D, Schulz1 B et al. (2005) From anisotropic photo-fluidity towards nanomanipulation in the optical near-field. Nature Mat 4:699–703
16. Su W, Han K, Luo Y et al. (2007) Formation and photoresponsive properties of giant microvesicles assembled from azobenzene-containing amphiphilic diblock copolymers. Macromol Chem Phys 208:955–963
17. Shang T, Smith K A, Hatton T A. (2003) Photoresponsive surfactants exhibiting unusually large, reversible surface tension changes under varying illumination conditions. Langmuir 19:10764–10773
18. Tong X, Wang G, Soldera A, Zhao Y. (2005) How can azobenzene block copolymer vesicles be dissociated and reformed by light? J Phys Chem B 109:20281–20287
19. Wang G, Tong X, Zhao Y. (2004) Preparation of azobenzene-containing amphiphilic diblock copolymers for light-responsive micellar aggregates. Macromolecules 37:8911–8917
20. Discher D E, Eisenberg A. (2002) Polymer vesicles. Science 297:967–973
21. Haag R. (2004) Supramolecular drug-delivery systems based on polymeric core-shell architecture. Angew Chem Int Ed 43:278–282
22. Kazmierski K, Hurduc N, Sauvet G, Chojnowski J. (2004) Polysiloxanes with chlorobenzyl groups as precursors of new organic-silicone materials. J Polym Sci Part A: Polym Chem 42:1682–1692
23. Materials Studio 4.0., Accelrys Software, Inc., San Diego (licensed to Nicolae Hurduc)
24. Ikeda T. (2003) Photomodulation of liquid crystal orientations for photonic applications. J Mater Chem 13:2037–2057
25. Shishido A, Tsutsumi O, Kanazawa A, Shiono T, Ikeda T, Tamai N. (1997) Distinct photochemical phase transition behavior of azobenzene liquid crystals evaluated by reflection-mode analysis. J Phys Chem B 101:2806–2810
26. Yager K, Tanchak O, Godbout C, Fritzsche H, Barrett C. (2006) Photomechanical effects in azo-polymers studied by neutron reflectometry. Macromolecules 39:9311–9319
27. Keniry M A, Owen E, Shafer R. (1997) The contribution of thymine-thymine interactions to the stability of folded dimeric quadruplexes. Nucleic Acids Res. 25:4389–4392
28. Lutz J-F, Thunemann A F, Nehring R. (2005) Preparation by controlled radical polymerization and self-assembly via base-recognition of synthetic polymers bearing complementary nucleobases. J Polym Sci Part A: Polym Chem 43:4805–4818

29. Wang G, Tong X, Zhao Y. (2004) Preparation of azobenzene-containing amphiphilic diblock copolymers for light-responsive micellar aggregates. Macromolecules 37:8911–8917
30. Jiang J, Tong X, Morris D, Zhao Y. (2006) Toward photocontrolled release using light-dissociable bloc copolymer micelles. Macromolecules 39:4633–4640
31. Zhao Y. (2007) Rational design of light-controllable polymer micelles. Chem Record 7:286–294
32. Lin Y, Alexandridis P. (2002) Self-Assembly of an Amphiphilic Siloxane Graft Copolymer in Water. J Phys Chem B 106:10845–10853
33. Kalyanasundaram K, Thomas J K. (1977) Environmental effects on vibronic band intensities in pyrene monomer fluorescence and their application in studies of micellar systems. J Am Chem Soc 99:2039–2044
34. Kuiper J M, Engberts J B. (2004) H-Aggregation of azobenzene-substituted amphiphiles in vesicular membranes. Langmuir 20:1152–1160
35. Tong X, Cui L, Zhao Y. (2004) Confinement effects on photoalignment, photochemical phase transition and thermochromic behavior of liquid crystalline azobenzene-containing diblock copolymers. Macromolecules 37:3101–3112

To evaluate the best ratio between acceptor and donor groups, we carried out a UV-Vis analysis of chloroform solutions of the complex, with a concentration of donor and/or acceptor groups of 0.05 N. The spectra of the two primary products and of the charge transfer complex are presented on Fig. 4.

The Job's method [18, 26] consists in plotting, for different wavelengths, the values of UV absorbance at various acceptor/donor molar fractions, while keeping constant the total molar concentration of the solution [donor] + [acceptor] (here 0.05 N). The molar fraction ratio at the maximum absorbance point corresponds to the stoichiometry of the charge transfer complex. Figure 5 shows the Job plot for charge transfer complexation between α,ω-(3,5-dinitrobenzoate) PDMS and free 9H-carbazole-9-ethanol. The curves in Fig. 5 exhibit a maximum when the donor and the acceptor are in equimolar ratio, suggesting the formation of a 1:1 mole/mole complex, thus confirming the assumption made previously. [26] However, the fact that Job's curves are not symmetrical also prove that a larger complex is able to

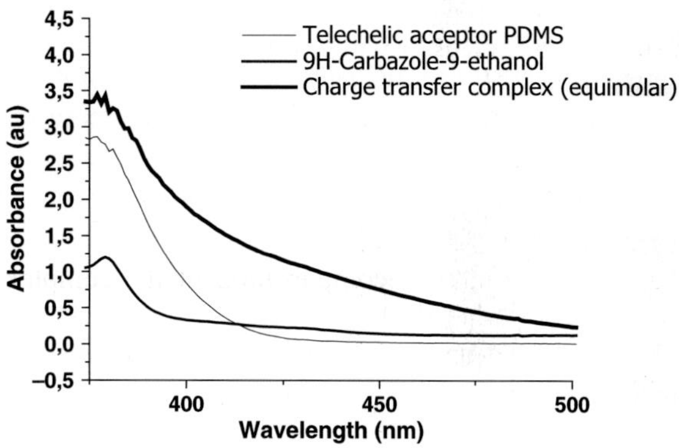

Fig. 4 CUV-vis spectra of 9H-carbazole-9-ethanol, α,ω-(3,5-dinitrobenzoate) PDMS and charge transfer complex in stoechiometric conditions

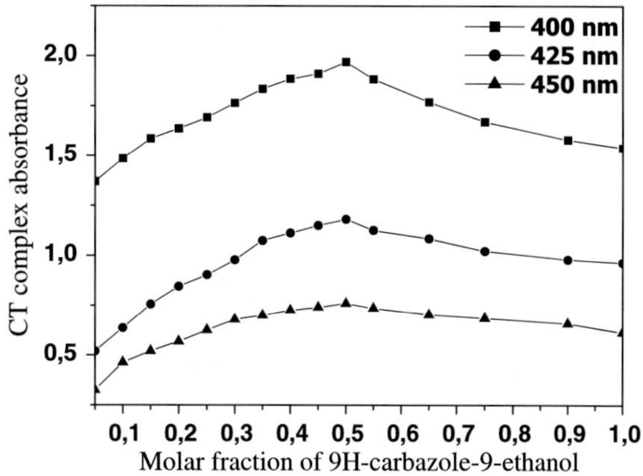

Fig. 5 UV-Vis Job plot for chloroform solutions of α,ω-(3,5-dinitrobenzoate) PDMS/9H-carbazole-9-ethanol between 400 and 450 nm

form in solution, thus explaining the viscosification of the charge transfer complex while removing the solvent.

3.2 *PCzEMA-b-PDMS-b-PCzEMA Triblock Copolymer Synthesis and Complexation with Dinitrobenzoate PDMS*

The monomer 2-(N-carbazolyl)ethyl methacrylate (CzEMA) was first prepared by reacting 9H-carbazole-9-ethanol with methacryloyl chloride, using an excess of triethylamine, in dichloromethane at 0°C. The yields of reaction were not very high (typically 60%) due to the presence of the nitrogen atom on the monomer, which can trap a proton during the reaction. This protonated monomer is eliminated while making several aqueous washes to remove the excess of salts in the product. The purity of the final monomer was demonstrated by ^{1}H-NMR (Fig. 6).

Besides, a di-bromo PDMS macroinitiator was prepared according to two reaction steps given in Scheme 2: first, an esterification of a α,ω-(3-hydroxypropyl) PDMS by bromoisobutyril bromide in dichloromethane in the presence of triethylamine (yields > 99%); second, a redistribution of this macroinitiator in the presence of D_4 and a strong acid resin catalyst to increase the average molar mass of the

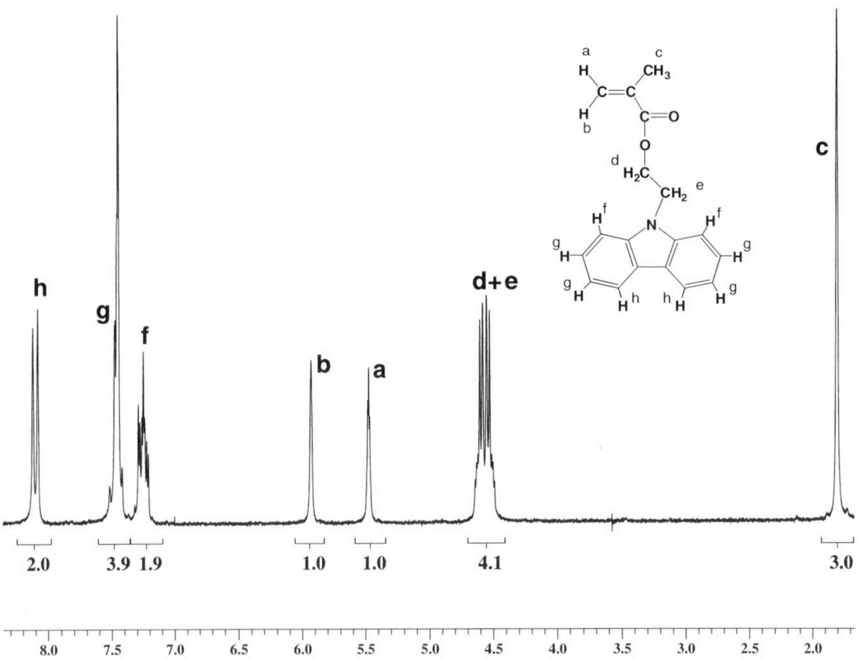

Fig. 6 ^{1}H-NMR of 2-(N-carbazolyl)ethyl methacrylate

Thermoreversible Crosslinking of Silicones Using Acceptor-Donor Interactions

Emmanuel Pouget, François Ganachaud, and Bernard Boutevin

Abstract Partial charge transfer between donor and acceptor groups has been used as a tool to physically structure silicone materials. In a first part, interactions between a single donor molecule, i.e. 9H-carbazole-9-ethanol, and a poly(dimethylsiloxane) functionalized on both chain-ends by acceptor groups, i.e. 3,5-dinitrobenzoate, were studied by DSC and UV-Vis analysis. The thermoreversibility of the interactions was likely demonstrated by UV-Vis analyses, which also showed a most favourable association of the complex for a 1/1 stoechiometry. The second part of this work deals with the synthesis of a triblock copolymer poly (2-(N-carbazolyl) ethyl methacrylate)-*b*-poly(dimethylsiloxane)-*b*-poly(2-(N-carbazolyl) ethyl methacrylate) by atom transfer radical polymerization in the presence of copper bromide and 1,1,4,7,10,10-hexamethyltriethylene tetramine in THF. The association of this triblock copolymer with the telechelic acceptor–based poly(dimethylsiloxane) used in the model study has been checked by rheology, which showed on heating and cooling a thermoreversible elastic/viscoelastic transition.

1 Introduction

Current silicone elastomers produced in the industry are made of long chemically crosslinked PDMS chains in which silica fillers are introduced to increase the mechanical properties of the material, thus preventing its recycling. Recently, Wacker Company commercialized a non-chemically crosslinked silicone elastomer

F. Ganachaud
Institut Charles Gerhardt – UMR 5253 CNRS, Ingénierie et Architectures Macromoléculaires, Ecole Nationale Supérieure de Chimie de Montpellier, 8, rue de l'école normale, 34296 Montpellier cedex 5, France
e-mail: ganachau@enscm.fr

F. Ganachaud et al. (eds.), *Silicon Based Polymers*,

called Geniomer®, [1–4] a multiblock copolymer made of urea-PDMS sequences. Supramolecular interactions [5–8] between urea or urethane groups, based on strong plural hydrogen bonding, has already been shown to structure silicones for quite a long time. For instance, Yilgor et al. and others [9–16] demonstrated that the strength of the hydrogen bonds had a direct influence on mechanical and thermal properties of the material, silicone-urethane systems being less efficient than silicone – urea systems. Also, the phase segregation between hard segments, namely crystalline urea or urethane, and soft poly(dimethylsiloxane) segments generates a thermoplastic elastomer, without the need of adding silica particles. Colombani et al. [17] studied slightly different supramolecular PDMS, introducing bisurethanes as side groups. They observed also a great improvement of mechanical properties, ascribed to an organization of the material as detected by Small Angle X-Ray diffraction. They also showed that, between 100 and 140°C, the material is easy to reprocess whereas at higher temperatures, degradation reactions produced isocyanate and amine groups back, which, by anarchic reactions, chemically crosslinked the material.

In this context, we ought to develop physically crosslinked silicone materials, starting from PDMS functionalized by carbazole (electron donor) and dinitrobenzene (electron acceptor) groups. The partial electron transfer between donor and acceptor groups has been reviewed by Simionescu et al. [18], and its application to the compatibilization and enhancement of mechanical properties of functionalized polymers described in several research articles [19–24]. Particularly, the glass transition temperature (T_g) and the temperature of decomplexation (T_m) were the highest while mixing a (poly(2-(3,5-dinitrobenzoyloxy)ethyl methacrylate) (acceptor polymer) and a (poly[2-(9-ethyl) carbazolyl methyl methacrylate]) (donor polymer) of low molar masses. [21] To our knowledge, only one report of Zentel et al. [25] studied some PDMS modified by acceptor/donor groups. A series of poly(dimethylsiloxane)s (PDMS) was functionalized either by electron donor (N-ethyl carbazole) or electron acceptor (3,5-dinitrobenzene) groups, statistically arranged along the polymer chain or grafted at the chain-ends (telechelic polymers). Blending of the donor- and acceptor-modified statistical copolymers led to a strong increase of the melt viscosity as a consequence of complex formation. In contrast to this, no significant enhancement of the viscosity was observed in blends of telechelic polymers, since only a linear chain extension was believed to occur in this case.

In this paper, we first present a model study on blending a α,ω-3,5-dinitrobenzoate PDMS and free 9H-carbazole-9-ethanol, in order to check whether the recently proposed 1:1 stoechiometry between carbazole and dinitrobenzoate molecules indeed applies (Scheme 1). [26] Then, we describe the preparation of triblock copolymers, poly[2-(N-carbazolyl)ethyl methacrylate]-*b*-PDMS-*b*-poly[2-(N-carbazolyl)ethyl methacrylate] (P(CzEMA)-*b*-PDMS-*b*-P(CzEMA)), using Atom Transfer Radical Polymerization (ATRP), and their blending with the acceptor-functionalized PDMS. In both studies, the physical association and thermal reversibility of these were followed by different techniques, including UV-Vis spectroscopy, DSC or Rheology.

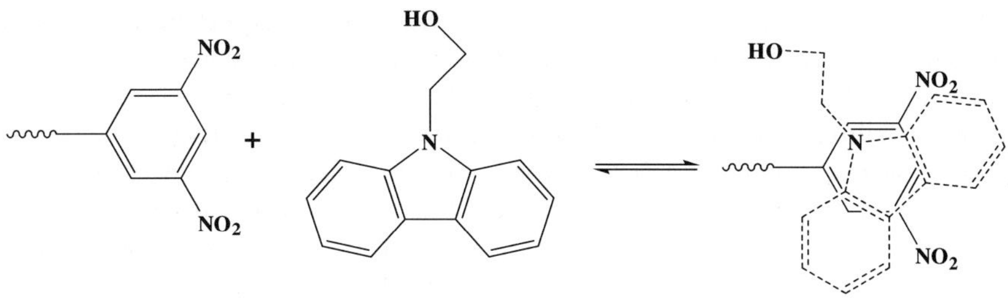

Scheme 1 Proposed geometry of the complex between a dinitrobenzoate and 9H-carbazole molecules, as given in ref. [26]

2 Experimental Part

Materials. The α,ω-(3-hydroxypropyl)polydimethylsiloxane was a gift of Rhodia Silicones. All other chemicals were purchased from Aldrich, unless stated. 3,5-dinitrobenzoyl chloride (98%), 9H-carbazole-9-ethanol (95%), dichloromethane (for analysis), octamethylcyclotetrasiloxane (D_4, Shin-Etsu, 99%), TONSIL EXE0096 (Süd-Chemie), 2-bromoisobutyryl bromide (98%), HMTETA (1,1,4,7,10,10-hexamethyltriethylene tetramine, 98%), copper bromide (98%) and triethylamine (98%) were used as received. Methacryloyl chloride (97%) was distilled before use. All reactions were carried out under inert nitrogen atmosphere.

Synthesis of the α,ω-(3,5-dinitrobenzoate)PDMS. Typically, 10 g of α,ω-(3-hydroxypropyl)polydimethylsiloxane ($M_{n,SEC} \approx 970\,g.mol^{-1}$, 10.3 mmol) were reacted with 5.229 g of 3,5-dinitrobenzoyl chloride ($M = 230.56\,g.mol^{-1}$, 22.6 mmol) in 50 mL dichloromethane at 0°C during 1 h and then at room temperature overnight. The solution of the 3,5-dinitrobenzoyl chloride was added dropwise onto the solution of the PDMS containing 2.29 g of triethylamine ($M = 101.19\,g.mol^{-1}$, 22.6 mmol) used to trap hydrochloric acid formed during the reaction. After filtration of the triethylammonium chloride, an extraction of the excess of 3,5-dinitrobenzoyl chloride by a basic aqueous solution was carried out. After drying over $MgSO_4$, the product was recovered by removal of dichloromethane under vacuum. The reaction is quantitative (>98%).

Synthesis of the α,ω-bromoPDMS macroinitiator. Typically, 10 g of α,ω-(3-hydroxypropyl)polydimethylsiloxane ($M_{n,SEC} \approx 970\,g.mol^{-1}$, 10.3 mmol) were reacted with 5.21 g of 2-bromoisobutyryl bromide ($M = 229.9\,g.mol^{-1}$, 22.6 mmol) in the presence of 2.29 g of triethylamine ($M = 101.19\,g.mol^{-1}$, 22.6 mmol) in 50 mL dichloromethane at 0°C during 3 h and then at room temperature overnight. The 2-bromoisobutyryl bromide was added drop-by-drop onto the solution of triethylamine and PDMS. After filtration of the triethylammonium bromide salt formed during the reaction, a basic extraction was carried out to remove the excess of 2-bromoisobutyryl bromide. After drying, the pure di-bromo telechelic PDMS macroinitiator was recovered by removal of dichloromethane under vacuum.

Redistribution of the PDMS macroinitiator with octamethylcyclotetrasiloxane (D_4). This reaction allowed to increase the molecular weight of the PDMS macroinitiator. Typically, the PDMS macroinitiator synthesized during the previous step ($M_n = 1270\,g.mol^{-1}$, 1.6 g, 1.26 mmol) was introduced in a round-bottom flask equipped with a magnetic stirrer. 23 g of D_4 ($M = 296.6\,g.mol^{-1}$, 77 mmol) and 0.23 g of TONSIL (a sulfonic acid modified clay) were added to the flask. The solution was stirred and heated at 80°C during 72 h. A pure product was recovered by filtration of the catalyst and precipitation of the polymer in excess methanol. Results: $M_{n,th} = 19400\,g.mol^{-1}$, $M_{n,RMN} = 21100\,g.mol^{-1}$, $M_{n,SEC} = 20900\,g.mol^{-1}$, PDI = 1.49, yield: 85%.

Synthesis of 2-(N-carbazolyl)ethyl methacrylate. 9H-carbazole-9-ethanol ($M = 211.3\,g.mol^{-1}$, 10 g, 47.3 mmol) was added dropwise to a solution of freshly distilled methacryloyl chloride ($M = 104.5\,g.mol^{-1}$, 14.8 g, 141.2 mmol) and triethylamine ($M = 101.19\,g.mol^{-1}$, 5.05 g, 50 mmol) in 100 mL dichloromethane at 0°C during 1 h. Then the solution was stirred overnight at room temperature. After filtration, the solution was extracted with a slightly basic aqueous solution to remove the excess of methacryloyl chloride. After drying over $MgSO_4$, the pure product was recovered by removal of dichloromethane under vacuum at room temperature (yield: 60%).

Atom Transfer Radical Polymerization of 2-(N-carbazolyl)ethyl methacrylate. The conditions of polymerization were adapted from the study published by Zeng et al. [27] The ATRP of CzEMA (277 $g.mol^{-1}$, 0.3437 g, 1.24 mmol) was carried out at 60°C under argon atmosphere in THF, in the presence of copper bromide (142.7 $g.mol^{-1}$, 13.7 mg, 96 μmol) and HMTETA (229.2 $g.mol^{-1}$, 22 mg, 96 μmol). The solution of the PDMS macroinitiator ($Mn = 20900\,g.mol^{-1}$, 1 g, $4.79\ 10^{-5}$ mol) dissolved in 1 mL of THF was bubbled under argon during 15 min, and then introduced in the round-bottom flask containing the solution of CuBr, HMTETA and CzEMA in 3 mL THF, previously freeze-thawed three times. After 7 h at 60°C, the monomer conversion was quantitative. The polymer was precipitated in methanol, filtrated, dissolved in a little amount of THF and passed through a silica column to remove the copper species. THF was then removed under vacuum before characterization.

Characterizations. Size Exclusion Chromatography (SEC) apparatus was composed of a SpectroSeries P100 pump equipped with a Shodex Rise-61 refractometer detector and two 300 mm columns thermostated at 30°C (mixed-D PL-gel 5μm columns from Polymer Laboratories: $2 \times 10^2 - 4 \times 10^5\,g.mol^{-1}$ molecular weight range). Toluene was used as the eluent at a flow rate of $0.8\,mL.min^{-1}$. Calibration was performed with polystyrene standards from Polymer Laboratories. ^1H NMR analyses were performed in $CDCl_3$ on a Bruker Advance 250 MHz. UV-Vis analyses were performed on a spectrometer UV-Vis mono-beam HP8453. Rheology experiments were carried out on a TA instruments AR1000 apparatus between 15°C and 205°C. Every 10°C, after 1 minute of equilibration, 10 measurements were made

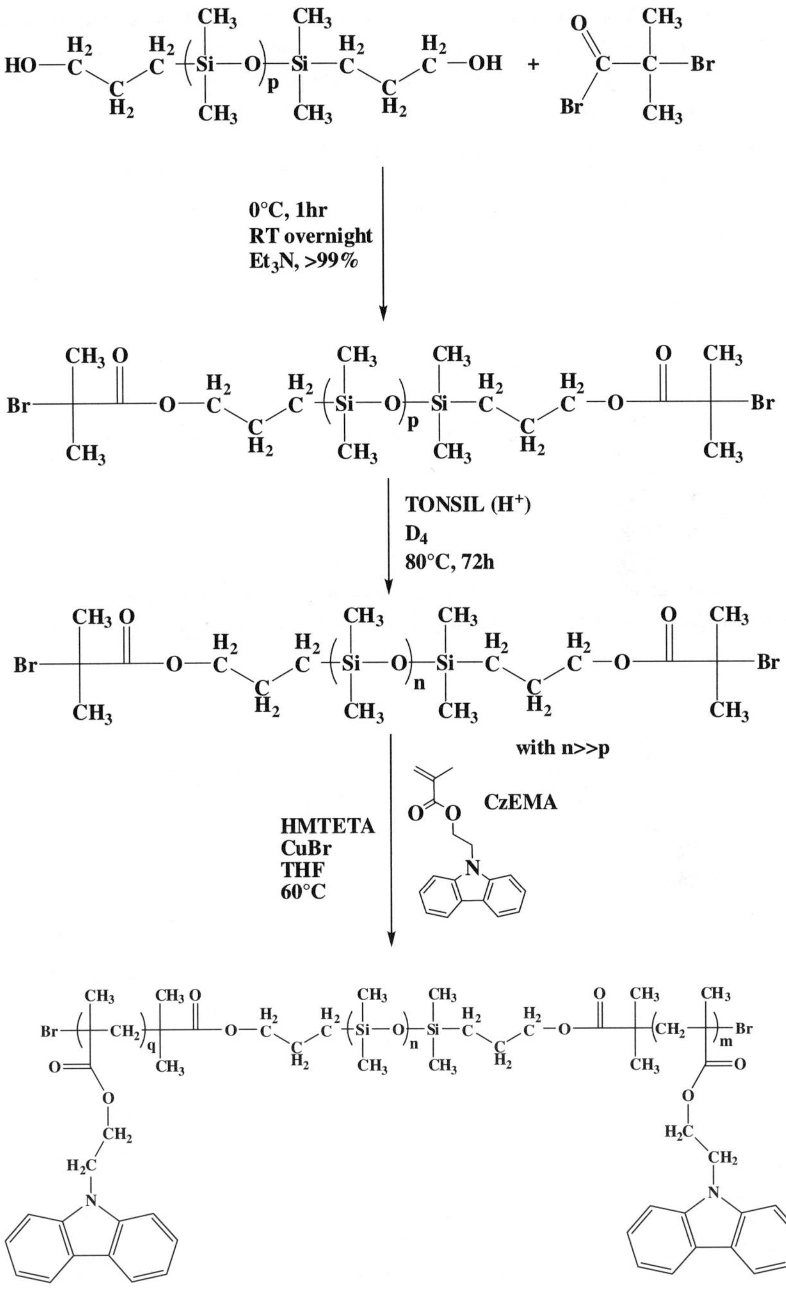

Scheme 2 General scheme for the synthesis of the PCzEMA-b-PDMS-b-PCzEMA triblock copolymer. HMTETA is 1,1,4,7,10,10-hexamethyltriethylene tetramine

functionalized PDMS. After removal of the catalyst by filtration of the excess of D_4 by precipitation of the polymer, this latter has been analyzed by ^{1}H-NMR (Fig. 7) and size exclusion chromatography. The molecular weights obtained by NMR ($M_{n,NMR}$ = 21100 g.mol^{-1}) and size exclusion chromatography ($M_{n,SEC}$ = 20900 g.mol^{-1}) were found very close, which confirmed that the functionality of the polymer is retained during the redistribution reaction. Note that the macroinitiator could not be prepared the other way round, since the functionalization of long PDMS chains is unlikely.

The polymerization of CzEMA by ATRP, starting from the PDMS macroinitiator, was adapted from the protocol described by Zeng et al. [27] applied to the ATRP of 2-(dimethylamino)ethyl methacrylate, which has a structure quite similar to the CzEMA, i.e. in THF at 60°C using HMTETA and CuBr. The desired triblock copolymer PCzEMA-*b*-PDMS-*b*-PCzEMA was obtained in very good yields (monomer conversion >99%), and the final product analyzed by SEC (not shown) and ^{1}H-NMR (Fig. 8).

By SEC, we observed that the full distribution of PDMS chains shifted to large molar masses, thus proving the good initiation efficiency of such macroinitator. The final polydispersity index (M_w/M_n = 1.25) is smaller than the polydispersity index of the PDMS macroinitiator (1.49), another evidence that all PDMS chains were functionalized. The final molecular weight obtained by SEC for the triblock copolymer

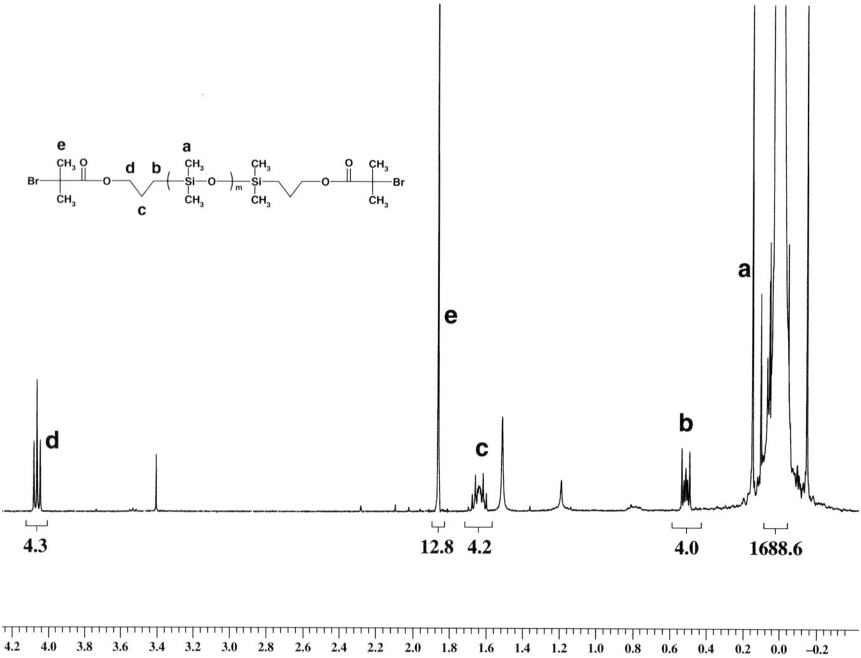

Fig. 7 ^{1}H-NMR of the di-bromo PDMS macroinitiator for the ATRP of CzEMA

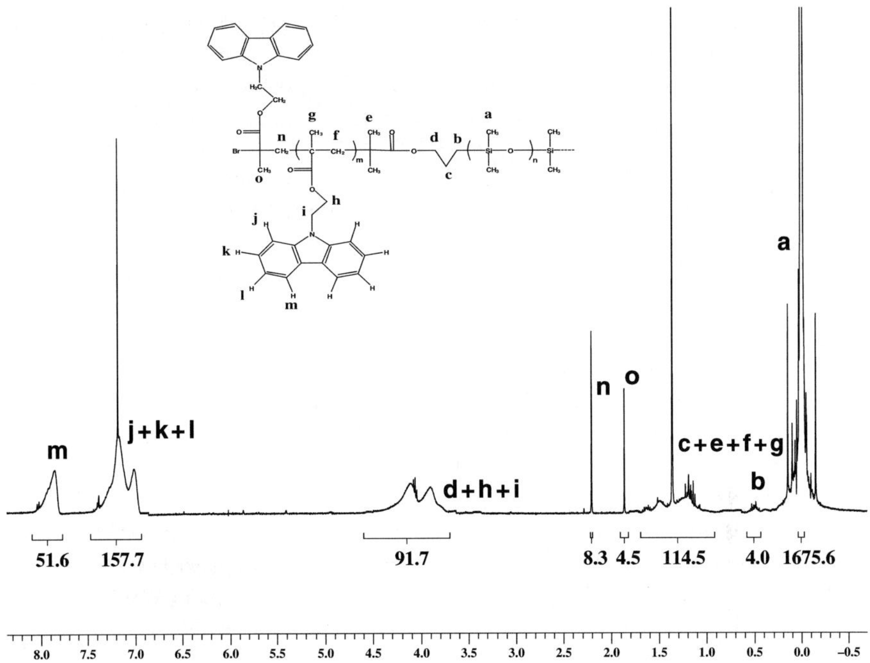

Fig. 8 ^{1}H-NMR of the PCzEMA-*b*-PDMS-*b*-PCzEMA triblock copolymer

($M_{n,SEC}$ = 31000 g.mol^{-1}) was higher than the theoretical molecular weight at the conversion (28000 g.mol^{-1} at 99%), since molecular weights were calculated on a polystyrene standard calibration.

This triblock copolymer was associated in dichloromethane, with the α,ω-(3,5-dinitrobenzoate) PDMS used in the model study, at a ratio between acceptor and donor groups strictly equal to 1. After solvent removal, a soft and lightly sticky material was obtained, again with a bright orange colour. No demixions were observed even after several cycles of heating and cooling. Since the DSC did not detect any complexation transitions, we directed our interests towards rheological experiments to unravel the thermal properties of our materials. The complex was submitted to one heating step and one cooling step cycle between 25 and 220°C, and measurements at different frequencies produced the master curves presented in Fig. 9.

Three domains were observed on these master curves, which overlapped perfectly for heating and cooling cycles. On heating, the material is viscous at high temperatures (>105°C). Between 105°C and 45°C, the material is elastic whereas under 45°C, the material is glassy. Such behaviour is typical of a thermoplastic elastomer, where here the transformation from an elastic solid to a viscous liquid is due to the decomplexation between acceptor and donor groups. Since a small number of interactions are exerted between acceptor and donor groups, the elastic domain is not very wide; besides, the glass transition zone is very low (<45°C) due to the presence of short PDMS chains acting as plasticizing agents. The fact that the viscous modulus G” did

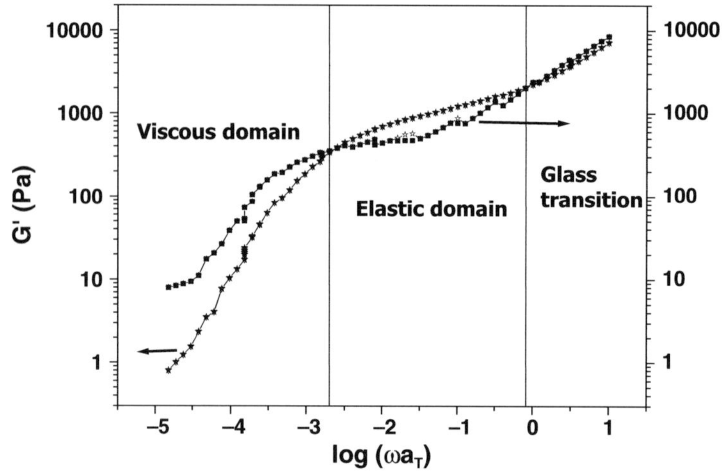

Fig. 9 Master curves obtained while heating (empty symbols) and cooling (full symbols) the material created by the association of α,ω-3,5-dinitrobenzoate PDMS with the PCzEMA-*b*-PDMS-*b*-PCzEMA triblock copolymer. Reference temperature: 25°C, (■) elastic modulus G', (*) viscous modulus G''

not show a minimum let one think that the elasticity of the material is due to the organization, rather than the entanglement, of the supramolecular chains. On cooling, the characteristic temperatures of the domains were found slightly different, as generally observed in DSC: the elastic domain is present between 95 and 30°C. This behaviour was reproducible upon heating and cooling few times, although some degradation reactions tended to occur at higher temperatures.

4 Conclusion

In conclusion, this study showed the possibility to thermoreversibly crosslink silicone elastomers using acceptor – donor interactions. The interactions are highly cooperative since a donor solid organic molecule, i.e. 9H-carbazole-9-ethanol, and a PDMS containing acceptor groups (3,5-dinitrobenzoyl groups) at the chain-ends are perfectly mixed together. A blend between this modified PDMS and a PCzEMA-*b*-PDMS-*b*-PCzEMA triblock was found stable on heating and cooling. The thermoreversibility of the interactions has been proved by rheological analysis. The preparation of blends between the triblock copolymer prepared in this study and PDMS containing many acceptor groups would be a solution to enlarge the elastic domain.

Acknowledgments The authors gratefully acknowledge Dr. Martin In for fruitful discussions on rheology experiments and Pascale Guiffrey for UV-Vis analysis. Rhodia silicones company is also acknowledged for the gift of the hydroxypropyl PDMS.

References

1. Hohberg T, Schaefer O, Mueller J et al. (2003) Textiles coated or finished with polysiloxane-polyurea-polyurethane block copolymers (Wacker-Chemie G.m.b.H., Germany) EP1336683.
2. Kuepfer J, Schaefer O. (2004) Polysiloxane-polyurea-polyurethane block copolymers and their preparation (Wacker-Chemie G.m.b.H., Germany) US2004254325.
3. Scheim U, Ziche W. (2006) Crosslinkable siloxane urea copolymers for molding compositions (Wacker-Chemie G.m.b.H., Germany) WO2006010486.
4. Schaefer O, Weis J, Delica S et al. (2004) Thermoplastic silicone elastomers. Polym. Prepr. (Am. Chem. Soc., Div. Polym. Chem.) 45:714–715.
5. Brunsveld L, Folmer BJB, Meijer EW et al. (2001) Supramolecular Polymers. Chem. Rev. (Washington, D. C.) 101:4071–4097.
6. Binder WH, Zirbs R. (2007) Supramolecular polymers and networks with hydrogen bonds in the main- and side-chain. Adv. Polym. Sci. 207:1–78.
7. Bouteiller L. (2007) Assembly via hydrogen bonds of low molar mass compounds into supramolecular polymers. Adv. Polym. Sci. 207:79–112.
8. ten Brinke G, Ruokolainen J, Ikkala O. (2007) Supramolecular materials based on hydrogen-bonded polymers. Adv. Polym. Sci. 207:113–177.
9. Kozakiewicz J. (1998) Siloxane oligomer diols as potential intermediates for novel durable coatings. Surf. Coat. Int. 81:435–439.
10. Kozakiewicz J. (1996) Polysiloxaneurethanes: new polymers for potential coating applications. Prog. Org. Coat. 27:123–131.
11. Yilgor E, Tulpar A, Kara S et al. (2000) High strength silicone-urethane copolymers: synthesis and properties. ACS Symp. Ser. 729:395–407.
12. Tyagi D, Wilkes GL, Yilgor I et al. (1982) Siloxane-urea segmented copolymers. 2. Investigation of mechanical behavior. Polym. Bull. (Berlin) 8:543–550.
13. Yilgor I, Riffle JS, Wilkes GL et al. (1982) Siloxane-urea segmented copolymers. 1. Synthesis and characterization of model polymers from MDI and α,ω-bis(aminopropyl)polydimethylsil oxane. Polym. Bull. (Berlin) 8:535–542.
14. Yilgor E, Yilgor I. (2001) Hydrogen bonding: a critical parameter in designing silicone copolymers. Polymer, 42:7953–7959.
15. Yilgor E, Ekin Atilla G, Ekin A et al. (2003) Isopropyl alcohol: an unusual, powerful, green' solvent for the preparation of silicone-urea copolymers with high urea contents. Polymer 44:7787–7793.
16. Sheth JP, Aneja A, Wilkes GL et al. (2004) Influence of system variables on the morphological and dynamic mechanical behavior of polydimethylsiloxane based segmented polyurethane and polyurea copolymers: a comparative perspective. Polymer 45:6919–6932.
17. Colombani O, Barioz C, Bouteiller L et al. (2005) Attempt toward 1D cross-linked thermoplastic elastomers: structure and mechanical properties of a new system Macromolecules 38:1752–1759.
18. Simionescu CI, Grigoras M. (1991) Macromolecular donor-acceptor complexes. Prog. Polym. Sci. 16:907–976.
19. Rodriguez-Parada JM, Percec V. (1986) Interchain electron donor-acceptor complexes: a model to study polymer-polymer miscibility? Macromolecules 19:55–64.
20. Simmons A, Natansohn A. (1991) Solid-state NMR study of charge-transfer interactions in polymer blends. Macromolecules 24:3651–3661.
21. Pugh C, Rodriguez-Parada JM, Percec V. (1986) The influence of molecular weight of the donor polymer on the solid-state behavior of interchain EDA complexes. J. Polym. Sci. Part A: Polym. Chem. 24:747–758.
22. Schneider HA, Cantow HJ, Massen U et al. (1982) Donor-acceptor complexation in macromolecular systems. 2. Synthesis and viscoelastic properties of donor-acceptor complexed poly(methyl methacrylate)s and poly(butyl methacrylate)s. Polym. Bull. (Berlin, Germany) 7:263–270.

23. Schneider HA, Cantow HJ, Percec V. (1982) Donor-acceptor complexation in macromolecular systems. 1. Viscoelastic properties of polydonor-polyacceptor blends and of corresponding copolymers. Polym. Bull. (Berlin, Germany) 6:617–621.
24. Schneider HA, Cantow HJ, Percec V. (1982) Viscoelastic properties of poly-donor-poly-acceptor blends and of corresponding copolymers. Polym. Prepr. (Am. Chem. Soc., Div. Polym. Chem.) 23:203–204.
25. Zentel R, Wu J, Cantow HJ. (1985) Influence of electron-donor-acceptor complex formation on the melt viscosity of some poly(dimethylsiloxane)s. Makromol. Chem. 186:1763–1772.
26. Cojocariu G, Natansohn A. (2003) Perturbation of Charge-Transfer Complexes in Aqueous Solutions of PEG-DNB by Adding N-Ethylcarbazole and Li+ Cations. J. Phys. Chem. 107:5658–5665.
27. Zeng F, Shen Y, Zhu S, Pelton R. (2000) Synthesis and Characterization of Comb-Branched Polyelectrolytes. 1. Preparation of Cationic Macromonomer of 2-(Dimethylamino)ethyl Methacrylate by Atom Transfer Radical Polymerization. Macromolecules 33:1628–1635.

Star-shape Poly(vinylmethyl-*co*-dimethyl)siloxanes with Carbosilane Core – Synthesis and Application

Anna Kowalewska and Bogumiła Delczyk

Abstract New polymeric supports, that can be used for preparation of novel catalytic systems, were obtained by grafting poly(vinylmethyl-*co*-dimethyl)siloxane arms onto multifunctional carbosilane moieties which belong to the class of exceptionally sterically hindered tris(silyl)methanes (T_{Si}). Three types of T_{Si} molecules were applied: 3-functional $HC(SiMe_2Br)_3$ (type A), 9-functional $HC[SiMe_2(CH_2)_5\text{-}C\ (SiMe_2Br)_3]_3$ (type B) and 4-functional T_{Si}-derivative $[SiMe_2C(SiMe_2Br)_2]_2$ (type C). The periphery-functionalized carbosilane-siloxane materials offer uniformly distributed and accessible sites for coordination of active catalytic species. New catalytic systems were thus prepared by coordination of platinum to vinyl moieties of the reported polymers, and used in hydrosilylation of vinyltrimethylsilane with 1,1,3,3-tetramethyldisiloxane.

1 Introduction

An easy recovery of a catalyst from a mixture of reagents/products as well as its simple handling and recycling are important problems in chemical synthesis. Consequently, new recoverable catalysts attract increasing attention and the use of polymeric supports became a common practice. Polystyrene [1] is one of the most popular polymer supports due to its availability, facile functionalization and chemical inertness. However, such organic polymers usually show a solvent swelling dependent performance, which impacts the catalytic activity of the supported species. Polysiloxanes, due to unusually high flexibility of the polymer chain and low barrier

A. Kowalewska
Centre of Molecular and Macromolecular Studies, Polish Academy of Sciences,
Sienkiewicza 112, 90–363 Łódź, Poland
e-mail: anko@cbmm.lodz.pl

F. Ganachaud et al. (eds.), *Silicon Based Polymers*,

of energy of skeletal bonds rotation, are well soluble in organic solvents. Active species attached to polysiloxane chains can be thus readily available to reactants in a catalyzed system. For example, silica hybrids of (vinylmethyl-*co*-dimethyl)siloxane polymers of well defined structure and various topologies were applied for immobilization of Pt(II) complex, by ligand displacement at $PtCl_2(PhCN)_2$ with vinyl groups of the hybrids. [2]

We would like to report the synthesis of a star-shape poly(vinylmethyl-*co*-dimethyl)siloxane polymers functionalized in their exterior, which makes them especially suitable for application as catalytic supports. Similarly to catalysts bound to periphery-functionalized dendrimers [3] they offer regularly distributed and available catalytic sites.

The polysiloxane arms were grafted onto carbosilane cores of an unique architecture (Scheme 1). Such sterically overloaded tris(trimethylsilyl)methane and related structures have been the subject of an extensive study in organometallic chemistry. [4, 5] More recently those moieties were applied for modification of polymers. [6–9] It was shown that T_{Si}-type ligands can provide steric protection to macromolecular systems, resulting in important changes of properties that reflected a substantial decrease of polymer chain mobility (the glass transition temperature increase and thermal resistance enhancement). T_{Si}-type moieties were also applied as scaffolds for platinum catalysts, either as "free" carbosilane dendrimers with alternating silicon-carbon bonds [10] or dendrons grafted on linear polystyrenes. [11]

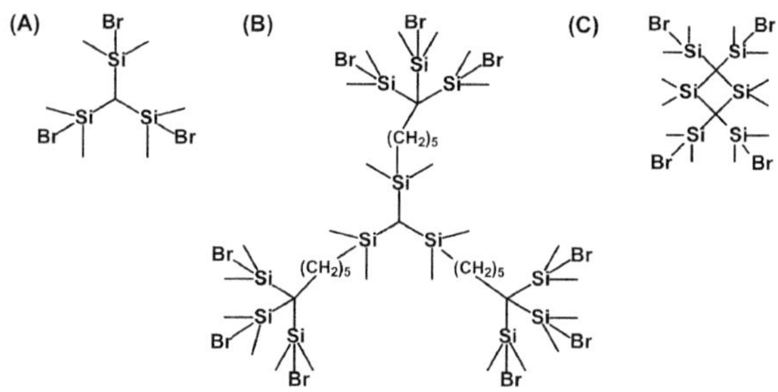

Scheme 1 Structure of T_{Si}-carbosilane cores

2 Experimental

2.1 Analysis and General Methodology

NMR spectra were recorded, in $CDCl_3$ with DRX-500 MHz spectrometer (TMS as the reference). Size exclusion chromatography (SEC) was performed using a LDC Analytical refractoMonitor IV instrument [RI detector, two columns SDV 8×300 (5 µm and 104 Å porosity) and SDV 8×300 (5 µm and 100 Å porosity), eluent – toluene (0.7 ml/min)]. Parallel measurements were effected with Wyatt Optilab 903 apparatus

[LKB 2150 HPLC pumps, dual detector MALS (Multiangle Laser Light Scattering)/ RI, two (TSK G4000HLX and G2000HLX) columns, eluent – CH_2Cl_2 (0.8 ml/min)]. Molecular masses were derived from a calibration curve based on polystyrene standards. Phase transitions of polymers were studied by differential scanning calorimetry (DSC) technique (DuPont 2000 thermal analysis system). Thermograms were taken for samples (sealed in aluminium pans) quenched rapidly from the melt (room temperature) and then heated at the rate of 10 K/min (selected samples were heated also at 20 K/min) from 118 K to 323 K. The sample was kept at 323 K for 3 min to destroy any thermal history. Subsequently, it was quenched to 118 K and heated again at 10 K/min. Each thermal treatment was repeated at least 2 times for each sample. The transition temperatures were taken as corresponding to the maximum of the enthalpic peak. Thermogravimetric measurements were performed by use of a Hi-Res TGA 2950 Thermogravimetric Analyzer (TA Instruments) in nitrogen atmosphere (heating rate 10 K/min, resolution 3, sensitivity 3). MALDI TOF spectra were recorded with Voyager-Elite (Matrix Assisted Laser Desorption/Ionization Time of Flight) mass spectrometer (NaI or AgTFA were used as the salt and DT or 5-MSCA as the matrix). Viscosity measurements were performed at room temperature (298 K) with a Brookfield viscometer (model DV-II+) with S40 cone rotating at 5 RPM. Observations of ultra-thin films of selected samples of polymers with coordinated platinum were performed with a transmission electron microscope (TESLA BS500). The thin films were prepared by placing a droplet of the Pt-containing polymer solution in CH_2Cl_2 directly on a TEM copper support grid covered with a carbon layer. The solvent evaporated slowly in air at room temperature to leave thin polymer film. IR spectra were recorded using a FT-IR ATI Mattson spectrometer.

The anionic ring-opening copolymerization of hexamethyl-cyclotrisiloxane (D_3) with 2,4,6-trivinyl-2,4,6-trimethyl-cyclotrisiloxane (V_3) was performed under dry N_2 atmosphere in a glass ampoule equipped with a teflon stopcock. All other reactions, except the preparation of Pt(0)-[poly(vinylmethyl-*co*-dimethyl)siloxane]-carbosilane complexes, were performed using standard Schlenk's or syringe techniques under an atmosphere of argon.

2.2 Reagents

Vinyltrimethylsilane (97%), trimethylchlorosilane (98%), chlorodimethylsilane (97%), bromoform (96%), 5-bromo-1-pentene (95%), nBuLi (2.5 M solution in hexanes), MeLi (1.6 M solution in diethyl ether), *n*-decane (puriss. p.a., standard for GC, ≥99.8%) and hexachloroplatinic(IV) acid hydrate were purchased from Aldrich. Platinum divinyltetramethyldisiloxane complex (Karstedt's catalyst, 3% solution in xylenes), hexamethyl-cyclotrisiloxane (95%), vinylmethyldichlorosilane (97%) and 1,1,3,3-tetramethyldisiloxane (97%) were bought from ABCR. Bromine (puriss) was bought from Fluka. Triethylamine (pure for analysis) and zinc oxide (pure) was purchased from Chempur. Solvents (tetrahydrofurane, diethyl ether, methylene chloride, pentane, ethyl acetate) were supplied by POCh (Polish Chemical Reagents).

Tris(dimethylsilyl)methane $HC(SiMe_2H)_3$ was prepared according to the literature procedure. [12] 2,4,6-Trivinyl-2,4,6-trimethyl-cyclotrisiloxane was synthesized using Takiguchi method. [13] The cyclosiloxane monomers were kept melted over CaH_2 for several days in an ampoule equipped with a teflon stopcock, and distilled (separately) under vacuum into the polymerization apparatus. All solvents were carefully dried according to literature procedures [14] and distilled prior to their use. Other reagents were used as received.

2.3 Experimental Procedures

2.3.1 Synthesis of Tris[(5-bromopentyl)dimethylsilyl]Methane

5-Bromo-1-pentene (1.66 ml, 0.014 mol) was stirred with Karstedt's catalyst ($[Pt]/[CH_2=CH]=1\times10^{-4}$) at room temperature for 30 min. A solution of tris(dimethylsilyl) methane (0.724 g, 0.0038 mol) in toluene (10 ml) was added drop-wise and the reaction mixture ($[CH_2=CH]/[Si\text{-}H]=1.23$) was stirred for 30 minutes at room temperature. Then, in order to achieve the total conversion of Si-H bonds, the mixture was stirred for 28 days at 343 K. Progress of the reaction was followed by ^1H NMR. Once the addition of Si-H to $CH_2=CH$ was completed, the volatiles were removed under reduced pressure, and the product was dried at RT under high vacuum to leave 2.1 g of tris[(5-bromopentyl)dimethylsilyl]methane (Y = 86.7%).

$HC[SiMe_2CH_2(a)CH_2(b)CH_2(c)CH_2(d)CH_2Br]_3$
NMR: ^1H (δ ppm): –0.79 (s) HC (1H), 0.05 (s) $SiMe_2$ (18H), 0.58 (m) CH_2 (a) (6H), 1.25 (m) CH_2 (b) (6H), 1.45 (m) CH_2 (c) (6H), 1.88 (m) CH_2 (d) (6H), 3.38 (t) CH_2Br (6H)
^{13}C(δ ppm): 3.2 HC, 0.9 $SiMe_2$, 18.5 CH_2 (a), 23.4 CH_2 (b), 33.8 CH_2 (c), 32.5 CH_2 (d), 31.9 CH_2Br
^{29}Si (δ ppm): 0.2 $SiMe_2$
DSC: Tg = 195 K

2.3.2 Synthesis of Tris[1,1,1-tri(dimethylsilyl)hexyl-dimethylsilyl]Methane

4 ml of MeLi in diethyl ether (1.6 M solution, 0.0064 mol) was placed in a Schlenk's flask. Volatiles were removed under vacuum and a solution of $HC(SiMe_2H)_3$ (1.20 g, 0.0063 mol) in THF (15 ml) was added. The solution became turbid after 10 min of stirring at room temperature. It was stirred at 343 K for additional 2 h and then cooled to room temperature. $LiC(SiMe_2H)_3$/THF was added drop-wise at room temperature to a solution of $HC(SiMe_2CH_2(CH_2)_3CH_2Br]_3$ (0.9 g, 0.0014 mol) in THF (15 ml). After 1 hour of stirring, the mixture became transparent and colourless. The mixture was then stirred for additional 24 h at 343 K. Volatiles were removed under reduced pressure, and the product was twice dissolved in a small amount of CH_2Cl_2

and precipitated into methanol, separated and dried (for 16h under vacuum at room temperature) to leave 0.92g of a viscous solid – tris[1,1,1-tri(dimethylsilyl) hexyl-dimethylsilyl]methane (Y=68.0%).

$HC[SiMe_2CH_2(a)CH_2(b)CH_2(c)CH_2(d)CH_2(e)C(SiMe_2H)_3]_3$
NMR: ^1H (δ ppm): –0.5 (s) HC (1H), 0.09 (s) $SiMe_2$ (18H), 0.17 (d) $SiMe_2H$ (54H) J(H-H)=3.43Hz, 0.5 (m) CH_2 (a) (6H), 1.25 (m) CH_2 (c+d) (12H), 1.5 (m) CH_2 (b+e) (12H), 4.0 (m) SiH (9H) J(H-H)=3.43Hz
^{13}C(δ ppm): –3.0 $SiMe_2H$, 1.1 $SiMe_2$, 18.8 CH_2 (a), 24.0 CH_2 (b), 35.1 CH_2 (c), 29.0 CH_2 (d), 30.3 CH_2 (e)
^{29}Si (δ ppm): 0.3 $SiMe_2$, -11.9 $SiMe_2H$
DSC: Tg=233K

2.3.3. Synthesis of 1,1,3,3-tetramethyl-2,2,4,4-tetrakis(dimethylsilyl)-1, 3-disilacyclobutane

$LiC(SiMe_2H)_3$ (0.008mol) has been prepared (see the procedure 2.3.2 above) as 0.15M solution in THF. $LiC(SiMe_2H)_3$/THF was added slowly at room temperature to a solution of $HC(SiMe_2Br)_3$ (0.0025mol) in THF (10ml). After 30 minutes of stirring, the mixture became transparent. It was stirred overnight at room temperature. Afterwards the volatiles were removed under reduced pressure. The solid residue was dissolved in pentane (50ml) and the precipitate (LiBr) was removed by filtration. 1.1g of crude product was obtained. After recrystallization in MeOH 0.3g (Y=31.8%) of pure 1,1,3,3-tetramethyl-2,2,4,4-tetrakis(dimethylsily)l-1,3-disilacyclobutane was obtained.

$[SiMe_2C(SiMe_2H)_2]_2$
NMR: ^1H (δ ppm): 0.17 (d) $SiMe_2H$ (24H) J(H-H)=3.43Hz, 0.51 (s) $SiMe_2$ (12H), 4.44 (m) SiH (4H) J(H-H)=3.43Hz
^{13}C(δ ppm): 0.38 CSi_4, 1.56 $SiMe_2H$, 3.90 $SiMe_2$
^{29}Si (δ ppm): 2.5 $SiMe_2$, –17.7 $SiMe_2H$
TGA: sublimation at 478K under the analytic conditions.
MS (CI, 70eV) m/z: 375 $[M-2H]^+$ (100%), 361 $[M-MeH]^+$ (10%), 317 $[M-Me_2SiH_2]^+$ (3%), 187 (1%), 73 (1%)

2.3.4. General Procedure of Bromination of Carbosilanes with Si-H Groups

A solution of the hydride precursor of a carbosilane core (A, B or C) (0.128mmol of SiH) in CH_2Cl_2 (5ml) was prepared. 0.1ml of 1.34M solution of Br_2 in CH_2Cl_2 was added dropwise at room temperature. The reaction is exothermic and discolouration of Br_2 solution occurs immediately (at the end of reaction the solution is pale orange). The mixture was stirred for an additional 1 hour at room temperature. Subsequently, the volatiles were pumped off under vacuum to leave white solids (quantitative yields), that were used immediately for siloxane arm grafting. Samples were taken to confirm the complete conversion of Si-H to Si-Br by ^1H NMR.

Core A – $HC(SiMe_2Br)_3$ [12]
NMR: ^1H (δ ppm): 0.58 (s) HC (1H), 0.86 (s) $SiMe_2Br$ (18H)
Core B – $HC[SiMe_2CH_2(a)CH_2(b)CH_2(c)CH_2(d)CH_2(e)C(SiMe_2Br)_3]_3$
NMR: ^1H (δ ppm): –0.55 (s) HC (1H), 0.09 (s) $SiMe_2$ (18H), 0.92 (s) $SiMe_2Br$ (54H), 0.5 (m) CH_2 (a) (6H), 1.28 (m) CH_2 (c+d) (12H), 1.8 (m) CH_2 (b+e) (12H)
Core C – $[SiMe_2C(SiMe_2Br)_2]_2$
NMR: 1H (δ ppm): 0.52 (s) $SiMe_2$ (12H), 0.88 (s) $SiMe_2Br$ (24H)

2.3.5 General Procedure of Poly(vinylmethyl-*co*-dimethyl)siloxane Arms Preparation and Their Grafting onto Brominated Carbosilane Cores

Given amounts of hexamethyl-cyclotrisiloxane, 2,4,6-trivinyl-2,4,6-trimethyl-cyclotrisiloxane, and THF were distilled under vacuum into a glass ampoule fitted with a magnetic stirrer, that was then filled with dry nitrogen and thermostated at given temperature. *n*-Decane (internal standard for GC, 7.5% wt for monomers) and a calculated amount of n-BuLi were added with a syringe. The reaction mixture was stirred at the set temperature. The reaction course was controlled chromatographically. Samples were taken out of the reaction mixture at given intervals to be quenched in a mixture of Me_3SiCl/Et_3N (molar ratio $Me_3SiCl/Et_3N=1$), centrifuged and analyzed by GC. The reaction was carried out to the desired conversion of monomers. Once the expected monomer conversion was reached, the reaction mixture was added in excess to a solution of a brominated carbosilane core in CH_2Cl_2 ([Si-Br] = 2.0 M). The solution of the living polymer obtained by AROP was distributed simultaneously into four separate Schlenk's flasks containing appropriate solutions of a carbosilane core (A, B or C) or the mixture Me_3SiCl/Et_3N (in order to estimate the final structure and molecular weight of the arm alone). The reaction mixtures containing cores A, B or C were stirred for 15 h at room temperature. Then, the mixture Me_3SiCl/Et_3N was added in excess in order to quench the unreacted siloxane chains. The resulting mixtures were stirred for 1 h at room temperature. The crude mixtures containing stars and the precursor arm were purified by precipitation into a large amount of MeOH, after the volatiles removal under reduced pressure and dilution with small amounts of CH_2Cl_2. The precipitation was repeated three times and the purified material was dried under high vacuum to the constant weight.

$Me_3SiO\text{-}(SiMe_2O)_m\text{-}(SiViMeO)_n\text{-}CH_2(a)\text{-}CH_2(b)\text{-}CH_2(c)\text{-}CH_3$
NMR: ^1H (δ ppm): 0.07 (s) $Me_3SiO+SiMe_2O$, 0.11 (s) SiViMeO, 0.50 CH_2 (a), 1.28 CH_2 (b+c), 0.88 CH_3, 5.9 (m) -CH=CH_2
^{13}C(δ ppm): –1.7 SiViMeO, –0.8 Me_3SiO, 1.8 $CSiMe_2O$, 13.8 CH_3, 18.0 CH_2 (a), 25.5 CH_2 (b), 26.4 CH_2 (c), 136.8 -CH=, 133.1 =CH_2
^{29}Si (δ ppm): 7.3 Me_3SiO, –11.0 BuViMeSiO, –22.0 $SiMe_2O$, –35.1 SiViMeO

$HC[SiMe_2O\text{-}(SiMe_2O)_m\text{-}(SiViMeO)_n\text{-}CH_2(a)\text{-}CH_2(b)\text{-}CH_2(c)\text{-}CH_3]_3$
NMR: ^1H (δ ppm): 0.08 (s) $CSiMe_2O$, 0.17 (s) $SiMe_2O$, 0.24 (s) SiViMeO, 0.50 CH_2 (a), 1.28 CH_2 (b+c), 0.88 CH_3, 5.9 (m) -CH=CH_2
^{13}C(δ ppm): –1.7 SiViMeO, –0.8 $SiMe_2O$, 1.7 $CSiMe_2O$, 13.8 CH_3, 18.0 CH_2 (a), 25.5 CH_2 (b), 26.4 CH_2 (c), 136.7 -CH=, 132.8 =CH_2
^{29}Si (δ ppm): 7.5 $CSiMe_2$, –11.0 BuViMeSiO, –21.9 $SiMe_2O$, –35.0 SiViMeO

HC{$SiMe_2CH_2$(a)CH_2(b)CH_2(c)CH_2(d)CH_2(e)C[$SiMe_2O$-$(SiMe_2O)_m$-$(SiViMeO)_n$-CH_2(f)-CH_2(g)-CH_2(h)-$CH_3]_3\}_3$
NMR: [1]H (δ ppm): 0.09 (s) $SiMe_2$, 0.10 (s) $SiMe_2O$+$CSiMe_2O$, 0.18 (s) SiViMeO, 0.52 (m) CH_2 (f), 0.6 (m) CH_2 (a), 0.93 (m) CH_3, 1.35 (m) CH_2 (c+d+g+h), 1.60 (m) CH_2 (b+e), 5.9 (m) -CH=CH_2
[13]C(δ ppm): –1.7 SiViMeO, –0.7 $SiMe_2O$, 1.1 $SiMe_2$, 1.5 $CSiMe_2O$, 13.8 CH_3, 18.0 CH_2 (f), 18.6 CH_2 (a), 22.5 CH_2 (g), 24.0 CH_2 (b), 26.4 CH_2 (h), 29.0 CH_2 (d), 30.3 CH_2 (e), 35.2 CH_2 (c), 136.7 -CH=, 132.8=CH_2
[29]Si (δ ppm): 7.5 $CSiMe_2O$, –2.5 $SiMe_2CH_2$, –11.0 BuViMeSiO, –21.9 $SiMe_2O$, – 35.0 SiViMeO

{$SiMe_2$C[$SiMe_2O$-$(SiMe_2O)_m$-$(SiViMeO)_n$-CH_2(a)-CH_2(b)-CH_2(c)-$CH_3]_2\}_2$
NMR: [1]H (δ ppm): 0.07 (d) $SiMe_2O$, 0.11 (s) SiViMeO, 0.16 (s) $CSiMe_2O$, 0.27 (s) $SiMe_2$, 0.50 CH_2 (a), 1.28 CH_2 (b+c), 0.88 CH_3, 5.9 (m) -CH=CH_2
[13]C(δ ppm): –1.6 SiViMeO, –0.9 $SiMe_2O$, 1.7 $CSiMe_2O$, 3.90 $SiMe_2$, 13.8 CH_3, 18.0 CH_2 (a), 25.5 CH_2 (b), 26.4 CH_2 (c), 136.7 -CH=, 132.8=CH_2
[29]Si (δ ppm): 7.4 $CSiMe_2O$, 3.0 $SiMe_2$, –10.8 BuViMeSiO, –21.6 $SiMe_2O$, –34.9 SiViMeO

2.3.6 Exemplary Procedure of Formation of Pt(0)-[poly(vinylmethyl-*co*-dimethyl) siloxane]-Carbosilane Complexes

A solution of C-R-2 (0.45 g, [-CH=CH_2]=0.21 mol/dm^3) in toluene was prepared in a flask, equipped with a magnetic stirrer. H_2O ([-CH=CH_2]/[H_2O]=0.52), and then solid $H_2PtCl_6x6H_2O$ were added ([-CH=CH_2]/[Pt]=3). The mixture was stirred at 343 K for 24 hours. Once cooled down to room temperature, the solution was admixed with $NaHCO_3$ ([Na]/[Pt]=7.3). After filtration, solvents were removed and the product was dried under vacuum to leave brownish viscous paste (0.39 g, Y=78.5%), very well soluble in toluene or CH_2Cl_2 (stable, clear brown solutions).

Pt(0)-{$SiMe_2$C[$SiMe_2O$-$(SiMe_2O)_m$-$(SiViMeO)_n$-CH_2(a)-CH_2(b)-CH_2(c)-$CH_3]_2\}_2$ complex
NMR: [1]H (δ ppm): 0.07 (s) $SiMe_2O$, 0.09 (s) SiViMeO, 0.17 (s) $CSiMe_2O$, 0.28 $SiMe_2$, 0.52 (m) CH_2(a), 0.90 (m) CH_2 (b), 1.31 (m) CH_2 (c), 1.60 (m) CH_3, 3.85 (m) -CH=CH_2···Pt

2.3.7 Exemplary Procedure of Kinetic Studies on Hydrosilylation of Vinyltrimethylsilane with 1,1,3, 3-Tetramethyldisiloxane

A solution of vinyltrimethylsilane and 1,1,3,3-tetramethyldisiloxane ([-CH=CH_2]/[SiH]=1.0) in toluene ([-CH=CH_2]=0.55 mol/dm^3) was prepared and its reference FT-IR spectrum was recorded (the concentration of SiH in sample "0" was adjusted by addition of toluene in order to match the amount of solvent in the subsequent kinetic samples). Platinum catalyst C-R-2 as a solution in toluene ([Pt]=0.33 M) was added to the reaction mixture ([Pt]/[-CH=CH_2]=1.7×10^{-3}). The reaction was carried out at room temperature. The conversion of SiH in the reaction mixture was followed by FT-IR. Once the reaction was completed, the mixture was analyzed with ^1H NMR to examine the selectivity of α/β addition to the double bond during the reaction.

$Me_3SiCH_2(a)CH_2(b)SiMe_2OSiMe_2CH_2(b)CH_2(a)SiMe_3$
NMR: ^1H (δ ppm): –0.03 (s) $SiMe_3$ (18H), 0.04 (s) $SiMe_2$ (12H), 0.39 (m) CH_2 (a+b ß addition) (8H), 1.00 (d) CH_3 [α addition (traces)]

3 Results and Discussion

3.1 Preparation of Carbosilane Cores

Bromination of tris(dimethylsilyl)methane yielded $HC(SiMe_2Br)_3$ (core A). [12] Hydride precursors of carbosilane cores B and C: tris[1,1,1-tri(dimethylsilyl)hexyl-dimethylsilyl]methane $\{HC[SiMe_2(CH_2)_5C(SiMe_2H)_3]_3\}$ and 1,1,3,3-tetramethyl-2,2,4,4-tetrakis-(dimethylsilyl)-1,3-disilacyclobutane $\{[SiMe_2C(SiMe_2H)_2]_2\}$ were also prepared utilizing $HC(SiMe_2H)_3$.

Core B was prepared stepwise. In the first step tris[(5-bromopentyl)dime thylsilyl]methane was obtained by hydrosilylation of 5-bromo-1-pentene by tris(dimethylsilyl)methane in toluene, catalyzed by Karstedt's catalyst at 343 K (Scheme 2). 67% of Si-H groups reacted with 5-bromo-1-pentene within 24 h, since addition of the third part of Si-H groups to $–CH=CH_2$ proceeded slowly due to the steric hindrance. $HC[SiMe_2(CH_2)_4CH_2Br]_3$ was used as a substrate for the synthesis of tris[1,1,1-tri(dimethylsilyl)hexyl-dimethylsilyl]methane in a reaction with an organolithium reagent – $[LiC(SiMe_2H)_3]$. $LiC(SiMe_2H)_3$ can be prepared with quantitative yield (97%) using $HC(SiMe_2H)_3$ and $[(CH_3)_2CH]_2NLi$. [15] The same results were obtained utilizing MeLi as the metallating agent. Whereas the cross-coupling of $LiC(SiMe_2H)_3$ with $HC[SiMe_2(CH_2)_4CH_2Br]_3$

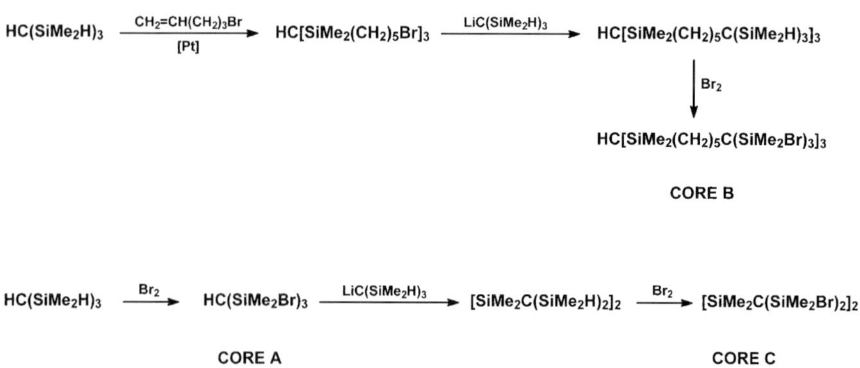

Scheme 2 Synthesis of carbosilane cores A, B and C

provided the desired product, an analogous reaction with $HC(SiMe_2Br)_3$ gave 1,1,3,3-tetramethyl-2,2,4,4-tetrakis(dimethylsilyl)-1,3-disilacyclobutane (the hydride precursor of core C). It was reported that $LiC(SiMe_2H)_3$ reacts in a similar way with $SiCl_4$ at ambient temperatures, although its cross-coupling with R_3SiCl (Me_3SiCl, $MeSiHCl_2$, $HSiCl_3$ or $MeSiCl_3$) at 195 K yields carbosilanes of $R_3SiC(SiMe_2H)_3$ type. [15] The proposed mechanism of the formation of $[SiMe_2C(SiMe_2H)_2]_2$ involves hydrogen-halogen exchange, elimination of LiCl and generation of a reactive silene, $(HMe_2Si)_2C{=}SiMe_2$, that dimerizes to give the disilacyclobutane.

The reactive cores B and C were prepared by bromination of their hydride precursors. Due to their high moisture susceptibility, all Si-Br containing products were not isolated, but used directly in the siloxane arms grafting step. To confirm the availability of all Si-Br groups in the core B, which appeared as the most sterically demanding one, a low molecular weight poly(dimethylsiloxane) star with core B was prepared (Mn of an average arm ~1000 D) and its structure was analyzed and proved by NMR.

Cores A and C, as well as the hydride precursor of C, are crystalline. T_{Si}-cores of II-nd generation, in spite of their relatively small molecular weight and spherical shape, exhibit some features characteristic of amorphous phases (the hydride precursor of core B has its glass transition (Tg) at 233 K, and its parent compound $HC[SiMe_2(CH_2)_5Br]_3$ at 195 K).

3.2 *Preparation of Poly(vinylmethyl-*co*-dimethyl)siloxane Arms*

Simultaneous copolymerization of monomers can lead to materials of diverse architectures. We aimed to prepare copolymeric siloxane arms of blocky structure, so that vinyl groups, interacting with platinum atoms, were placed most accessibly after grafting on the core – at the part of the siloxane arm opposite to the carbosilane centre. Anionic ring opening polymerization (AROP) seemed to be the method of choice, since it makes possible the preparation of polysiloxanes with controlled molecular weight and polydispersity, as well as arrangement of functional groups along the main chain. Side reactions on siloxane chain can be avoided under carefully chosen reaction conditions. [16, 17] Vinylmethylsiloxane-*co*-dimethylsiloxane polymer arms were thus obtained by *n*-BuLi catalyzed AROP of hexamethyl-cyclotrisiloxane (D_3) with 2,4,6-trivinyl-2,4,6-trimethyl-cyclotrisiloxane (V_3) (Scheme 3). Kinetic studies of anionic ring opening polymerization of D_3 and V_3 performed in THF using *n*-BuLi as the initiator [18], proved that propagating

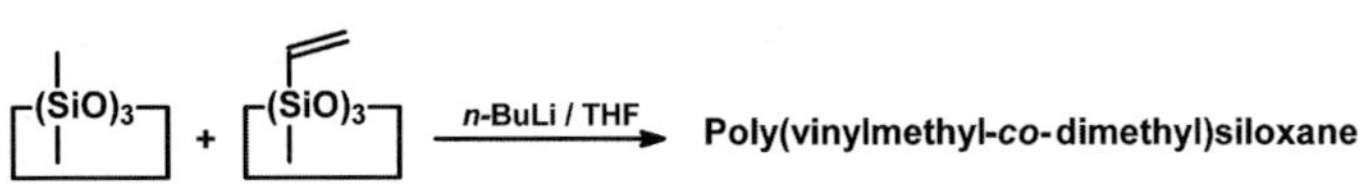

Scheme 3 Preparation of co-oligosiloxane arms

species preferentially add V_3 [the apparent reactivity ratios (RT) of monomers $rD_3=0.011$ and $rV_3=49.3$]. We could thus easily copolymerize simultaneously V_3 and D_3 at room temperature to obtain quasi-block copolymeric siloxane arms having vinylmethylsiloxane segment placed opposite to the reactive chain end (Fig. 1). A series of copolymers differing by molecular weight and molar ratio of siloxane units [D]/[V] was obtained. One of the polymers (R-1) was prepared in a one pot process at low temperature (268 ± 1 K) to assure the block structure. Decrease of the reaction temperature impedes the reactivity of monomers and restrains the insertion of dimethylsiloxane units into the growing polymer chain composed of vinylmethylsiloxane segments (Fig. 2). However, the considerable difference in the monomer reactivity prevents formation of gradient polymer, even at room temperature. ^{29}Si NMR (reverse gate) studies showed that the materials obtained in the simultaneous polymerization of a mixture of monomers have virtually block structure (Fig. 3). The same results were obtained for various $[D_0]/[V_0]$ and [SiO]/[*n*BuLi] (Table 1).

Molecular weight of poly(vinylmethyl-*co*-dimethyl)siloxane arms was determined by SEC and ^{1}H NMR spectroscopy (Table 1). The SEC values obtained in

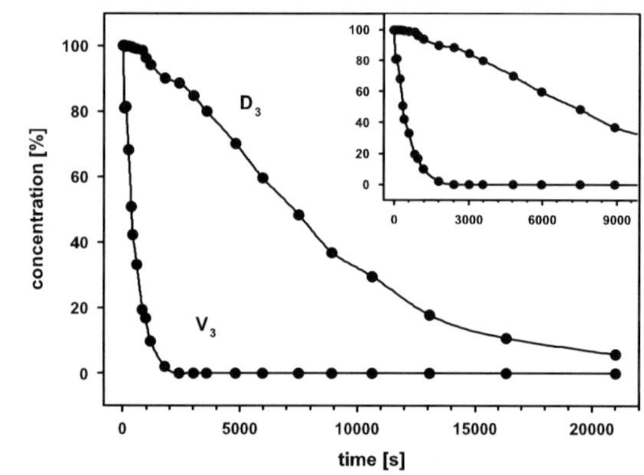

Fig. 1 Kinetics of copolymerization of V_3 and D_3 at room temperature ($[D_3]_0:[V_3]_0:[BuLi]_0=2:1:0.053$) (polymer R-3)

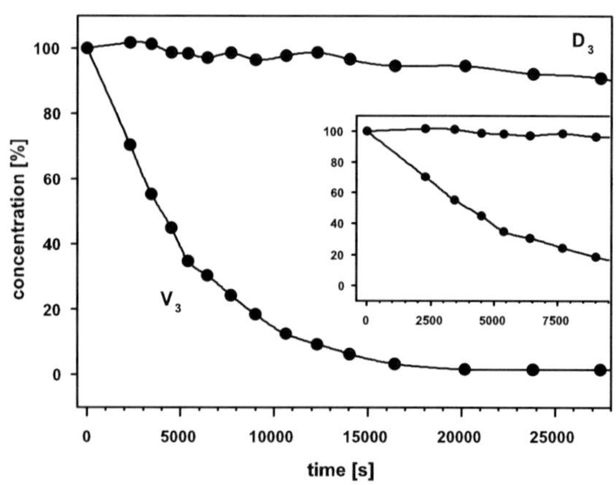

Fig. 2 Kinetics of copolymerization of V_3 and D_3 at -5°C ($[D_3]_0:[V_3]_0:[BuLi]_0=1:1:0.035$) (polymer R-1)

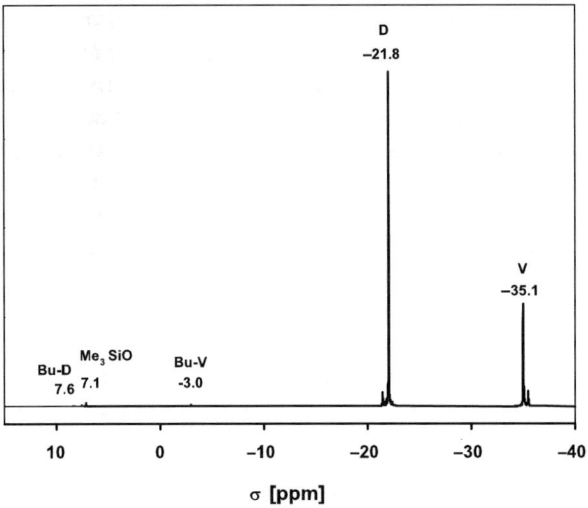

Fig. 3 ^{29}Si NMR spectrum of polymer R-3 obtained by AROP (Fig. 1)

Table 1 Preparation of copolymeric siloxane arms

run	T [K]	a_0	[SiO]/ [Li]	X_D [%]	X_V [%]	M_t	a_1	b_1	c_1	d_1	v [mPas]	ΔH_c [J/g]	ΔH_m [J/g]
R-1	268	1.0	171.1	20	100	2900	0.10	8200	1000	2.2	–	–	–
R-2	295	7.2	83.5	94	100	6300	7.59	6600	15600	1.3	20.9	18.5	20.4
R-3	297	2.0	174.3	94	100	13600	1.93	9300	16100	1.3	–	13.7	12.0
R-4	295	11.4	277.3	95	100	21000	11.82	18500	38300	1.3	340.1	4.8	33.6

T – reaction temperature
a_0 – [D]/[V] in the starting mixture
X_D – conversion of hexamethyl-cyclotrisiloxane at quenching
X_V – conversion of 2,4,6-trivinyl-2,4,6-trimethyl-cyclotrisiloxane at quenching
M_t – the theoretical value calculated from the ratio [SiO]/[nBuLi]
a_1 – [D]/[V] in the product estimated by ^{1}H NMR
b_1 – M of the product estimated by ^{1}H NMR
c_1 – M_n of the product estimated by SEC in toluene
d_1 – M_w/M_n of the product estimated by SEC in toluene
v – melt viscosity at 248 K
ΔH_c – enthalpy of crystallization (10 K/min)
ΔH_m – enthalpy of melting (10 K/min)

CH_2Cl_2 were not always accurate, due to insufficient differences in RI between the solvent and copolymeric fractions (especially for polymers with substantial amount of vinylmethylsiloxane segment). Measurements in toluene gave more reliable results. However, large differences between M_{NMR} and M_{SEC} were observed. M_{NMR} was found to be close to those calculated on the basis of the reaction stoichiometry and the conversion of monomers at quenching. It seems that the relationship between hydrodynamic volumes of polystyrene standards and the studied block poly(vinylmethyl-*co*-dimethyl)siloxanes (and also their star-shape analogs) may slightly deviate from the well defined one known for linear homopolysiloxanes.

Poly(vinylmethyl-*co*-dimethyl)siloxane arms were proved to be semicrystalline. DSC analysis of block copolymers showed a characteristic cold crystallization transition at 188 K and two exothermic peaks at 229 K and 235 K, typical for melting of long PDMS blocks. Arms R-2 and R-3 undergo mostly "cold" crystallization during DSC analysis (minor difference in the enthalpy of crystallization ΔH_c and melting ΔH_m – Table 1), but R-4 is partly crystalline ($\Delta H_m - \Delta H_c = 29$ J/g). For all copolysiloxane arms the heat of crystallization and fusion recorded for a sample is the same in subsequent DSC runs, which proves a facile chain alignment. The polymer R-1 containing the largest amount of vinyl groups does not show crystallization or melting because PDMS segments are too short.

3.3 *Multi-Arm Co-Polysiloxane Stars*

3.3.1 Preparation

Multiarm star-shape systems can be obtained in several ways, e.g.: by coupling of linear macromolecules to a multifunctional core, by addition of linear macromolecules with polymerizable end-groups to difunctional monomers and by growing linear chains from multifunctional initiator. [19] Coupling of living anionic macromolecules can be difficult and can lead, due to steric hindrance, to incomplete conversion of core reactive groups. Still, this method can be successful for star molecules with low number of arms. [20,21] Because of $rV_3 >>> rD_3$, it seems that the third method is the preferable way for preparation of macromolecules grafted with block copolysiloxane arms, and having vinylmethylsiloxane block in their outer sphere.

Once the expected conversion of monomers was reached in the cyclosiloxane AROP step, the living polymer was quenched by addition of a freshly prepared brominated T_{Si}-core, with deficiency of Si-Br groups, relative to polymer chain ends. The AROP of R-1, used for the preparation of A-R-1, was stopped before the ratio [D]/[V] corresponding to the stoichiometric one was obtained. The mixture of a carbosilane core and Li-terminated siloxane chains was stirred overnight at room temperature. Then Me_3SiCl/Et_3N was added to the reaction mixture to quench all reactive chain ends left after core-grafting. No chain scission was detected under the applied conditions. GC analyses of the reaction mixture performed after 15 h of stirring at room temperature did not show any new peaks that could indicate the presence of cyclic siloxanes, being the result of side reactions on polysiloxane chain. Polymers were then purified by precipitation several times into methanol and dried under high vacuum at room temperature. The raw product fractionalization allowed for separation of the highest molecular weight species, although their yield was rather poor (Table 2).

Table 2 Preparation of star-shape siloxane/carbosilane polymers

arm	core	a_2	c_2	d_2	v [mPas]	ΔH_c [J/g]	ΔH_m [J/g]	Y [%]
R-1	A	0.74	10800	1.6	68.0	–	–	24.5
R-2	B	9.96	12900	1.6	36.6	3.1	3.0	15.6
R-2	C	9.09	13200	1.6	21.4	8.0	7.7	33.4
R-3	A	1.87	18800	1.3	20.9	3.0	2.8	33.9
R-4	B	12.85	11400	1.6	26.2	14.7	14.3	13.4
R-4	C	12.59	–*	–*	15.7	3.2	3.6	37.4

a_2 – [D]/[V] in the star-shape polymer estimated by ^{1}H NMR
c_2 – M_n of the star-shape polymer estimated by SEC in toluene
d_2 – M_w/M_n of the star-shape polymer estimated by SEC in toluene
v – melt viscosity at 248 K
ΔH_c – enthalpy of crystallization (1st run, 10 K/min)
ΔH_m – enthalpy of melting (1st run, 10 K/min)
Y – the reaction yield
* – not measurable

3.3.2 Characterization

The carbosilane framework introduces extraneous component to the polysiloxane system. That is why the analysis of such materials is rather difficult. The evaluation of the number of arms in star copolymers was complex. We were not able to obtain MALDI TOF spectra with any combination of salt/matrix available. SEC in CH_2Cl_2 did not give rational results since the presence of T_{Si}-cores changes substantially RI of star-shape systems, compared to their linear precursors. Better results were obtained with toluene as the SEC solvent. Still, the hydrodynamic volume (hv) of the molecule becomes unpredictable on addition of T_{Si}-core and thus we had stars of the retention time "higher", "equal to" and "smaller" than the parent arms (Fig. 4). SEC analysis with MALLS detector can be often used for number of arms determination of the star-shaped polymers. The method has some limitations [only polymers of high molecular mass (10 nm radius of gyration is the MALLS detector limit) and uniform architecture can be analyzed]. [22] The analysis of poly(vinylmethyl-*co*-dimethyl)siloxane-carbosilane materials did not give good enough results. NMR studies were finally carried out to prove the polymer structure. The comparison of ^{1}H and ^{29}Si NMR (reverse gate) spectra was done, and the intensities of peaks corresponding to carbosilane Si-Me groups and *n*Bu-V chain ends (assuming that there are virtually no *n*Bu-D in block copolymers) were compared.

The branched architecture has great influence on the packing of molecular chains. In general, dendrimers have smaller hydrodynamic radius and the melt and solution viscosity of a hyperbranched polymer is expected to be lower than that of a parent linear polymer. Viscosity measurements performed with a cone viscometer confirmed the decrease of viscosity of star-shape polymers compared to the respective high molecular weight arms (polymers B-R-4 and C-R-4, Tables 1 and 2). This observation is consistent with the decrease of hydrodynamic volume observed for

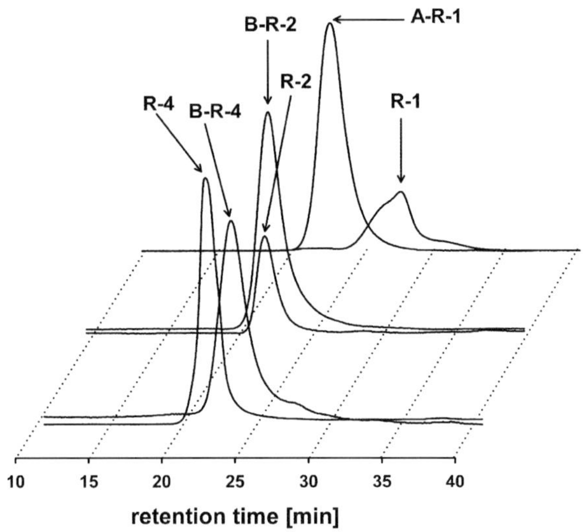

Fig. 4 SEC (in toluene) of siloxane arms parent to respective carbosilane-core stars of various multiplicity, molecular weight and structure

high M_n stars (Fig. 4). The copolymers (B-R-2, C-R-2) obtained with low molecular weight polysiloxane arms retained the range of viscosity of their precursors.

The core seems to control thermal properties of star-shape polysiloxanes. Siloxane stars grafted onto T_{Si} core are structurally close to the parent linear polymers (similar temperatures of devitrification). Glass transition is an indicator of the change in the heat capacity, as the polymer goes from the glassy state to the rubbery one. [23, 24] Its presence is a proof of an amorphous content in polymeric material and the change of specific heat ΔCp can be its measure. An apparent change of specific heat Cp at glass transition can be observed between linear siloxane arms and the respective star-shape polymers. The steep of heat flow at the glass transition and thus ΔCp depends on the type of carbosilane core for B-R-2 and C-R-2. The increase of M_n of siloxane arms prompts large increase of ΔCp for both cores B and C.

We observed that the heat of fusion of block-arm stars is much smaller than that of the respective arm alone. It implies an expected decrease of ordering due to the interference of T_{Si}-core. Samples of long PDMS block (B-R-4 and C-R-4) do not show the peak corresponding to the secondary melting (235 K, Fig. 5). Their maximum crystallization temperature increases. It may suggest faster crystallization of poly(vinylmethyl-*co*-dimethyl)siloxane arms from the glassy state (>145K) due to more facile alignment of PDMS block in linear polymers than in their analogs grafted onto carbosilane cores.

The type of carbosilane core and the length of siloxane arms play an important role in this process. For high M_n siloxane arms the more flexible core (B) causes large ($\Delta H_m = 14$ J/g for B-R-4 versus $\Delta H_m = 34$ J/g for R-4), but slightly less pronounced decrease of transition enthalpy than compact ones [A, C ($\Delta H_m = 4$ J/g for C-R-4)] (Fig. 5). It implies easier orientation of polymer chains on heating. The opposite was found for lower M_n polymers ($\Delta H_m = 3$ J/g for B-R-2 and $\Delta H_m = 8$ J/g for C-R-2). The traces recorded in the course of DSC experiment for siloxane stars of compact carbosilane core and shorter siloxane arms (C-R-2) do not show any difference

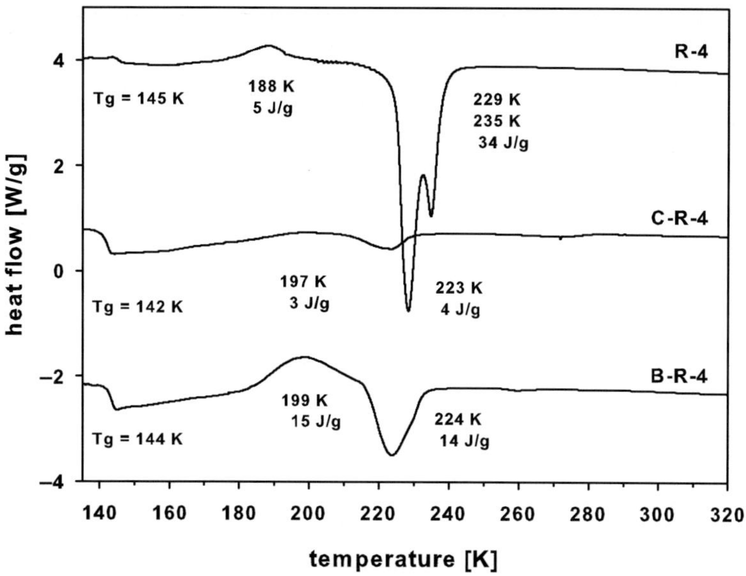

Fig. 5 DSC traces of siloxane arm R-4 and T_{Si}-siloxane star polymers B-R-4 and C-R-4 at 10 K/min

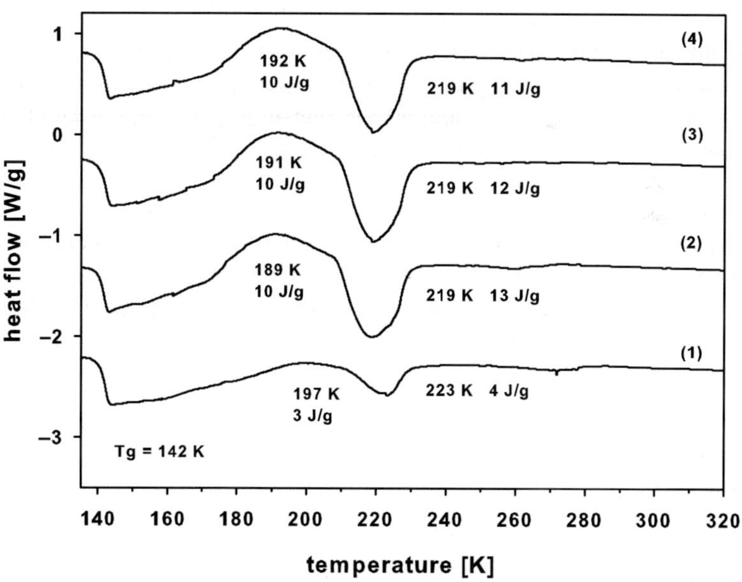

Fig. 6 DSC traces of subsequent heating runs recorded for C-R-4 at 10 K/min

between the first heating and the subsequent runs. Those with the same core but longer PDMS chain (C-R-4, Fig. 6) tend to line up during repeated heating/cooling. A significant difference in the enthalpy of crystallization/melting can be observed

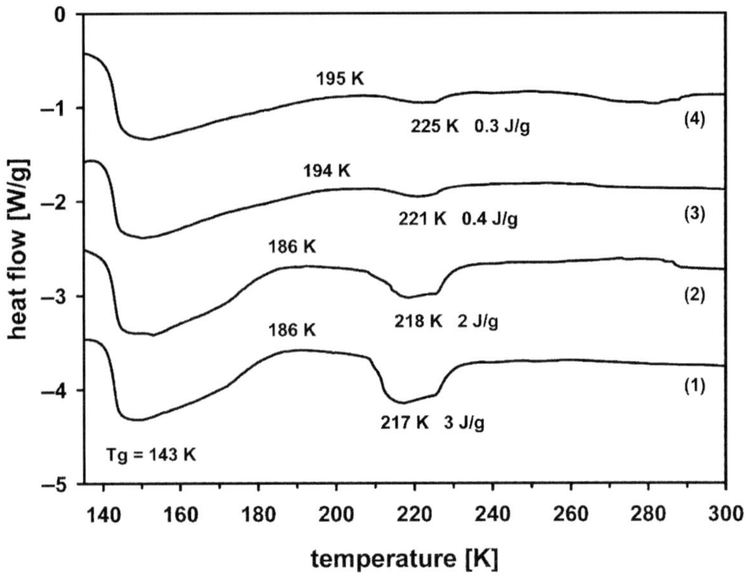

Fig. 7 DSC traces of subsequent heating runs recorded for C-R-4 at 20 K/min

between the first and the subsequent DSC runs. Transitions are much broader than those characteristic for shorter arm stars. For the corresponding sample B-R-4, no relevant chain organization on heating was observed.

Samples B-R-2, C-R-2 and C-R-4 were also analyzed for their phase transitions under fast heating (20 K/min). As expected, the heat of cold crystallization and fusion, evolved during the first DSC run, is slightly smaller than those recorded at 10 K/min. As under slow heating regime, any chain organization was observed during repeated heating/cooling of C-R-2 at 20 K/min. Surprisingly, and contrary to the experimental results obtained for C-R-4 at slow heating, the subsequent heating runs at 20 K/min caused a decrease of order in its sample (Fig. 7).

3.4 Platinum Catalysts Based on TSi-siloxane Stars – Preparation and Catalytic Activity

Platinum was bound to -CH=CH_2 groups of star-shape copolymeric supports in reaction with $H_2PtCl_6x6H_2O$, performed according to the literature procedure. [25] Polymers A-R-1, B-R-2 and C-R-2 were utilized in this reaction. The molar ratio [H_2PtCl_6]/[-CH=CH_2]=3 was used, according to the known stoichiometry of [Pt]/ [-CH=CH_2] in Pt-2,4-divinyl-tetramethyldisiloxane complex (Karstedt's catalyst). [26] The samples with Pt coordinated to vinyl groups resemble a greenish paste after neutralization with $NaHCO_3$ and drying. After dilution with toluene or CH_2Cl_2, transparent pale brown solutions were obtained. For samples with R-2 arms a complete coordination of Pt to all vinyl groups in block copolymers was confirmed by ^{1}H NMR {disappearance of "free" vinyl groups and formation of a set of multiplets in the range 3.4–3.95 ppm (the chemical shift typical for vinylsiloxanes with Pt coordinated to -CH=CH_2 [27] and their Rh analogs [28])}.

Disappearance of the crystalline phase and 16 K increase of glass transition temperature during DSC measurements were observed for these copolymers (Fig. 8). Most probably, since the coordination of Pt takes place, and it involves vinyl groups that can belong to various siloxane chains, a sort of physical cross-linking occurs and hinders the mobility of polymeric arms. This phenomenon does not cause a decrease of activity of catalysts based on block siloxanes. The -CH=CH_2 groups are few and belong to flexible siloxane chains, which assures the catalyst solubility and thus good catalytic performance. Similar polymeric catalysts based on $(ViMe_2Si)_3CCH_2$ dendronized polystyrene [11] suffered much more, concerning the supported catalyst solubility, on such inter/intra-chain coordination. In the case of copolymer A-R-1, due to the large amount of the potential cross-linking sites, most of our attempts to coordinate Pt to all vinyl groups ended up in formation of insoluble materials. The only soluble product, obtained with A-R-1 contained only a small number of active metal sites (8% out of all -CH=CH_2 was involved in coordination to Pt). For this sample Tg increased by 2 K. Figure 9 shows TEM images of platinum supported on poly(vinylmethyl-*co*-dimethyl)siloxane-carbosilane star-shape polymer. Platinum particles organized into small assemblies (ca. 1–2 μm) uniformly distributed over the sample can be seen.

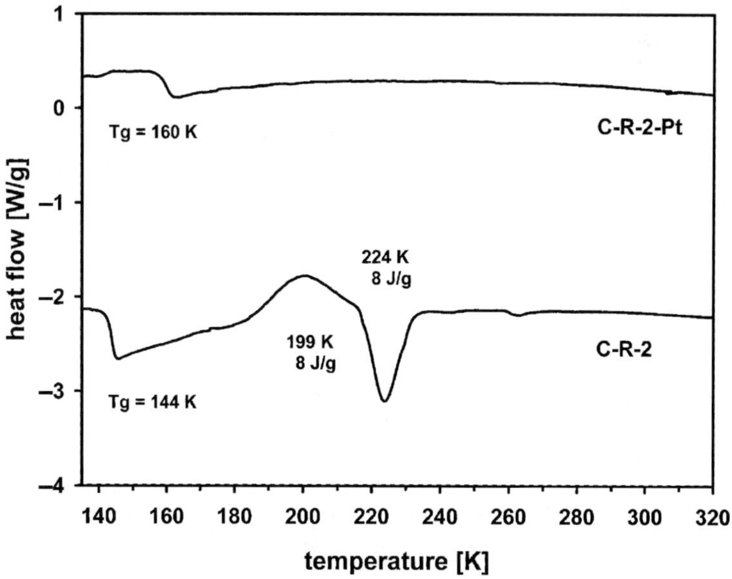

Fig. 8 DSC traces of platinum catalyst C-R-2-Pt and its precursor C-R-2 at 10 K/min

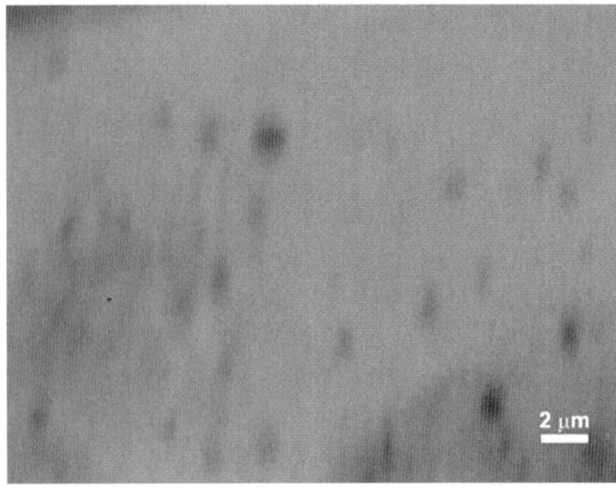

Fig. 9 TEM image of C-R-2-Pt catalyst

The catalysts were tested in a model reaction of hydrosilylation of trimethylvinylsilane with 1,1,3,3-tetramethyldisiloxane (Scheme 4). A solution of both reagents in toluene ([Si-H] = 0.55 mol/dm^3, [Si-H]/[-CH=CH$_2$] = 1) was prepared. A solution in toluene of the immobilized platinum catalyst was added to the reaction mixture at room temperature ([Pt]/[-CH=CH$_2$] = 1.7×10^{-3}). The amount of platinum in the catalyst, assuming that Pt coordinates to 3 -CH=CH$_2$ moieties, was calculated on the basis of the relative decrease of vinyl groups in ^{1}H NMR spectrum [their number (related to Me groups) was compared to those in the precursor].

The composition of the reaction mixture was studied by FT-IR and the progress of Si-H bonds conversion was followed [disappearance of the characteristic Si-H band at 2100 cm^{-1} (Fig. 10)]. Once the reaction was completed the mixture was

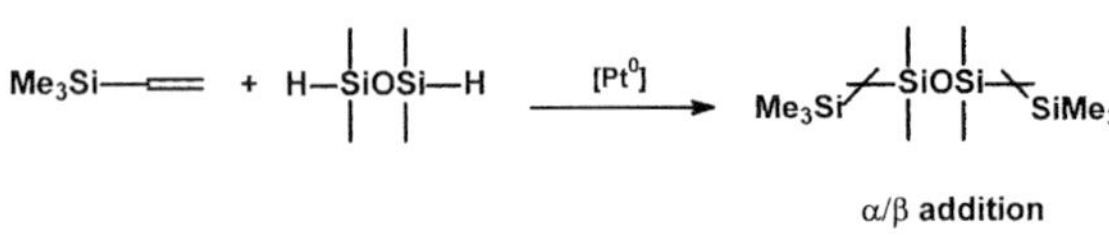

Scheme 4 Model hydrosilylation reaction catalyzed with platinum catalyst supported on T_{Si}-siloxane star-shape polymers

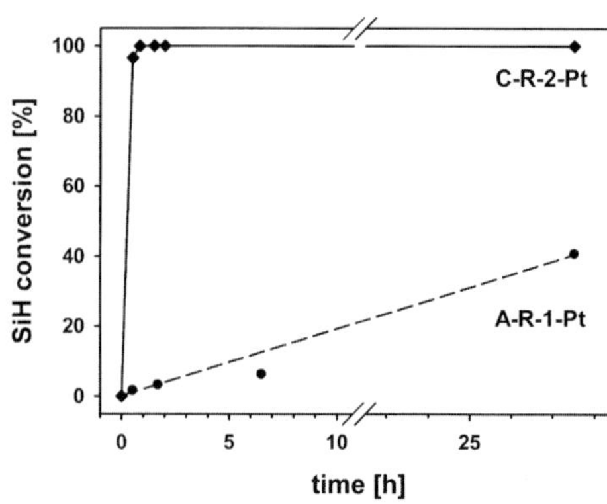

Fig. 10 Conversion of SiH groups in C-R-2-Pt and A-R-1-Pt catalyzed hydrosilylation of trimethylvinylsilane with 1,1,3,3-tetramethyldisiloxane ([Si-H] = 0.55 mol/dm^3, [Si-H]/[-CH=CH$_2$] = 1, [Pt]/[-CH=CH$_2$] = 1.7×10^{-3}, room temperature)

analyzed by ^{1}H NMR to examine the selectivity of the α/β addition of Si-H to the double bond. The ratio was found to be 0.06, which is comparable to the selectivity of Karstedt's catalyst (0.03) under the same conditions. [11] The activity of the platinum catalyst supported on A-R-1 star was found to be much poorer than that for its C-R-2-Pt counterpart. The model reaction carried out under the same conditions is much slower. It may be caused by a decrease of accessibility of Pt-containing sites.

5 Conclusions

We have described a new system of polymeric supports based on multifunctional, exceptionally sterically hindered carbosilane moieties, grafted with block poly(vinylmethyl-*co*-dimethyl)siloxane arms. They offer uniformly periphery-distributed active sites (-CH=CH_2 moieties) and can be used for preparation of novel catalysts. Platinum was thus attached to the polymers via coordination to vinyl groups. The materials used in hydrosilylation of vinylsilanes can be considered as an alternative for traditional platinum catalysts. The utility of the catalysts seems to be dependent on the ratio [D]/[V] in the copolysiloxane arm. Those with too high amount of vinyl groups suffer from poor solubility and catalytic performance due to excessive inter/intra-chain coordinative cross-linking.

Acknowledgements We thank the Polish Ministry for Science and Information Technology for support within Grant PBZ-KBN-118/T09/02. TEM measurements were performed in Polymer Physics Department at Centre of Molecular and Macromolecular Studies (the help from colleagues working at Polymer Physics Department is most kindly acknowledged).

References

1. McNamara CA, Dixon MJ, Bradley M (2002) Recoverable catalysts and reagents using recyclable polystyrene-based supports. Chem Rev 102:3275–3300
2. Rózga-Wijas K, Chojnowski J, Fortuniak W, Ścibiorek M, Michalska Z, Rogalski Ł (2003) Branched functionalised polysiloxane–silica hybrids for immobilisation of catalysts. J Mater Chem 13:2301–2310
3. Van Heerbeek R, Kamer PCJ, van Leeuwen PWNM, Reek JNH (2002) Dendrimers as support for recoverable catalysts and reagents. Chem Rev 102:3717–3756
4. Eaborn C, Smith JD (2001) Organometallic compounds containing tris(trimethylsilyl)methyl or related ligands. J Chem Soc Dalton Trans 1541–1552
5. Eaborn C (2001) Unusual mechanistic pathways. The novel chemistry of compounds with tris(trimethylsilyl)methyl or related ligands on silicon. J Chem Soc Dalton Trans 3397–3406
6. Kowalewska A, Stańczyk WA, Boileau S, Lestel L, Smith JD (1999) Novel polymer systems with very bulky organosilicon side chain substituents. Polymer 40:813–818
7. Kowalewska A, Stańczyk WA (2003) Highly thermally resistant UV-curable poly(siloxane)s bearing bulky substituents. Chem Mater 15:2991–2997
8. Safa KD, Babazadeh M, Namazi H, Mahkam M, Asadi MG (2004) Synthesis and characterization of new polymer systems containing very bulky tris(trimethylsilyl)methyl substituents as side chains. European Polym J 40:459–466

9. Kowalewska A, Kupcik J, Pola J, Stańczyk WA (2008) Laser irradiation of oligosiloxane copolymer thin films functionalized with side chain bulky carbosilane moieties. Polymer. doi:10.1016/j.polymer.2007.12.022
10. Kowalewska A, Stańczyk WA (2006) New dendrimeric systems based on sterically hindered carbosilane units – synthesis and application. ARKIVOC v:110–115
11. Kowalewska A (2008) Dendronized Polystyrene Supports for New Catalytic Systems. J Organomet Chem. doi:10.1016/j.jorganchem.2008.03.021
12. Eaborn C, Hitchcock PB, Lickiss PD (1983) Some derivatives of tris(dimethylsilyl)methane. A novel bicyclic tris(disiloxane) with a manxane structure. J Organomet Chem 252:281–288
13. Takiguchi T, Sakurai M, Kishi T, Ichimura J, Izuka Y (1960) Notes – Preparation of Hexaphenylcyclotrisiloxane by the Reaction of Hexaphenyldichlorosilane with Zinc Oxide. J Org Chem 25:310–311
14. Perrin DD, Armarego WLF (1980) Purification of Laboratory Chemicals. Pergamon Press, Oxford
15. Hawrelak EJ, Ladipo FT, Sata D, Braddock-Wilking J (1999) Synthesis, Characterization, and Reactivity of $[LiC(SiMe_2H)_3]$ 2THF: Formation of 1,1,3,3-Tetramethyl-2,2,4,4-tetrakis(dimethylsilyl)-1,3-disilacyclobutane, $[Me_2SiC(SiMe_2H)_2]_2$. Organometallics 18:1804–1807
16. Kendrick TC, Parbhoo BM, White JW (1989) Polymerization of Cyclosiloxanes. In: Eastmond GC, Ledwith A, Russo S, Sigwalt P, editors. Comprehensive Polymer Chemistry. vol. 4. Oxford: Pergamon Press, pp. 459–523
17. Chojnowski J, Cypryk M (2000) Synthesis of Linear Polysiloxanes. In: Jones RG, Ando W, Chojnowski J, editors. Silicon-Containing Polymers. Dordrecht: Kluwer Academic Publishers, pp. 3–41
18. Cypryk M, Delczyk B Controlled synthesis of vinylmethylsiloxane-dimethylsiloxane gradient copolymers by anionic ROP of cyclotrisiloxanes. A kinetic study. (submitted)
19. Vasilenko NG, Rebrov EA, Muzafarov AM, Eßwein B, Striegel B, Möller M (1998) Preparation of multi-arm star polymers with polylithiated carbosilane dendrimers. Macromol Chem Phys 199:889–895
20. Roovers J, Bywater S (1972) Preparation and characterization of four-branched star polystyrene. 5:384–388
21. Roovers J, Hadjichristidis N, Fetters LJ (1983) Analysis and dilute solution properties of 12- and 18-arm-star polystyrenes. Macromolecules 16:214–220
22. Biela T, Duda A, Pash H, Rode K (2005) Star-shaped poly(L-lactide)s with variable numbers of hydroxyl groups at polyester arms chain-ends and directly attached to the star-shaped core – controlled synthesis and characterization. J Polym Sci: Part A: Polym Chem 43:6116–6133
23. Marti E, Kaiserberg E, Moukhina E (2006) Heat capacity functions of polystyrene in glassy and in liquid amorphous state and glass transition DSC and TMDSC study. J Therm Anal Cal 85:505–525
24. Wunderlich B (2007) Glass transition as a key to identifying solid phases. J Appl Polym Chem 105:49–59
25. Chandra G, Lo PY, Hitchcock PB, Lappert MF (1987) A convenient and novel route to bis(η-alkyne)platinum(0) and other platinum(0) complexes from Speier's hydrosilylation catalyst $H_2PtCl_6xH_20$. X-ray structure of $[Pt\{(\eta\text{-}CH_2{=}CHSiMe_2)_2O\}(P\text{-}t\text{-}Bu_3)]$. Organometallics 6:191–192
26. Hitchcock PB, Lappert MF, Warhurst NJW (1991) Synthesis and Structure of a rac-Tris(divinyldisiloxane) diplatinum(0) Complex and its Reaction with Maleic Anhydride. Angew Chem, Int Ed Engl 30:438–440
27. Lewis LN, Colborn RE, Grade H, Bryant GL, Sumpter ChA, Scott RA (1995) Mechanism of Formation of Platinum(0) Complexes Containing Silicon-Vinyl Ligands. Organometallics 14:2202–2213
28. Michalska ZM, Rogalski Ł, Rózga-Wijas K, Chojnowski J, Fortuniak W, Ścibiorek M (2004) Synthesis and catalytic activity of the transition metal complex catalysts supported on the branched functionalized polysiloxanes grafted on silica J Mol Catal A: Chem, 208:187–194

Copolycondensation of Functional Silanes and Siloxanes in Solution Using tris(pentafluorophenyl)borane as a Catalyst in a View to Generate Hybrid Silicones

Claire Longuet and François Ganachaud

Abstract This paper proposes new ways of preparation of hybrid silicones, i.e. an alternated multiblock sequence of silicone and alkyl spacers, via a polycondensation process catalyzed by the tris(pentafluorophenyl)borane, a water-tolerant Lewis acid, between methoxy and hydrogeno functionalized silanes and siloxanes at room temperature and in the open air. The protocol was first developed with model molecules which led to polydimethylsiloxane (PDMS) chains, in order to seize the best experimental conditions. Several factors were studied such as the contents of each reactants, the nature of the solvent or the rate of addition. The best conditions were then adapted to the synthesis of hybrid silicones, condensing alkylated oligocarbosiloxanes with methoxy or hydrogeno chain-ends and complementary small molecules. A systematic limitation in final molar masses of hybrid silicones was observed and explained by the formation of macrocycles, which cannot redistribute or condense further while formed.

1 Introduction

Tris(pentafluorophenyl)borane [$B(C_6F_5)_3$], is a powerful and selective Lewis acid catalyst used in many reactions in organic chemistry [1–4]. Parks and Piers [5] found that $B(C_6F_5)_3$ catalyzes the hydrosilylation of carbonyl compounds. The silylation of alcohols with the formation of H_2 as the only by-product [6] and the cleavage of silyl ether and ether bonds catalyzed by $B(C_6F_5)_3$ [7, 8] provide an

F. Ganachaud
Institut Charles Gerhardt – UMR 5253 CNRS, Ingénierie et Architectures Macromoléculaires, Ecole Nationale Supérieure de Chimie de Montpellier, 8, rue de l'école normale, 34296 Montpellier cedex 5, France
e-mail: ganachau@enscm.fr

F. Ganachaud et al. (eds.), *Silicon Based Polymers*,

easy and convenient route for the reduction of primary, secondary and tertiary alcohols as well as some ethers.

Cross-coupling reactions of organosilanols with hydrosilanes or aminosilanes are particularly convenient to synthesize polysiloxanes. These reactions occur rapidly and quantitatively under mild conditions, making them a useful method in organosilicon chemistry for the generation of siloxane bonds [9–11]. This method of forming siloxanes could be very attractive since the starting materials for this process are often cheap and easy to handle. The by-product of this condensation is a neutral and harmless aliphatic hydrocarbon. Chojnowski et al. [12] described the reaction of an alkoxytriorganosiloxane with a hydridotriorganosiloxane in the presence of $B(C_6F_5)_3$. UV analysis of the catalyst properties revealed that $B(C_6F_5)_3$ was present almost exclusively as the free species in the reaction system under these conditions. A mechanism drawing a parallel between kinetics and spectral data assumed the formation of an intermediate oxonium complex with a salt-like structure formed by both reactants and catalyst. The complex may decompose to regenerate the starting substrates, exchange reaction products, or redox reaction products. The yield of polymer was relatively low, due to the unavoidable formation of cyclotetrasiloxane as a by-product, and side reactions, like homocondensation of silanol groups and disproportionation process, which may however not be significant in this system [13]. Moreover, the comparison of results with various metal-based catalysts shows that $B(C_6F_5)_3$ has higher reactivity and stereoselectivity in the cross dehydrocoupling reaction [13].

Besides, the introduction of (-$(CH_2)_n$-) long spacer groups and (-pC_6H_4-) rigid structure increased the yields of polymer, though the reactivity of hydrosilane groups decreased by introducing these groups [13]. Recently, Chojnowski and Rubinsztajn highlighted that $B(C_6F_5)_3$ cleaved -(Me_2)SiOSiO(Me_2)- groups if at least one of their silicon atoms is bonded to hydrogen [14]. All these reactions involving the siloxane bond cleavage are proposed to proceed via a metathetic mechanism involving the intermediary of a trisilyloxonium ion. We also recently showed that the condensation between dimethoxydimethylsilane (DMDMS) and 1,1,3,3-tetramethyldisiloxane (L_2H) could be catalyzed by $B(C_6F_5)_3$ directly in aqueous suspension, although the main mechanisms of chain growth were D_3 polymerization and silanol co-condensation in these humid conditions [15]. All our attempts to condense precursor silicone bricks failed in these conditions.

The purpose of this work is then to study similar $B(C_6F_5)_3$-catalyzed polycondensation of hydrido and methoxy functionalized silanes and siloxanes in solution, in a view to prepare hybrid silicones (see general reaction pathway in Scheme 1). We considered three routes, first a model reaction to prepare PDMS and two others to synthesize different hybrid silicones, starting from hydrido and methoxy functionalized bricks.

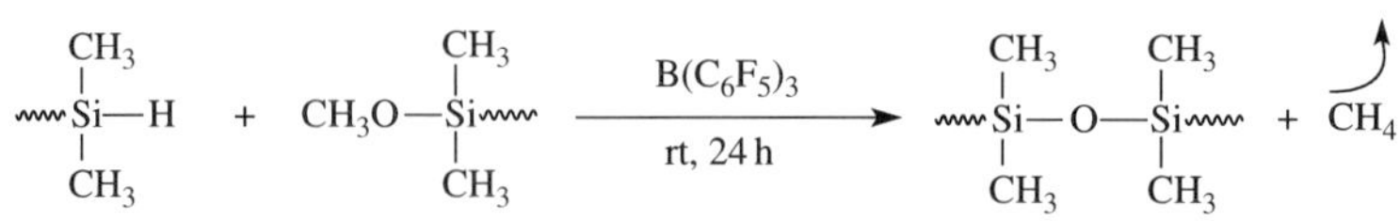

Scheme 1 General procedure of silicone polymer synthesis by cross-coupling reaction

2 Experimental Part

2.1 Materials

Tris(pentafluorophenyl)borane ($B(C_6F_5)_3$, purity 97%) was obtained from Lancaster, dimethoxydimethylsilane (DMDMS) from PCR, 1,1,3,3-tetramethyldisiloxane (L_2H, purity 97%) from ABCR, the methoxydimethylsilane (MDMS) from Fluorochem and 1,7-octadiene (purity 98%) from Aldrich. All of them were used without any further purification. All solvents were purchased from Aldrich and anhydrous toluene was stored over sodium. During the various experiments we used two batches of catalyst. Since the first one contained a fair amount of water, a small content of Brij 98® (from Aldrich), a non-ionic surfactant, was added, so that the ethylene glycol functions competed with water to complex $B(C_6F_5)_3$ and to increase its activity.

2.2 Methods

^1H and ^{29}Si NMR spectra were recorded on a BRUKER WP 250 spectrometer with TMS as the reference. All spectra were measured in $CDCl_3$ and the chemical shifts (δ) are given in ppm.

Conventional Size Exclusion Chromatography, SEC, was performed on a Spectra Physics apparatus with two PL gel columns (5 μm particle size, 300 mm length, one with 50 Å and one with 100 Å pore size) and a Styragel HR2 column (7.8 mm internal diameter × 300 mm length). The detection was achieved with a SP8430 differential refractometer. The toluene was eluted at a flow rate of 0.8 mL.min^{-1}.

Triple detection SEC was performed on a TDA300-EXD purchased from Viscotek Corporation (Crowthorne, Berkshire, UK). The separation used two ultrahydrogel linear columns (GMHHR-H column) and one guard column (HHR-H Guard Column) purchased from Viscotek. The mobile phase was toluene and the flow rate was set at 0.8 mL.min^{-1}, without a need for a flow marker. Data acquisition and calculations were performed using the Viscotek OmniSEC software version 4.1. Polymer samples in toluene were prepared at concentrations ranging from 1 to 3mg.mL^{-1}. The temperature for the SEC column set and the detector chamber was 35°C to ensure high chromatographic efficiency, stable baselines, and consistent results. The three detectors were connected in a serial arrangement such that light scattering detector was placed directly downstream of the SEC column set, followed by the refractometer and viscometer. Narrowly disperse polystyrene standard (M_p = 64500 g.mol^{-1}) solution at concentration of 1.705 mg.mL^{-1} was first injected into the SEC with the triple detection system. Three measurements were achieved by the injection of the narrow polymer standard solution. Firstly, the inter-detector delay volumes among the three detectors were compared and corrected. Secondly, the calibration factors and constants for the three detectors were obtained. Thirdly, the band broadening effect due to serial connections of three detectors was effectively corrected. It has previously been reported that the triple detection SEC analyses could be very sensitive to the interdetector delay [16,17]. To ensure that our device was well calibrated, MHS parameters were recalculated for polystyrene standard giving $K = 36.10^{-5}$ dL.g^{-1}, $\alpha = 0.620$ at 35°C (molar mass 64500 g.mol^{-1}) and dn/dc = 0.0962. Intrinsic viscosity and absolute molar mass of the sample were recalculated using determinate dn/dc value for each sample.

MALDI-TOF mass spectrometry analyses were performed on an Ultraflex (Bruker Daltonik, Bremen, Germany) time-of-flight mass spectrometer, equipped with a 337 nm, 50 Hz N_2 laser, a delayed extraction, and a reflector. An accelerating voltage of 25 kV was used in reflected mode. The matrix, 2,5-dihydroxybenzoic acid was dissolved in THF at a concentration equal to 20 g.l^{-1}. The polymer solution (1 µl of 10 g.l^{-1} in THF) was mixed with 10 µl of matrix solution. CF_3COOAg was added as a cationizing agent. The final solution (1 µl) was deposited onto the target and dried in air at room temperature before irradiation. The mass spectra represent averages over 250 consecutive laser shots. External calibrations were performed with peptide calibration standard (Burker Daltonic, Bremen, Germany).

2.3 Hybrid Precursor Preparation

Dihydrido-oligocarbosiloxane **H** was prepared in a 200 mL Carius tube purged with argon several times prior to introduction of reactants. An example of hydrosilylation procedure is given here [18]: 1,7-octadiene (5 g, 45.5 mmol), L_2H (60.9 g, 455 mmol) and Karstedt catalyst (9 µL) were introduced through a syringe. The flask was sealed and the reaction was allowed to proceed during 24 h at 90°C (L_2H boiling point is 71°C at room pressure). After cooling to room temperature the flask was opened and the excess of L_2H was evaporated at 60°C under 10^{-2} Pa. ^1H NMR analysis: 0.1 ppm (**t**, $(CH_3)_2Si$), 0.55 ppm (**m**, $\underline{CH_2}$-SiMe$_2$), 1.3 ppm (**m**, $\underline{CH_2})_nCH_2$-Si), 4.75 ppm (**s**, HSi-), $DP_n = 2.3$.

The reaction to prepare the dimethoxy-oligocarbosiloxane precursor **M** was carried out in a 200 mL one neck flask equipped with a magnetic stirrer. The flask was charged with dimethylmethoxysilane (1 g, 11 mmol) and Karstedt catalyst (1 µL) via a glass syringe and then was purged with argon. The 1,7-octadiene (0.67 g, 5 mmol) was introduced dropwise via a syringe. The flask was left at room temperature and stirred 24 h. During the course of the reaction the ^1H NMR spectrum of the reaction mixture was registered at 2 h intervals to detect the end of reaction. ^1H NMR analysis: 0.1 ppm (**t**, $(CH_3)_2Si$), 0.55 ppm (**m**, $\underline{CH_2}$-SiMe$_2$), 1.3 ppm (**m**, $\underline{CH_2})_nCH_2$-Si), 3.5 ppm (**m**, CH_3OSi-).

2.4 Polycondensation

All polycondensations proceeded equally [13]: a flask, equipped with a magnetic stirrer, was charged with $B(C_6F_5)_3$ (0.005 mmol), DMDMS or precursor **M** and solvent under argon atmosphere. Later, L_2H or precursor **H** was added drop by drop and methane release was followed with a bubbler. During the course of the reaction the SEC chromatogram of the reaction mixture was recorded at 2 h intervals to detect the end of reaction. The reaction was typically stopped after 24 h. In the model condensation, the reaction conditions were: DMDMS (0.6g, 5 mmol) and L_2H (0.67 g, 5 mmol) added drop by drop. In hybrid condensation, the reaction conditions were: DMDMS (2. mmol) and precursor **H** (2.65 mmol) in the first case; precursor **M** (1.71 mmol) and L_2H (1.88 mmol) in the second case.

3 Results and Discussion

3.1 Model Polycondensation to Prepare PDMS

In the first part of this study, we checked several conditions of reaction using the simplest components, namely L_2H and DMDMS (Scheme 2). Model reaction was carried out at room temperature, using rather high contents of monomers in stoichiometric conditions, typically 5 mmol, in 5 mL of various good solvents of PDMS and 0.1 mol% (compared to monomer) of catalyst.

A typical SEC trace obtained in the course of reaction is given in Fig. 1. It shows that slow addition of L_2H on DMDMS forms linear PDMS chains and small cycles of typically 3–5 D units, the D_3 being consumed. The large content of small cycles,

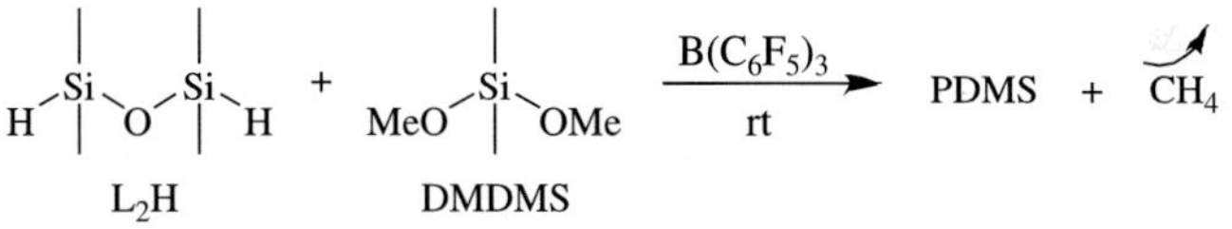

Scheme 2 Model polycondensation by cross-coupling reaction

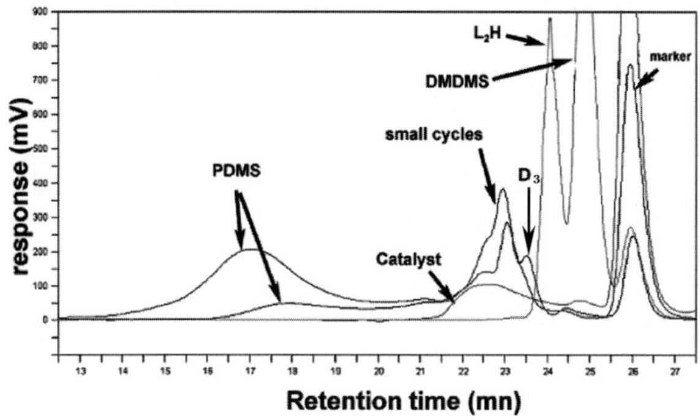

Fig. 1 Typical SEC trace obtained in this study (taken from final sample, Run 5, Table 1)

Table 1 Summary of various experiments of model reaction[a]

Run	L_2H excess (%)	Solvent	Addition time (h)	Polymer content (%)	$\overline{M}_n$ (g.mol^{-1})	ip	Cycle content (%)
1[b]	0	bulk	~2	85	1000	2.25	12
2[b]	0	CH_2Cl_2	~2	19	2000	2.67	80
3[b]	0	Et_2O	~2	46	1000	1.63	54
4[b]	0	H_2O	~1	69	4000	1.71	23
5[b]	0	Toluene	~2	59	1500	1.38	39
6[b]	0	Toluene	~1	57	16000	1.43	42
7	0	Toluene	~10	60	3000	1.67	40
8[d]	0	Toluene	<0.5	50	2700	1.51	34
9	0	Toluene	<0.5	24	7300	1.43	65
10[d]	0	Toluene	<0.5	33	3300	1.64	58
11	10	Toluene	<0.5	34	6000	1.27	65
12[d]	15	Toluene	<0.5	33	6200	1.39	62
13	30	Toluene	<0.5	21	13000	1.50	79
14	45	Toluene	~1	22	2200	1.32	78

[a] reaction conditions: 5 mmol 1,1,3,3-tetramethyldisiloxane (L_2H), 5 mmol dimethoxy-dimethylsilane (DMDMS), solvent (5 mL), and catalyst (0,1% molar compared to monomer), room temperature, 24 h; [b] addition of surfactant (Brij 98®) to dehydrate the catalyst; [c] runs stopped after 6 days; [d] Presence of high molar masses: run 8, polymer content = 15%, Mn = 58000 g.mol^{-1}; ip = 1.70; run 10: polymer content = 9%, M_n = 58000 g.mol^{-1}; ip = 1.44; run 12: polymer content = 5%, M_n = 75000 g.mol^{-1}, ip = 1.38

primarily D_4, generated here is surely a consequence of the low purity of the reactants, which were used as received, contrary to Rubinsztajn studies where reactants were thoroughly distillated and dried. [12].

Table 1 summarizes the various attempts carried out in this study as well as the final parameters of the condensation, i.e. conditions, number average molar mass $\overline{M}_n$, and cycle content at final conversion. In most instances, large contents of small cycles were generated, whereas polymer chains of low molar masses were synthesized. During certain experiments, a double distribution of the molar masses was observed (see Table 1), probably due to two types of simultaneous polymerization (vide infra).

Figure 2 and Table 1 show the content of polymer formed and the molar masses obtained using different solvents of different polarity. In solvents of moderate to high dielectric constants (typically $C_2H_5OC_2H_5$ and CH_2Cl_2), the content of cycles is too high for the experiment to be worth, whereas experiments in bulk or toluene reached a good conversion in polymer of low molar masses. The case of water as a solvent has already been treated in a separate study, and will not be further described here [15].

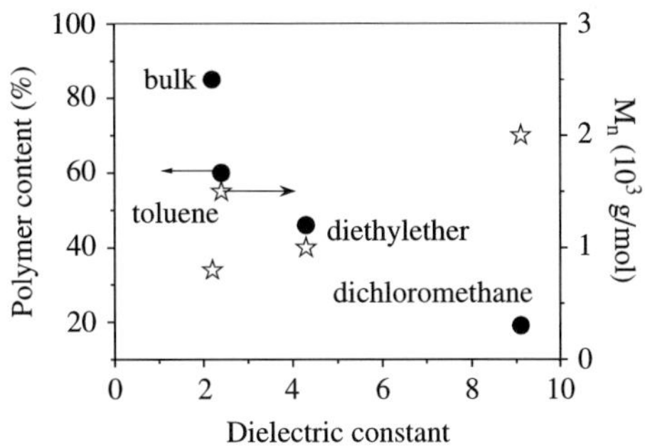

Fig. 2 Polymer conversion (●) and M_n (☆) obtained in the model reaction carried out in different solvents (see recipes Table 1)

Several experiments were also carried out using similar recipe (runs 8–10). Reproducibility was difficult to achieve in terms of molar masses and cycle contents, surely because reactants were used as received, i.e. neither distilled not dried, and also due to the difficulty to control the addition of L_2H thoroughly. This fact is particularly striking while looking at the evolution of the number average molar mass and polymer content for different L_2H period of time as reported in Fig. 3. Polymers of high molar masses were obtained only for low conversion, i.e. large contents of small cycles. There are no clear trend of the effect of addition time on final conversion and molar masses except for slow addition times, where final molar masses were systematically low.

Figure 4 shows the evolution of the polymer content as a function of the excess L_2H. The polymer content decreases to 10% when 45% excess of L_2H was introduced whereas molar masses were maximal at 30% excess. These results show that excess of L_2H does not really favor the generation of high polymer content, but that stoichiometry does not lead either to high molar masses, albeit in these unpurified conditions [19].

We further characterized a sample by MALDI-TOF and ^{29}Si NMR (run 5, Table 1). The MALDI TOF analysis shows the presence of multiple chain populations with different chain-ends, SiOH/SiOH (**A**), SiOH/SiOMe (**B**) and SiOMe/ SiOMe (**C**) (Fig. 5). A ^{29}Si NMR spectrum of the same sample is given in Fig. 6, where it shows the presence of D_4, D_5, D_6 in addition to the linear polymer; and the absence

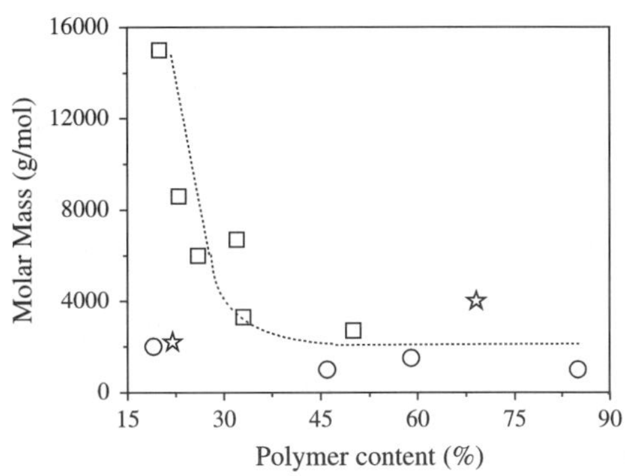

Fig. 3 Evolution of the molar masses according to polymer conversion for quick addition (~15 min, ❑), moderate one (~1 h, ❍) or slow addition (~2 h, ☆) in the model reaction (Table 1)

Fig. 4 Molecular weight, M_n (★), and polymer content (●) in the model reaction where excess L_2H were introduced in the reactor (runs 10–16, Table 1)

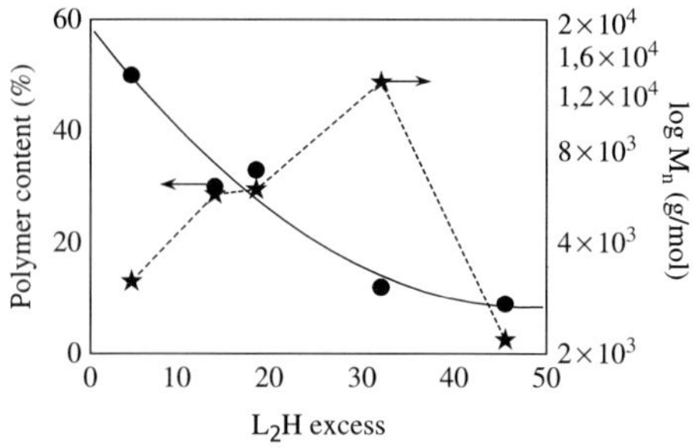

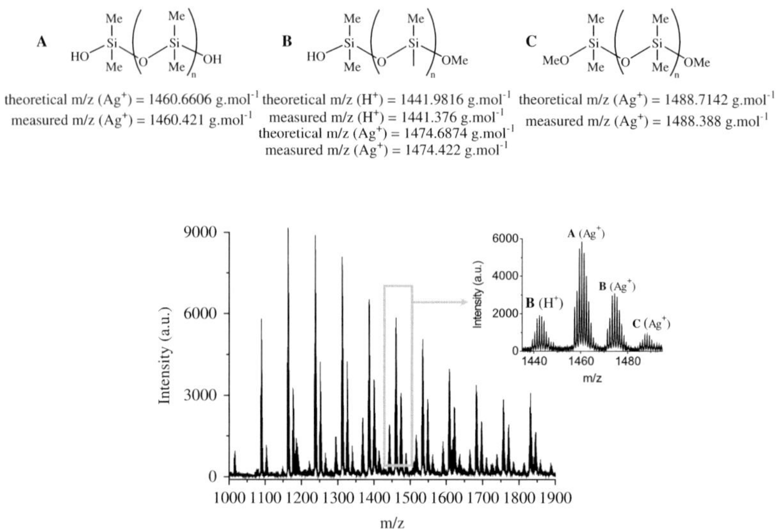

Fig. 5 MALDI TOF spectrum of final PDMS sample in run 5, Table 1

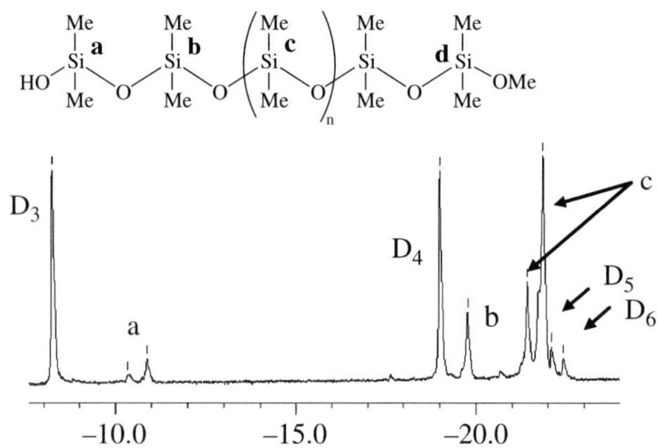

Fig. 6 ^{29}Si NMR spectrum of final PDMS sample in run 5, Table 1

of peaks corresponding to the initial reactants, L_2H at −5.2 ppm and DMDMS at −2.5 ppm. A significant content (about 30%) of SiOH groups, in addition to the expected SiOMe chain ends, are also observed. Clearly, the SiOH groups arise from hydrolysis, catalyzed by $B(C_6F_5)_3$, of SiH functions from water traces [15].

The difficulty of controlling the present condensation technique lies presumably to the three different synthetic pathways available to prepare the PDMS: the first one presented in Scheme 2, namely SiH/SiOMe condensation, the second one entailing a co-condensation of SiH and SiOH groups, and finally the cyclization of L_2H and DMDMS into D_3, then consumed by ring opening polymerization. Contrary to the study in water [15], D_3 was not found in great contents, which let one think that ROP may not be the main reaction scheme here. On the other hand, the polycondensation of silanol-terminated PDMS on SiOH or SiH moieties is likely to be catalyzed by $B(C_6F_5)_3$. We carried out a blank experiment where a disilanol functionalized telechelic PDMS (L_nOH) was let to react in toluene, containing the $B(C_6F_5)_3$ catalyst and possibly L_2H. The results are given in Fig. 7.

In the absence of L_2H, the condensation of silanol groups is slow and inefficient, since after three days the molar masses slightly shifted (note that the reactions quoted in Table 1 were carried out only during 24 h, unless otherwise stated). In the presence of L_2H, the reaction is much faster and larger molar masses are reached, which seems to indicate that the SiH/SiOH condensation is competitive with the SiH/SiOMe one. Such result may explain the double distribution found in some runs of Table 1, where both fast L_2H addition and fair content of water present the best conditions for this co-condensation reaction to occur.

3.2 Hybrid Silicone Syntheses

We then concentrated this study on the generation of hybrid silicones, starting from dihydrido oligocarbosiloxane bricks (precursor **H**) and DMDMS (Scheme 3) or

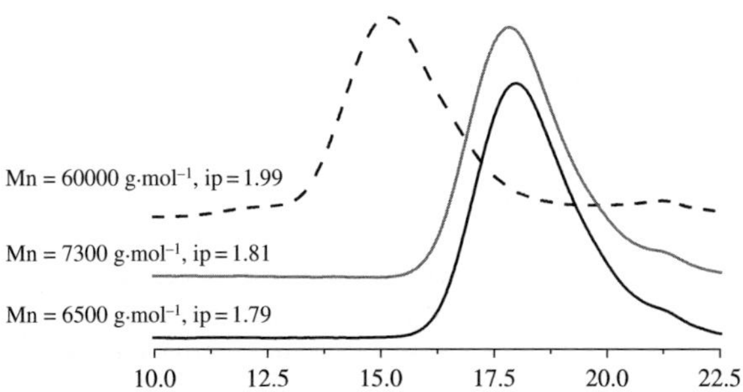

Fig. 7 Bis-silanol PDMS condensation chromatogram carried out during three days. Reaction conditions: L_nOH, 0.5 mmol, 2.1 g; $B(C_6F_5)_3$, 0.004 mmol, 2 mg; L_2H, 0.5 mmol, 0.067 g in 5 mL toluene. (*black line*) pristine L_nOH; (*grey line*) $L_nOH+B(C_6F_5)_3$, (*black dashed line*) $L_nOH+B(C_6F_5)_3+L_2H$

dimethoxy ones (precursor **M**) and L_2H (Scheme 4). Tables 2 and 3 present a summary of the reactions carried out during this study. Note that here, the production of small cycles is not likely, which is a clear advantage compared to the previous model system.

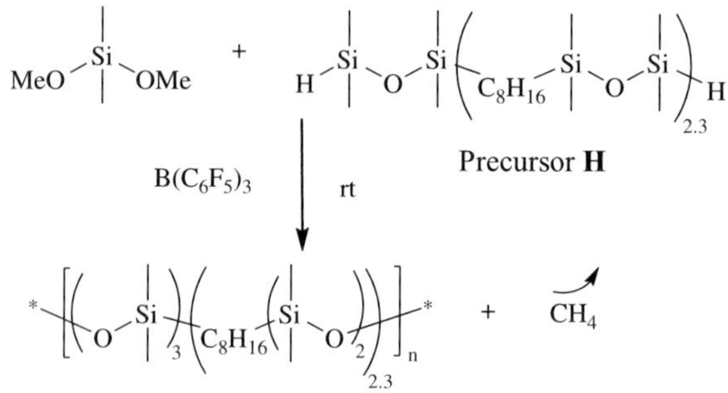

Scheme 3 Synthesis of hybrid silicones starting from DMDMS

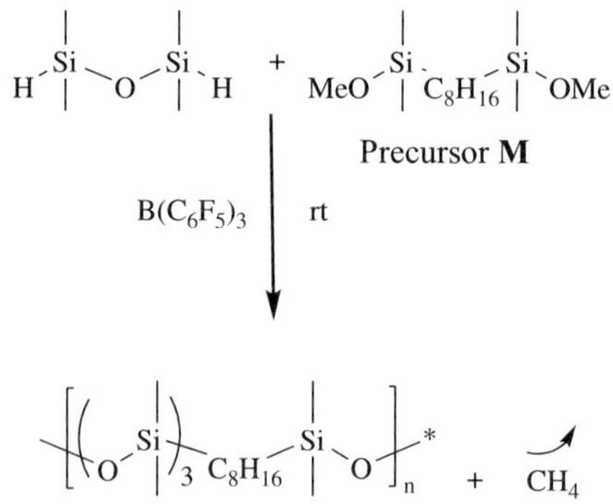

Scheme 4 Synthesis of hybrid silicones starting from L_2H

In the first system, we observed by looking at kinetics of the reaction, that the polymer content and the molar mass smoothly evolved concomitantly in 10 hours (not shown). The molar masses obtained were generally large, with Ips around 1.5, but with low precursor conversions. This is better shown in Fig. 8, where the polymer content and final molar masses are plotted against the content of precursor oligomer that did not reacted. Low conversion was systematically accompanied by the generation of large molar masses (at best 75000 g.mol^{-1}), a result that we never achieved before. When the major part of the precursor oligomer was consumed, low molar masses were observed. Such an effect could be correlated to the larger consumption of DMDMS compared to the low reactivity of the longer oligocarbosiloxane **M** (DP_n=2.3). Large molar masses could arise from long sequences of PDMS in the final hybrid copolymer through conventional acid catalyzed hydrolysis/condensation of DMDMS (vide infra).

Table 2 Summary of the various reactions using precursor **H** (Scheme 3)[a]

Run	Solvent (mL)	Polymer yield (%)	$\overline{M}_n$ (g.mol^{-1})	Ip	Precursor consumption (%)	DMDMS consumption (%)
15	5	85	8000	1.47	81	–
16	5	69	20000	1.91	61	95
17	5	11	49000	1.55	12	10
18	5	12	77000	1.57	17.5	80
19	5	6	63000	1.43	12.5	5
20	2.5	74	21000	1.67	71	90
21	2.5	86	24000	1.88	82.5	–

[a]Recipe: DMDMS (2 mmol) + precursor **H** (2.65 mmol) + $B(C_6F_5)_3$ (4.89 10^{-3} mmol), at room temperature, 24 h in toluene. The consumptions are given in percentages of introduced quantities for each reactant.

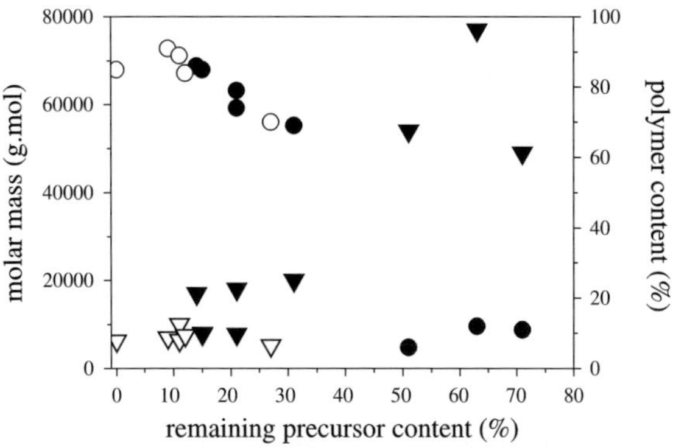

Fig. 8 Evolution of the molar mass (▼) and polymer content (○) of hybrid silicones prepared from precursor **H** (Table 2: full symbol) and precursor **M** (Table 3: empty symbol). Remaining content represents the fraction of unreacted precursor

Since the previous protocol led to low precursor conversions, we turned our attention to the second system, starting from precursor **M** and excess L_2H (Scheme 4). To confirm the validity of working off stoichiometry, we plotted in Fig. 9 the evolution of the molar mass and polymer content of an hybrid silicone prepared from precursor **M** and by fractionated addition of L_2H. The molar masses increase smoothly up to the end of the L_2H addition. In parallel, the polymer content grows more quickly up to a constant level.

Table 3 summarizes the various experiments carried out with precursor **M**. Compared to the previous system, it is now clear that most of the oligocarbosiloxane and DMDMS are consumed, and the reproducibility is good in terms of contents of polymers, their molar masses and Ips. Logically, the final molar masses are in this case relatively low compared to the system with precursor **H** (see also Fig. 8), which let one think that true alternated sequences are generated here.

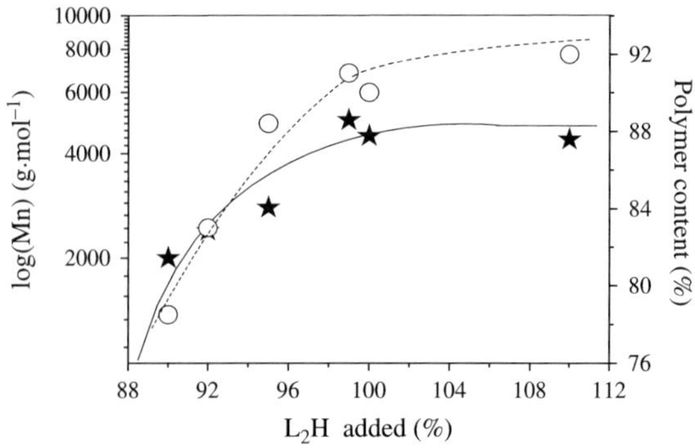

Fig. 9 Evolution of the polymer content (○) and molar mass (★) in the polycondensation of precursor **M** and increasing L_2H concentration (see other reactant contents in footnote, Table 3)

Table 3 Summary of the various reactions using precursor **M** (Scheme 4)[a]

Run	Solvent (mL)	Polymer yield (%)	$\bar{M}_n$ (g.mol^{-1})	Ip	Precursor consumption (%)	L_2 H consumption (%)
22[b]	2,5	70	5200	1.4	60	100
23	0,5	85	6300	1.42	100	100
24	0	91	7000	1.55	87	100
25	0,5	89	10000	1.55	84	100
26	1	89	64000	1.35	84	100
27	2	84	7500	1.49	82	88
28	0	88	7000	1.44	82	100

[a]Recipe: L_2H (1.88 mmol) + precursor **M** (1.71 mmol) + $B(C_6F_5)_3$ (4.71 10^{-3} mmol), room temperature, 24 h in toluene; [b)]stoichiometry test: L_2H (1.71 mmol) + precursor **M** (1.71 mmol) + $B(C_6F_5)_3$ (4.28 10^{-3} mmol). The consumptions of reagents were calculated in relation to the introduced mass.

The effect of solvent proportion on the reaction was also tested. Figure 10 shows a decrease of polymer content when the solvent volume increases, presumably since high dilution favors the generation of macrocycles, which are not able to further condense or redistribute.

To confirm the possible formation of macrocycles, otherwise saying the absence of chain-ends, ^{29}Si NMR was carried after the end of reaction. The spectrum (Fig. 11) shows signals for D and M (hybrid segment) units and D_3 arising from L_2H self-condensation. Neither methoxy nor silanol end-groups could be traced on this spectrum, nor remaining L_2H at −5.2 ppm or precursor **M**, around 15 ppm (see for instance [20]). The residual peak at +2.5 ppm could not be ascribed.

Another elegant method to characterize linear versus cyclic macromolecular chains is the SEC triple detection analysis. Absolute molar masses and intrinsic

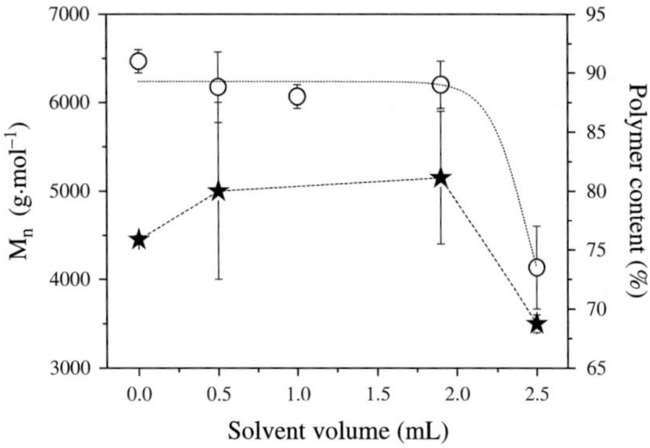

Fig. 10 Evolution of the polymer content (○), and M_n (★) of hybrid silicones prepared from precursor **M** and L_2H in increasing toluene dilution

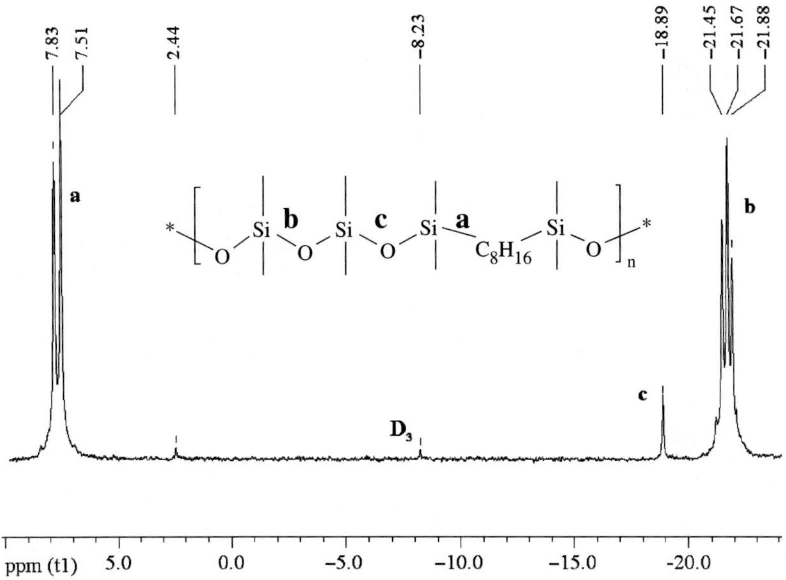

Fig. 11 ^{29}Si NMR spectrum of the hybrid silicone synthesized from L_2H and precursor **M** with $B(C_6F_5)_3$ (from run 26, Table 3)

viscosities, independent of calibration curves, can be determined if one knows the dn/dc value for each polymers (see experimental part). This is done by injecting known concentrations of polymers and looking at their refractive index variations. The dn/dc for PDMS and hybrid silicones in toluene at 35°C were calculated as 0.0764 and 0.0328, respectively. Mark Houwink Sakaruda plots for PDMS and hybrid silicones are drawn in Fig. 12.

Figure 12 shows the overlapping of the curves for trimethyl terminated commercial PDMS and those of the PDMS synthesized with the $B(C_6F_5)_3$. We can thus conclude that the PDMS chains prepared in this study exhibits a linear structure, a point that was confirmed both by MALDI-TOF and NMR studies already.

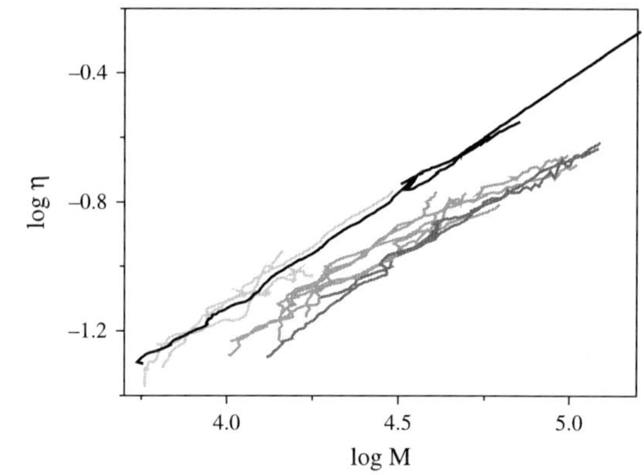

Fig. 12 Mark-Houwink-Sakaruda plots for commercial PDMS terminated by $SiMe_3$ (*black*), PDMS synthesized with $B(C_6F_5)_3$ (*grey*), hybrid silicones synthesized with $B(C_6F_5)_3$ (*green*: from precursor **H**/ magenta: from precursor **M**) (See also Plate 15 in the Colour Plate Section)

Besides, the hybrid silicones synthesized using the method in Scheme 3 produced scattered linear lines, whereas those obtained from precursor **M** mostly overlap on a single line, with a *y* intercept lower than 0.2 units compared to the PDMS one. This difference is typical of the theoretical lower hydrodynamic volumes for cyclic structures, compared to linear ones [21, 22]. From this analysis, it is reasonable to conclude that the formation of macrocycles prevents high molar masses to be reached, since the hybrid silicones are not capable of ring scission through redistribution reactions within the macromolecular chains, unless PDMS chains.

From the triple detection SEC, it is also possible to determine the alpha coefficient in the Mark Houwink equation, which indicates the nature of interactions between the polymer and the solvent: one can differentiate between a good solvent ($0.65 \leq \alpha \leq 0.85$), a bad solvent ($0.5 \leq \alpha \leq 0.65$), a theta solvent ($\alpha = 0.5$) or a non-solvent ($\alpha < 0.5$) [23]. The alpha coefficient could not be obtained for hybrid silicones obtained from precursor **H**, which varies between 0.59 and 0.80. The one calculated for the hybrid silicones obtained from precursor **M** was 0.62 ± 0.02, against 0.7 ± 0.01 for PDMS. Toluene is thus a good solvent for PDMS and a bad one for hybrid silicones. The fact that during the synthesis, hybrid silicones tend to form strained coils in toluene could explain the preferential cyclization of this latter. The generation of macrocycles could tentatively be prevented if the synthesis of hybrid silicone was carried out in a mixture of solvents, i.e. toluene, a good solvent of PDMS, and a co-solvent allowing the solubilization of the hydrogenated alkane part, such as DMF or THF.

4 Conclusive Remarks and Prospects

The polycondensation reaction between telechelic hydride and telechelic methoxy siloxane in the presence of very low levels of $B(C_6F_5)_3$ has been described. The compromise between large polymer content and reasonable molar masses is found by tailoring the addition rate and excess content of L_2H reagent. Rubinsztajn et al. showed that with silphenylene-co-PDMS copolymers, the reaction was completed over a period of 2–3 h, and high molecular weight (M_w = 10K–50 Kg.mol^{-1}) materials were obtained [11]. During hybrid silicon synthesis a conversion limit is observed, at best 90% of polymer is produced. The low conversion can be explained by the partial hydrolysis of SiH which are not available for the condensation with $SiOMe_2$. The PDMS conversion is also limited but we can explain such effect with the formation of small cycles. For the hybrid silicones, no small cycles are produced, but the formation of macrocycles stops the condensation process abruptly.

Next works will be devoted to the synthesis of hybrid silicones mixing both precursors together, in order to limit the generation of undesired small cycles from L_2H hydrolysis and to favor high conversions and high molar mass polymers. A preliminary experiment showed that, at the stoichiometry, up to 93% of a hybrid polymer of large molar mass (14000 g.mol^{-1}) and reasonable Ip (1.52) with less than 5% of each oligocarbosiloxane bricks remaining. We will also further characterize the microstructure of hybrid silicones generated so far, by DSC and 2D ^{1}H NMR, as well as check their thermal resistance, typically by TGA.

Acknowledgements Francine Guida-Pietrasanta, Amédée Ratsimihety and Bernard Boutevin are acknowledged for suggestions and support, and Christine Joly-Duhamel for triple detection SEC analysis. CL's Ph. D. funding was supported by a grant from the French Ministry of Research and Education.

References

1. W. E. Piers and T. Chivers, Chem. Soc. Rev., 26, 1997, 345.
2. W. E. Piers, Adv. Organomet. Chem., 52, 2004, 1.
3. K. Ishihara and H. Yamamoto, Eur. J. Org. Chem., 3, 1999, 527.
4. G. W. Coates, P. D. Hustad and S. Reinartz, Angew. Chem. Int. Ed., 41(13), 2002, 2236.
5. D. J. Parks and W. E. Piers, JACS., 118, 1996, 9440.
6. J. Blackwell, K. L. Foster, V. H. Beck and W. E. Piers, J. Org. Chem., 64, 1999, 4887.
7. V. Gevorgyan, M. Rubin, S. Benson, J. X. Liu and Y. Yamamoto, J. Org. Chem., 65, 2000, 6179.
8. V. Gevorgyan, J. X. Liu, M. Rubin, S. Benson and Y. Yamamoto, Tetrahedron. Lett., 40, 1999, 8919.
9. US 2004127668 (2004), GE Plastics, invs.: S. Rubinsztajn and J. A. Cella; Chem. Abstr. 2004, 534027.
10. S. Rubinsztajn and J. A. Cella, Polym. Preprints, 45, 2004, 635.
11. S. Rubinsztajn and J. A. Cella, Macromolecules, 38, 2005,1061.
12. J. Chojnowski, S. Rubinsztajn, J. A. Cella, W. Fortuniak, M. Cypryk, J. Kurjata and K. Kazmierski, Organomet., 24, 2005, 6077.
13. D. Zhou and K. Yusuke, Macromolecules, 38 (16) 2005, 6902.
14. J. Chojnowski and S. Rubinsztajn, Macromolecules, 39, 2006, 3802.
15. C. Longuet, C. Joly-Duhamel and F. Ganachaud, Macromol. Chem. Phys., 208(17), 2007, 1883.
16. M. D. Zammit and T. P. Davis, Polymer, 38, 1997, 4455.
17. M. D. Zammit, T. P. Davis and K. G. Suddaby, Polymer, 39, 1998, 5789.
18. US 5453528 (1996), Dow Corning Corp, invs.: B. Boutevin, F. Guida-Pietrasanta, A. Ratsimihety and G. Caporiccio, Chem. Abstr. 1995, 875028; US 5527933 (1996), Dow Corning Corp, B. Boutevin, F. Guida-Pietrasanta, A. Ratsimihety and G. Caporiccio, Chem. Abstr. 1996, 425682

19. L. Lestel, H. Cheradame and S. Boileau, Polymer, 31, 1990, 1154
20. K. R. Brzezinska, 1 K. B. Wagener,1 G. T. Burns, J. Polym. Sci.: Part A Polym. Chem., 37, 1999 849–856.
21. B. Lepoittevin, M. A. Dourges, M. Masure, P. Hémery, K. Baran and H. Cramail, Macromolecules, 33, 2000, 8218.
22. B. Lepoittevin, X. Perrot, M. Masure and P. Hémery, Macromolecules, 34, 2001, 425.
23. P. J. Flory, « Principles of Polymer Chemistry », Cornell University Press, 1953.

Hydrosilylation of Polymethylhydrogenosiloxanes in the Presence of Functional Molecules Such as Amines, Esters or Alcohols

Corinne Binet, Matthieu Dumont, Juliette Fitremann, Stéphane Gineste, Elisabeth Laurent, Jean-Daniel Marty, Monique Mauzac, Anne-Françoise Mingotaud, Waêl Moukarzel, Guillaume Palaprat, and Lacramioara Zadoina

Abstract This article presents an overview of the methods used in our laboratory to successfully perform hydrosilylation reactions on polymethylhydrogenosiloxane polymer in the presence of several functional molecules, such as esters, acids, amines, alcohols or carbamates. The used catalysts are platinum based homogeneous systems. In each case, experimental parameters such as choice of catalyst, catalyst/ Si-H functions ratio, temperature have been examined. Also, a detailed kinetic study was performed in the case of liquid crystal molecules in order to better characterize the system.

1 Introduction

Polysiloxanes presenting functional groups receive an increasing interest because of their particular physical and chemical properties. Indeed, polydimethylsiloxane (PDMS) is a very versatile polymer, which has the characteristics of presenting the lowest glass transition temperature (ca. −123°C), making it a very soft polymer. It is optically transparent above 230 nm, and it is non toxic for applications in biomedical science. Moreover, it is permeable to gases. These properties explain the very wide use of so called silicones in everyday life, extending from sealants or adhesives for construction, insulators for use in electronics, additives for personal care formulations, objects such as tubing or contact lenses in health care [1]. In order to tune the exact property of each silicone used to each application, it has become essential to dispose of a wide diversity of functionalized polysiloxanes. This explains why hydrosilylation has become a key reaction in this field since it

A.-F. Mingotaud
Lab. des Interactions Moléculaires et Réactivité Chimique et Photochimique, UMR 5623 CNRS, Université de Toulouse, 118 Rte de Narbonne, 31062 Toulouse cedex 9, France
e-mail: afmingo@chimie.ups-tlse.fr

F. Ganachaud et al. (eds.), *Silicon Based Polymers*,

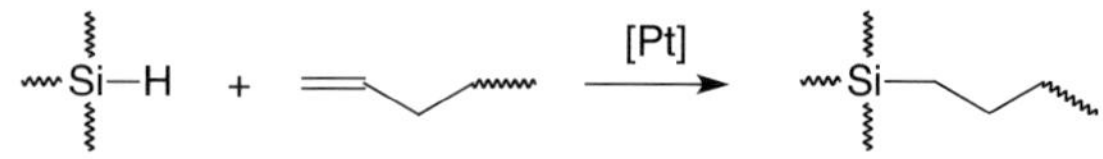

Fig. 1 Hydrosilylation reaction

enables the obtaining of various functionalized polysiloxanes starting from a simple polyhydrogenosiloxane chain [2, 3]. This importance is particularly shown by the number of references dealing with this reaction which amounts to ca. 5000, 70% of which are patents. Although this has been studied for a long time and a number of catalysts are now commercially available, a continuous research is still going on, because of several drawbacks. Hydrosilylation involves the reaction between a silicon hydride moiety with an alkene molecule. Depending on the catalyst and the alkene, side reactions can be observed such as migration of the double bond inside the carbon chain skeleton or reaction of functional groups with the catalyst, possibly leading to its poisoning or degradation of the functional group and/or of the silane functions [2–9]. Poisoning is typically the case when trying to perform hydrosilylation in the presence of amines or sulfur containing molecules. When carboxylic acids are present, these react with silane functions and lead to crosslinking. Among all metallic complexes catalyzing this reaction, those with platinum are the most widely used for the functionalization of polysiloxanes. Indeed, they present the advantage of minimizing side reactions in the presence of end alkenes and afford almost exclusively the desired adduct (Fig. 1).

Historically, hexachloroplatinic acid in alcohol, so-called Speier's catalyst, was the first homogeneous catalyst and remained the most used for a long time. This platinum (IV) complex has been replaced by other (II) or (0) species, some of which are presented in Scheme 1. Their advantages are a reactivity that can be tuned by the choice of solvent or inhibitor and the absence of hydrochloride formation. Moreover, they are now commercially available as homogeneous solutions, which ensures a perfect dissolution of the complex, and some of them are colorless.

$(Et_2S)_2PtCl_2$ Dichloro bis(diethylsulfide) platinum (II)
$DCPPtCl_2$ Dichloro(dicyclopentadienyl) platinum (II)

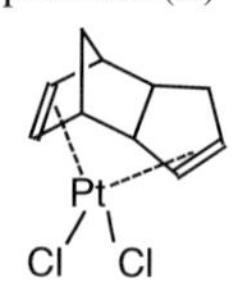

Pt(0) $(vinylMe_2Si\text{-}O\text{-}SiMe_2vinyl)_{1.5}$ Karstedt's catalyst, platinum divinyltetramthyldisiloxane complex

Scheme 1 Homogeneous platinum catalysts

This article presents an overview of our experience using these catalysts in the case of alkenes bearing functional groups, such as esters, acids, amines, alcohols or carbamates. It is divided in three parts. The first one describes the formation of liquid crystal polysiloxanes or elastomers by the grafting of liquid crystal moieties.

The second part deals with the grafting of amines or alcohols with protection/deprotection methods and the third one presents the grafting of alkenes bearing free hydroxyl or acid functions. Finally, a table summarizing all experiments is also presented.

2 Experimental Part

2.1 Materials

The syntheses of the functional alkenes have already been described in the literature [10, 11]. Polyhydrogenomethylsiloxane ($\overline{DP}_n$=80 (ABCR or Sigma Aldrich), poly(hydrogenomethyl-co-dimethyl siloxane) (synthesized according to an already published procedure [12]), α-methyl benzyl amine (Aldrich), diaminonaphthalene (Aldrich), diethyl 4-nitrobenzylphosphonate (DE4NBP) (Acros), cinchonidine (Aldrich) and 1,1,1-trifluoro-2-(trifluoromethyl)pent-4-en-2-ol (OHFl) (Fluorochem) were used as received. $DCPPtCl_2$ and Karstedt's catalyst were purchased from Gelest, $(Et_2S)_2PtCl_2$ from Aldrich.

2.2 Instruments

^1H and ^{29}Si NMR spectra were recorded on a Bruker Avance 300 spectrometer respectively at 300.13 and 75.6 MHz at room temperature. $CDCl_3$ and toluene were used as solvents. For ^{29}Si NMR, $Cr(acac)_3$ (0.03 M) was added in the tube and a delay between pulses of 20 s was set. IR spectra were recorded on a Perkin-Elmer IR FT 1760-X. The average molecular weight of the linear polymers was determined by size exclusion chromatography in THF (flow rate 1.0 mL.min^{-1}) on an apparatus equipped with a Waters refractive index detector, a Waters column pack (Ultrastyragel 10^4, 10^3, 100 Å) and a Minidawn Wyatt light scattering detector.

2.3 Hydrosilylation Reactions

Two typical examples of hydrosilylation processes are presented below.

Polymethylhydrogenosiloxane (1.3 mmol of SiH groups), 1.36 mmol 4-methoxy-4'-(3-butenyloxy)phenyl benzoate (105 mol% relative to the SiH units) and 0.067 mmol 4-undecenyloxybenzophenone (5 mol%) were mixed with 3 mL of degassed toluene under argon atmosphere. To this solution was added about 2.5 μmol of $DCPPtCl_2$ in 0.5 mL of degassed toluene. The resulting mixture was stirred for 65 h at 70°C under argon atmosphere. Toluene was evaporated and the resulting polymer was kept in the dark until its use in films. ^1H NMR δ, ppm: 0.23 (br s, Me-Si); 0.6–0.8 (m, -Si-CH_2-); 1.0–2.0

(m, CH_2); 3.8–4.0 (3 br s, CH_2-O and Me-O); 6.8–7.4 (m, aromatics); 7.5–7.9 (m, benzophenone aromatics); 8.0–8.3 (mesogen aromatics). ^{13}C NMR: 165.5 164.2 (C_{Ar}-O); 157.1 (C_{Ar}-O); 144.7 (C_{Ar}-OCO); 138.4 132.5 130.8 129.4 128.6 126.4 125.7 122.9 115.5 115.3 114.1(aromatics); 69.4 68.2 (CH_2-O); 55.7 (MeO-); 33.2 20.0 17.8 (CH_2); 0.3 (Me-Si). ^{29}Si NMR: −22.5 (O-Si -O).

Polymethylhydrogenosiloxane (1.5 mmol of SiH groups), 1.28 mmol 4-methoxy-4'-(3-butenyloxy)phenyl benzoate (85 mol%), 0.45 mmol OHFl (30 mol%) and 0.067 mmol 4-undecenyloxybenzophenone (5 mol%) were mixed with 3.8 mL of degassed toluene under argon atmosphere and protected from light. To this solution was added 40 µL of Karstedt's catalyst (0.2 mol% relative to the Si-H groups). The resulting mixture was stirred for 24 hours at room temperature under argon atmosphere. Toluene was evaporated and the resulting polymer was kept in the dark until its use in films. ^{1}H NMR: 0.13 (br s, Me-Si); 0.4–0.6 (m, -Si-CH_2-); 1.2–1.8 (m, CH_2); 3.8–3.9 (3 br s, CH_2-O and Me-O); 6.8–7.1 (m, aromatics); 7.4–7.9 (m, benzophenone aromatics); 8.0–8.3 (mesogen aromatics).

The reaction was followed by IR spectroscopy by the disappearance of the $\nu_{Si\text{-}H}$ band at 2160 cm^{-1}.

3 Results and Discussion

3.1 *Comparison of Catalysts in the Grafting of Liquid Crystal Side Groups*

In the field of so-called intelligent materials, the possibility of combining the liquid crystal properties to the mechanical strength of polymers has been developed and particularly studied using polysiloxanes based materials [13, 14]. Indeed, owing to the high flexibility of the siloxane chain, mesophases can easily self-organize while the polymer chain adapts itself to this order. This is why a lot of liquid crystalline polysiloxanes have been synthesized [15]. They can also lead quite easily to the formation of elastomers through crosslinking.

Two typical liquid crystal molecules, particularly studied in our laboratory and which can be linked to polyhydrogenosiloxane by hydrosilylation, are represented in Fig. 2 [11]. Basically, they are composed of three parts: an accessible double

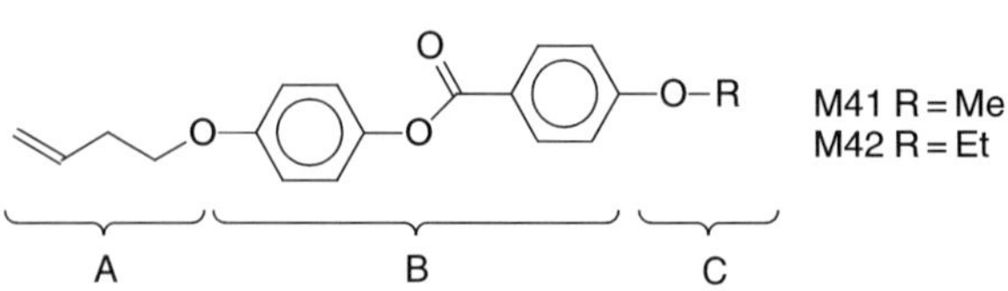

Fig. 2 Chemical structure of liquid crystal groups

bond for the hydrosilylation (A), an aromatic core for intermolecular interactions (B) and a small alkyl group to inhibit the formation of crystal order (C).

Although these molecules do not present any functions that are potentially poisonous to the platinum catalyst, the linking to polyhydrogenomethylsiloxane aimed at choosing the best experimental conditions to obtain a full reaction without any side reaction and a color as low as possible. Therefore, several hydrosilylations were followed by IR spectroscopy using both $DCPPtCl_2$ and Karstedt's catalyst. IR spectroscopy is particularly useful in this case owing to the very intense $\nu_{Si\text{-}H}$ band at 2160 cm^{-1}. Figure 3 shows the comparative kinetics of both catalysts (400 ppm, 60°C) for the reaction of M41 with a copolymer poly(methylhydrogeno-co-dimethyl siloxane) (59% of hydrogeno units) in toluene. These catalysts clearly behaved differently at the beginning of the reaction, $DCPPtCl_2$ being slower, although both led to the same result of ca. 80% after 2 hours. In some examples, an induction period as long as 20 min has been described in the literature with $DCPPtCl_2$ due to a slow appearance of the active species [6]. In our case, no such induction was observed.

The influence of temperature was also assessed on these two catalysts. The results are presented in Fig. 4 for M42 and show that decreasing the temperature to 40°C led to a very slow reaction with $DCPPtCl_2$ whereas the rate remained almost identical for Karstedt's catalyst.

In order to better characterize the system, a further kinetic study was carried out on these two catalysts. Hydrosilylation mechanism has been thoroughly studied in the literature. This is a complex system, since the mechanism depends altogether on the catalyst, the reactants and the experimental conditions. This also explains why for each new reactant, a whole new experimental set-up has to be developed. In most cases already described, the limiting step is the insertion of platinum in the Si-H bond, leading to an apparent rate of reaction independent of double bond concentration:

$$r = k_{app}[Si - H]^{\gamma}$$

In the case of hydrosilylation performed on a polymer, it has also been shown that the presence of several close hydrogenosilane functions may modify the mechanism,

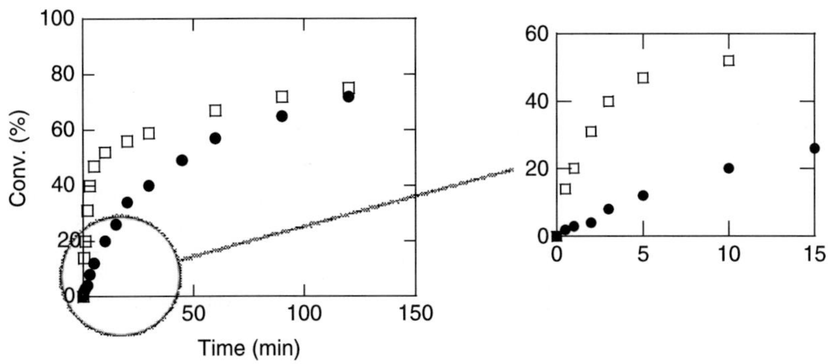

Fig. 3 Comparison of hydrosilylation kinetics with $DCPPtCl_2$ (•) and Karstedt's catalyst (❑). $[C{=}C]_0{=}[Si\text{-}H]_0{=}0.1M$. Right graph: zoom of the beginning of the reaction

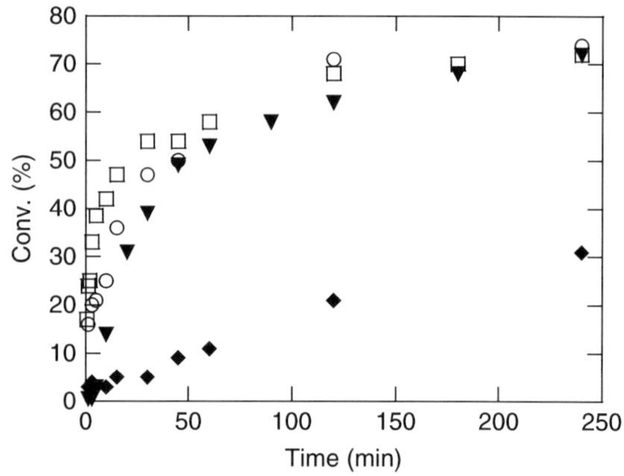

Fig. 4 Comparison of hydrosilylation kinetics with $DCPPtCl_2$ (◆ at 40°C, ▼ at 60°C) and Karstedt's catalyst (o at 40°C, ❑ at 60°C)

leading to fractional partial order for Si-H functions and a kinetic order of 1 for the alkene [16]. In other cases, kinetic orders higher than 1 may also be observed for Si-H functions [6]. Thus for both catalysts, the kinetic curves were treated according to a method described by Coqueret to determine first the global kinetic order, and in a second step, the partial kinetic orders:

$$r = -\frac{d[Si-H]}{dt} = -\frac{d[C=C]}{dt} = k[Si-H]^a[C=C]^b$$

where a and b are the partial kinetic orders and the global kinetic order is n = a + b

In the case where both reactants are in a stoichiometric ratio, it is possible to express the disappearance of silane moieties according to the following equation:

$$-\frac{d\alpha}{dt} = k[Si-H]_0^{n-1}\,\alpha^n \text{ where } \alpha = \frac{[Si-H]}{[Si-H]_0}$$

The integrated form of this equation can allow the determination of n by a graphical method.

$$F_n(\alpha) = \square\ \alpha^{-n}\,d\alpha$$

In a second step, based on this overall reaction order, $F_n(\alpha)$ can be expressed as a function of a and b, therefore leading to their determination.

This mathematical treatment enabled us to determine an overall kinetic order of 2 for $DCPPtCl_2$ and partial orders for silane and double bonds both equal to 1. Therefore, in this case, the reaction rate could be written as followed:

$$r = -\frac{d[Si-H]}{dt} = -\frac{d[C=C]}{dt} = k[Si-H]\,[C=C]$$

For Karstedt's catalyst, the same calculation showed that the overall kinetic order could be either 2 or 3, both curves $F_n(\alpha) = f(t)$ presenting similar shapes. We therefore could not express the reaction rate for this system.

This kinetic study enabled us to show that both $DCPPtCl_2$ and Karstedt's catalyst are satisfying catalysts for liquid crystal moieties, leading to ca. 80% of conversion at 60°C ($[C=C]_0=[Si\text{-}H]_0=0.1M$, Pt quantity 40 ppm) after 2 hours of reaction. In order to ensure a full conversion in silane, the time of reaction had to be increased to 24 h.

Beside the synthesis of linear polysiloxanes, hydrosilylation was also used to obtain liquid crystalline elastomers. This can be performed in two ways, either a one pot reaction involving a diolefin together with the functionalized mesogen or a two step process involving first the synthesis of a linear polysiloxane, bearing adequate functions for a further crosslinking. For example, a photosensitive group such as benzophenone could be introduced (Fig. 5, x and y are theoretical values). In order to ensure a full conversion in silane, the time of reaction was increased here to 65 h [17].

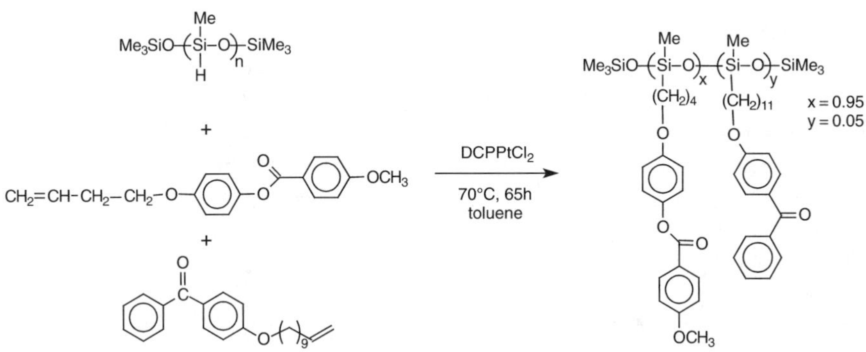

Fig. 5 Grafting of mesogens and benzophenone moieties

3.2 *Grafting of Functional Molecules with Protected Amines or Alcohols*

Some functional groups, especially protic groups such as alcohols, amines or carboxylic acids are hardly compatible with hydrosilylation. However, there is a high interest in introducing these groups in polysiloxane side chains, affording new properties, such as solubility in polar solvents, including water. This in turn enables a wider use of the material, which is essential for example in the case of medical applications. This is the reason why several types of amphiphilic polysiloxanes have been developed in the past and are commercially available. Polysiloxanes grafted with polyethyleneoxide, quaternary ammonium or pyrolidinone side-groups are some examples [18–21].

Another interest in the introduction of such groups lies in the opportunity to take advantage of the nucleophilicity (amines) of these side-groups for further functionalizing polysiloxanes. Conversely, methods for grafting a nucleophilic molecule on electrophilic side-groups on polysiloxanes are based essentially on the ring opening of epoxide (allylglycidylether) [22], succinic anhydride-functionalized polysiloxanes, isocyanate-functionalized polysiloxanes or chloride side-chains.

However, during hydrosilylation, protic groups can induce cross-linking side reactions (OH, COOH, NH_2), or can poison the catalyst (NH_2). In order to avoid reactions due to the protic groups, we adopted first a strategy using protective groups for amines or alcohols. When using protective groups, the deprotection giving back the free protic group is very often difficult, since it uses most of the time acids or bases in polar solvents, leading to the degradation of the polysiloxane backbone.

3.2.1 Polyaminopropylsiloxanes Obtained by Protection/Deprotection Method

Aminopolysiloxanes are commercially available, with 1 to 7 mol% of aminoalkyl side-chains and maximum molar masses around 8000 g.mol^{-1}. By polycondensation reactions, larger molecular weights and higher grafting contents cannot be obtained. In order to get a larger panel of molecular weights and grafting contents, the introduction of protected aminopropyl groups followed by deprotection has been described. We tried to reproduce one of these methods, by using N-terbutoxycarbonyl (tBoc) allylamine. Hydrosilylation of a poly(hydrogenomethyl-co-dimethyl)-polysiloxane with a mixture of dodecene and N-tBoc-allylamine in the presence of $(Et_2S)_2PtCl_2$ (1% mol/ Si-H, toluene, 70°C) afforded the grafted polysiloxane in good yields (Fig. 6). The reaction progress showed the disappearance of 90% of the silane functions after 40 min, while 24 h were necessary for their complete disappearance. After purification by precipitation in acetonitrile, the integrity of the polysiloxane backbone was checked by SEC in THF. By ^{29}Si NMR, only very tiny signals around −57 ppm (D^{OM}, -Si-O-SiCH$_3$(-OSi(CH$_3$)$_3$)-OSi-) and −67 ppm (T, -Si-O-SiCH$_3$(-OSi(R,CH$_3$,OSi-)-OSi-) were observed (Fig. 7), pointing out a slight rearrangement of the polysiloxane chain during hydrosilylation [23]. But neither macroscopical cross-linking nor functional group decomposition, as seen by ^{1}H and ^{13}C NMR, was observed. The platinum catalyst $(Et_2S)_2PtCl_2$ has been

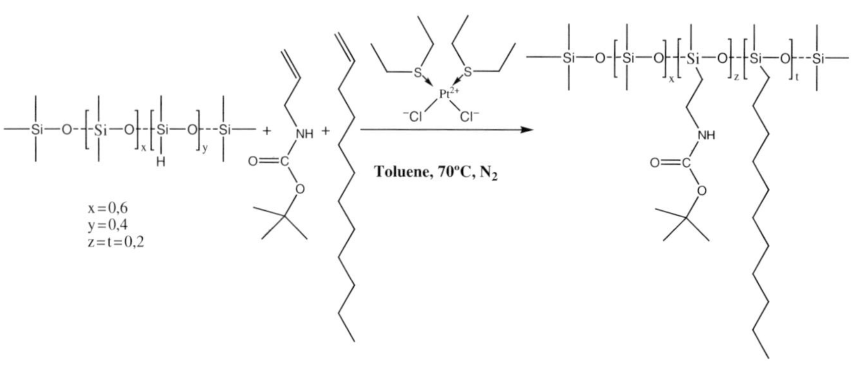

Fig. 6 Synthesis of poly(N-tBoc-aminopropylmethyl-co-dodecylmethyl-co-dimethyl)-siloxane

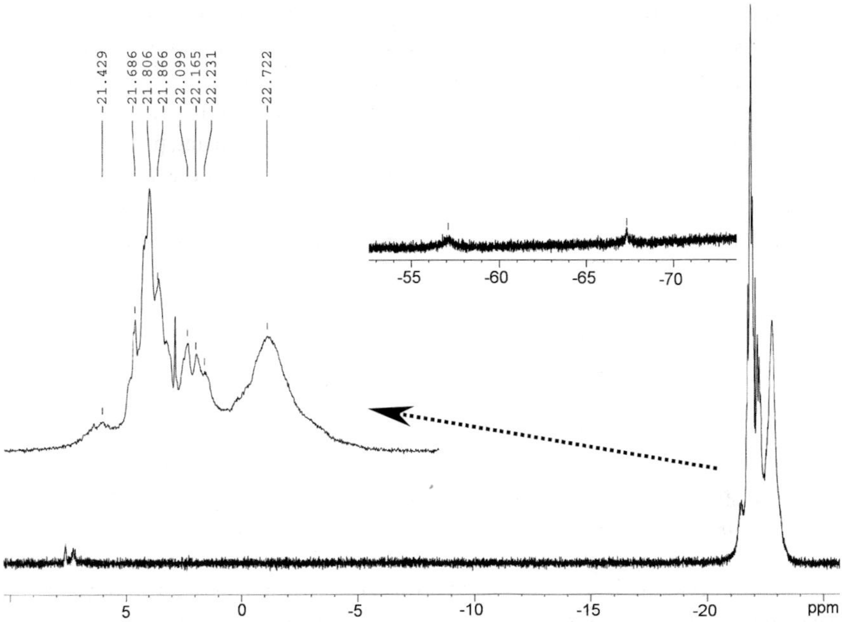

Fig. 7 ^{29}Si NMR spectrum of poly (N-tBoc-aminopropylmethyl-co-dodecylmethyl-co-dimethyl)-siloxane (signals centered at: 22.7 ppm: –Si-$(CH_2)_3$-NHtBoc; 22.1 ppm: -Si-dodecyl; 21.8 ppm: $Si(CH_3)_2$)

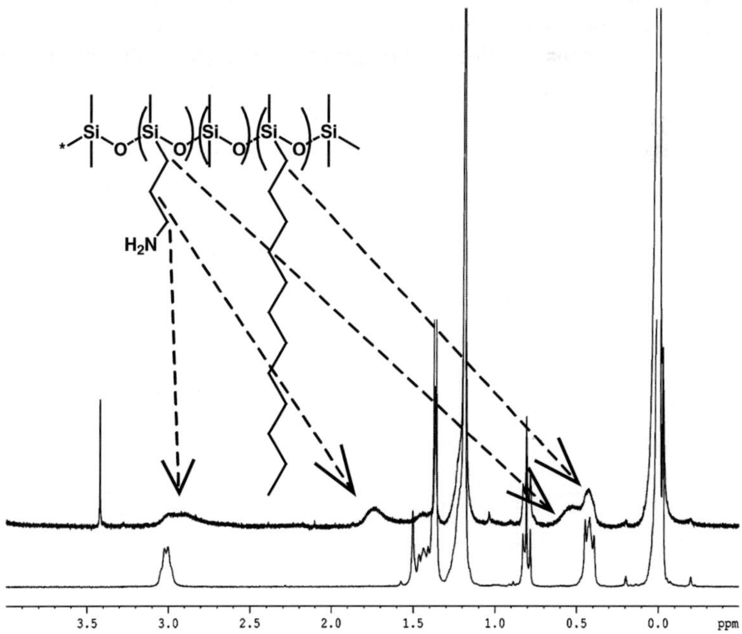

Fig. 8 Compared ^{1}H NMR spectra of poly(N-tBoc-aminopropylmethyl-co-dodecylmethyl-co-dimethyl)-siloxane *(lower)* and the deprotected aminopropyl polysiloxane *(upper)*

more especially selected according to systematic tests performed by Stadler et al. for grafting aldonamides by hydrosilylation. They demonstrated that reactions worked better by using this catalyst, when amide groups are involved [24].

Deprotection was performed first by using CF_3COOH, as described in ref [25]. In our hands, complete cleavage of the polysiloxane chain was observed. Another way to deprotect terminal tBoc groups has been described by Frechet et al., using trimethylsilyliodide (TMSI) in dichloromethane, applied to a hyperbranched poly(siloxysilane) with N-tBoc-aminopropyl end-groups [26]. At room temperature, once again a full cleavage of the chain was observed. However optimization of this method allowed us to keep the polysiloxane backbone intact. The reaction was then performed at −10°C for 6 min, in strictly anhydrous conditions and with 0.9–1.0 eq. of freshly provided TMSI. The reaction was quenched with anhydrous methanol and an excess of a resin with amine functions (e.g. Amberlyst A-21, thoroughly washed with anhydrous THF and dried under vacuum).

In these conditions, the deprotection rate varied between 90 and 99%, according to ^1H NMR (at 1.36 ppm for the tBu signal, Fig. 8), and the elution profile in SEC (THF), given in Fig. 9, showed similar profiles for both protected and unprotected polymers, a sign that the polysiloxane backbone was maintained. We cannot ensure that part of the polymer was not adsorbed on the columns or that some redistribution did not occur. However, it was observed in other aforementioned methods that some deprotection conditions led to the complete destruction of the polymer chain leading to the absence of peak in SEC in the polymer area. The peaks on the right in Fig. 9 have been attributed to the presence of soluble residues coming from the aminated resin used in the reaction.

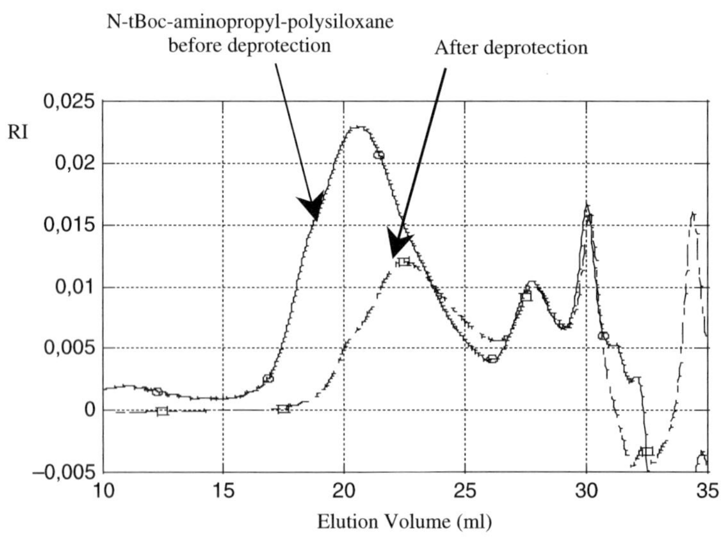

Fig. 9 Compared SEC elution profile (in THF) of poly(N-tBoc-aminopropylmethyl-co-dodecylmethyl-co-dimethyl)-siloxane before and after addition of TMSI in the reaction medium

Quenching with the aminated resin appeared absolutely necessary to avoid cleavage of the aminopropylpolysiloxane. After filtration, the aminopropyl-polysiloxane can be stored at −18°C in the CH_2Cl_2/MeOH solution without decomposition. Analysis by SEC can be made from THF solutions prepared after incomplete evaporation of the CH_2Cl_2/methanol mixture. Actually, they cannot be easily stored for long times and the polymers cannot be dried without degradation. Their sensitivity to TMSI appeared much higher compared with NBoc-functionalized hyperbranched poly(siloxysilanes) or NBoc-functionalized highly crosslinked siloxanes networks formed by sol-gel chemistry [27].

Further work would be useful on these points for making these aminopolysiloxanes more practical to use in further nucleophilic reactions or other applications.

3.2.2 Grafting of Molecules with Protected Alcohols

In some applications, it is also desirable to graft various molecules bearing OH groups or amine groups to the polysiloxane backbone. An example for a successful introduction of such a molecule has been shown by the grafting of cinchonidine. This molecule is well known to afford enantioselective reduction on metallic surfaces. The aim was first to use such a polymer to stabilize metal nanoparticles in organic solvents as well as to study the catalytic activities of such system.

Cinchonidine displays a tertiary amine, an aromatic amine and a free hydroxyl functionality. Direct hydrosilylation with unprotected cinchonidine (derivatized with a double bond for hydrosilylation), led to a cross-linked product. Thus, hydrosilylation was performed on PHMS with the trimethylsilyl derivative **1** (Fig. 10), in the presence of $(Et_2S)_2PtCl_2$ as catalyst (0.05%), for 6h at 80°C in toluene. The hydroxyl group was deprotected with methanol at 65°C during 120h. Size Exclusion Chromatography showed that the polysiloxane backbone was not degraded. A maximal grafting percent of 15% could be obtained, relative to the SiH units.

Hydroxyl groups have also been introduced through a ketal-dialkene **2**. The hydrosilylation was performed in the presence of $DCPPtCl_2$ at 60°C in toluene during 4h [10]. The ketal group could then be cleaved by treatment with p-toluenesulfonic acid in water/methanol. In this case however, since the final material was cross-linked

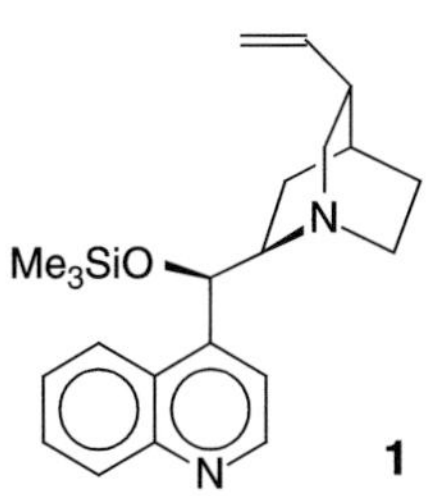

1

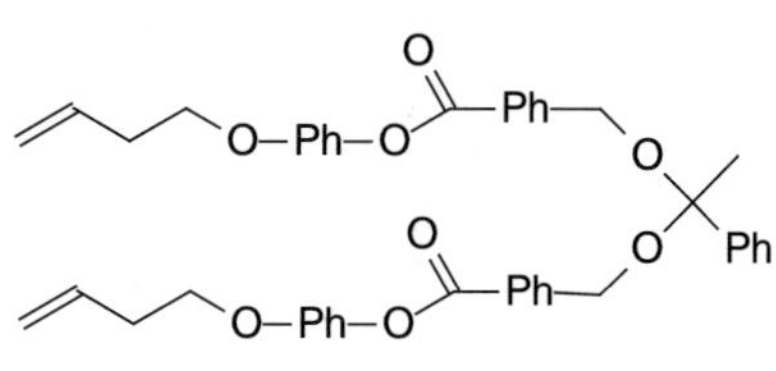

2

Fig. 10 Examples of functional alkenes used in hydrosilylation reactions

by hydrosilylation through the introduction of a dialkene in the same step, precise analyses of the polysiloxane skeleton could not be carried out.

3.3 Grafting of Free Alcohol or Acid Groups

The second strategy developed in our laboratory benefits from our research on molecular imprinted polymers. These are synthesized by polymerizing and crosslinking monomers in interaction with a chosen template. After its removal by washing steps, the material presents specific cavities that are complementary in shape and size to the template. We have recently developed new molecular imprinted polymers (MIPs) using liquid crystal elastomers [28–31]. In this field, the materials are basically synthesized in two steps, the first one attaching all substituents on the polysiloxane chain and the second one crosslinking the materials. The usual functional groups used for such synthesis are acids or alcohols since they constitute excellent hydrogen bond donors. By complexing them to the template molecule during the synthesis, hydrosilylation becomes possible in their presence. In this case, the template molecule acts as a kind of non covalent protective group. Figure 11 depicts two examples of this strategy, one including a carboxylic acid group and the other one a hydroxyl one [30, 32]. In these cases, either α-methylbenzylamine or diethyl 4-nitrobenzylphosphonate were used as templates.

At this linear polymer step, it is possible to precipitate and recover the polymer selectively, since the affinity constants of both complexes are not very high. Thus, this method has the advantage of affording polar polysiloxanes in a one step process. For the first reaction of Fig. 11, it is noteworthy that a stoichiometry between the functional groups and the template molecule is not mandatory. Indeed, we were able to carry out the first reaction of Fig. 12 with a ratio of acid/amine of 4. This result might be explained if a fast equilibrium exists between acids and amines leading to the protection of all acids. When no amine was added on the other hand, the reaction led to crosslinking for percentages of acid higher than 8%. This is usually attributed in the literature to side reactions of the acid group with the catalyst [9]. When the amine was replaced by the phosphonate as template, the protective group effect could be demonstrated. Indeed, when the percentage of acid group was 15%, a crosslinking was observed when the phosphonate was used as a ratio lower than stoichiometry, whereas no more crosslinking was observed when it was used in excess versus acid groups, although we do not know whether a chemical or physical crosslinking occurred in this case.

Using the second reaction of Fig. 11, we synthesized copolysiloxanes containing hydroxyl and liquid crystalline substituents (Fig. 12).

A change in hydrophilicity was observed in these polymers, the most hydrophobic precipitating in methanol whereas the polymer with 42 mol% of hydroxyl unit led to an emulsion in this same solvent. Although the synthesis of fully substituted hydroxyl polysiloxane was attempted, the reaction led in this case to a crosslinked material. In the literature, some examples of hydrosilylation in the presence of hydroxyl groups can be found. They often present only hydrosilylation on the polymer chain extremities, to minimize the possibility of crosslinking [33, 34].

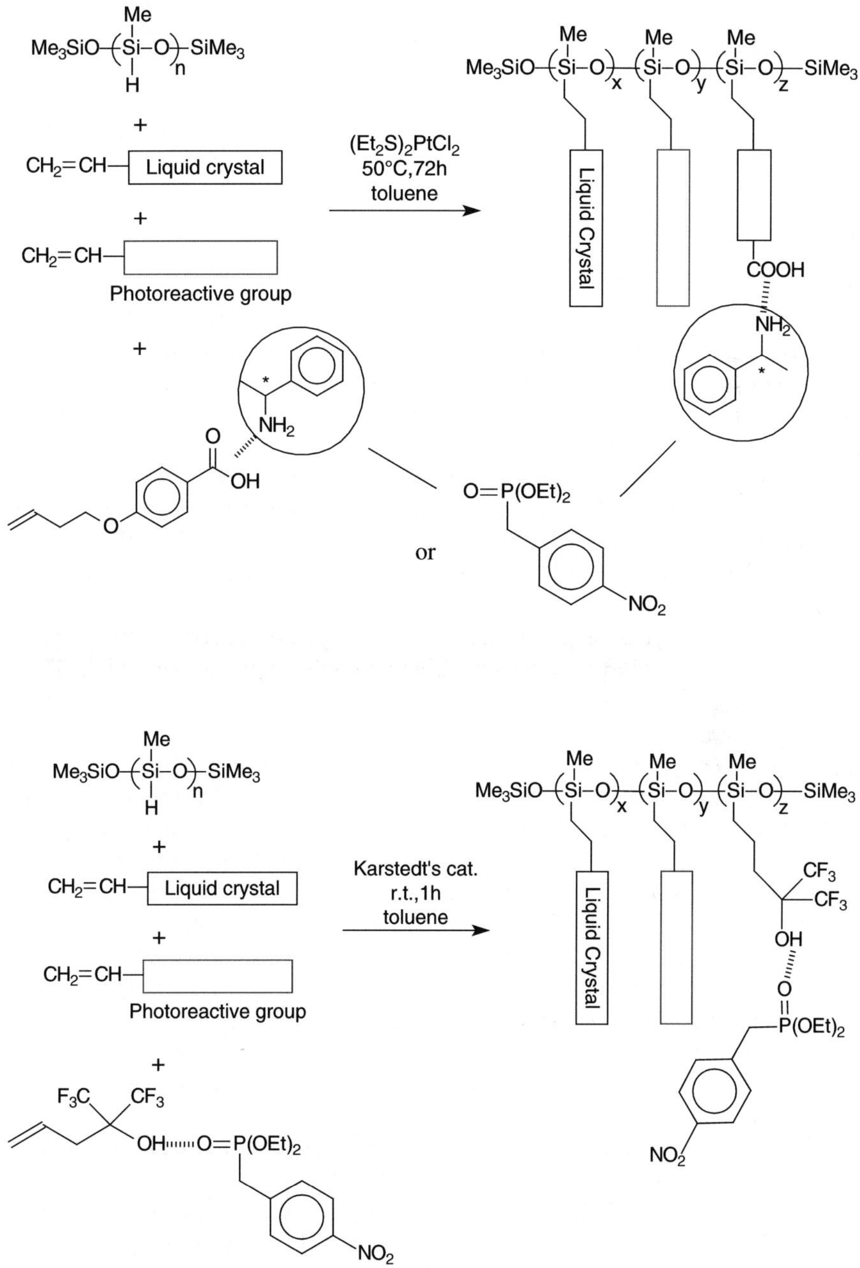

Fig. 11 One-step synthesis of functional polysiloxanes

Fig. 12 Structure of hydroxyl copolysiloxanes

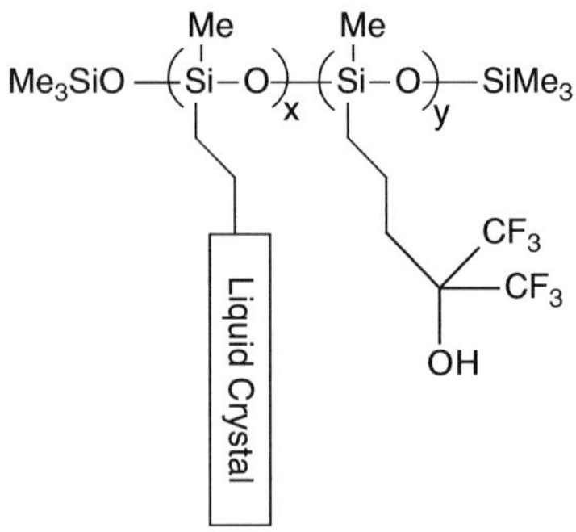

x=86% y=14%
x=75% y=25%
x=71% y=29%
x=58% y=42%

3.4 Summary

For all the reactions presented here, “OK” means that we have observed the total disappearance of SiH units and that no macroscopic crosslinking was present (Table 1).

Table 1 Summary of hydrosilylation reactions presented in this chapter

Alkenes	[=] (M)	[SiH] (M)	Cat.	[cat.] /SiH	Conditions	Results
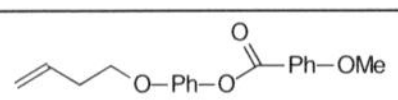 M41	0.11	0.1	$DCPPtCl_2$	0.05%	Toluene, Ar, 60°C, 24h	OK
M41 Docosadiene	0.14 0.15	0.2 Then 1.0	Karstedt	0.05%	Toluene, Ar, 60°C, 4h for the first phase	OK
Octene Benzophenone derivative	1.3 0.06	1.2	$DCPPtCl_2$	0.6%	Toluene, Ar, 65°C, 72h In the dark	3–10% SiH remaining
M41 Benzophenone derivative	0.35 0.017	0.33	$DCPPtCl_2$	0.2%	Toluene, Ar, 70°C, 65h In the dark	OK
M41 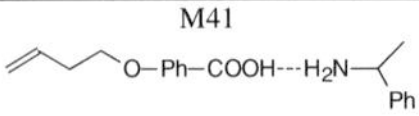 Benzophenone derivative	0.32 0.028 0.017	0.3	$(Et_2S)_2PtCl_2$	1.0%	Toluene, Ar, 60°C, 72h In the dark	OK
M41 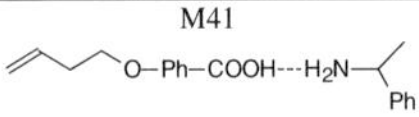 Benzophenone derivative			$DCPPtCl_2$		Toluene, Ar, 60°C, 72h In the dark	No reaction
M41 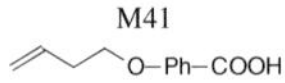			$DCPPtCl_2$ $(Et_2S)_2PtCl_2$			Crosslinking for percentages >8%
M41 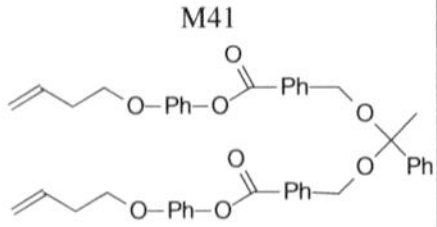	0.14 0.036	0.15	$DCPPtCl_2$	0.1%	Toluene, N_2, 60°C, 1h	OK
M41 docosadiene 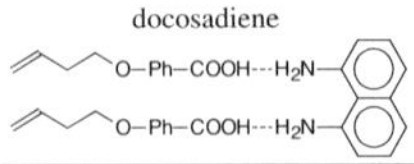			$DCPPtCl_2$	1.4%	N_2, 50°C, 96h	One step crosslinking
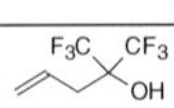 styrene	0.04 0.46	0.36	Karstedt	0.1%	Toluene, Ar, Tamb, 48h	2% SiH remaining

Table 1 (Continued)

M41 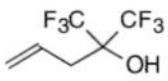	0.3 0.06	0.37	Karstedt	0.15%	Toluene, Ar, Tamb, 1h	OK
M41 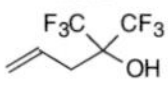			$DCPPtCl_2$ H_2PtCl_6 $(Et_2S)_2PtCl_2$			crosslinking
M41 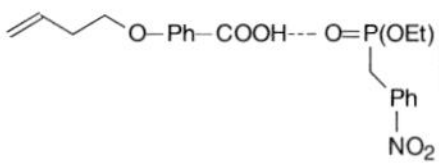	0.33 0.03	0.34	$(Et_2S)_2PtCl_2$	1.3%	Toluene, Ar, 50°C, 72h	OK
M41 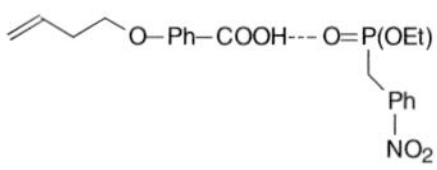	0.31 0.06	0.33	$(Et_2S)_2PtCl_2$	1.3%	Toluene, Ar, 50°C, 72h	Partial crosslinking
M41 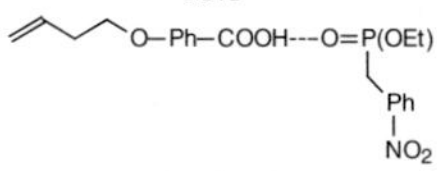 excess in phosphonate	0.31 0.06	0.33	$(Et_2S)_2PtCl_2$	1.3%	Toluene, Ar, 50°C, 72h	OK
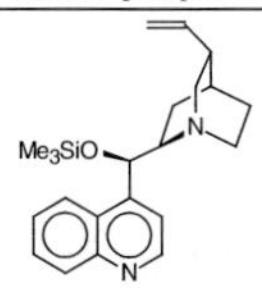	0.13	0.66	$(Et_2S)_2PtCl_2$	0.05%	Toluene, Ar, 80°C, 6h	OK
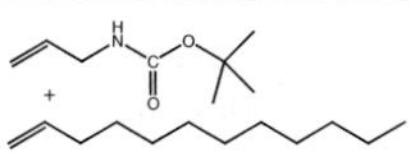	0.068 0.036	0.072	$(Et_2S)_2PtCl_2$	1%	Toluene, Ar, 70°C, 1-2h	OK

4 Conclusion

Hydrosilylation remains a particular reaction in chemistry, because no universal recipe can be applied. For each case, appropriate experimental conditions have to be determined. Our experience has only enabled us to establish general behaviours, especially for the most problematic ones. Among them, hydrosilylation with free carboxylic acids lead easily to crosslinking. For hydroxyl groups, the same problem was obtained with $DCPPtCl_2$, H_2PtCl_6 or $(Et_2S)_2PtCl_2$. Another discrepancy of behaviour was noted for carboxylic acids complexed to amines. Whereas hydrosilylation was possible in the presence of diaminonaphthalene using $DCPPtCl_2$, no reaction could be obtained in the presence of α-methyl benzylamine. The use of $(Et_2S)_2PtCl_2$ was mandatory in this case to obtain the grafting. Another point that remains unclear in the literature is the need of oxygen during the reaction. Some authors recommend that the reaction medium be fully deoxygenated, whereas others mention that oxygen plays a role in the catalytic cycle and is therefore mandatory [35]. In our case, we did not notice any important difference in the presence or in the absence of oxygen.

From a general standpoint, for molecules without any nitrogen atom or protic groups, either $DCPPtCl_2$, Karstedt's catalyst or $(Et_2S)_2PtCl_2$ can be used. For hydroxyl containing molecules, we have found that the use of Karstedt's catalyst can be efficient at room temperature. For amides or analogues of amides, $(Et_2S)_2PtCl_2$ leads to the desired polymer. Finally, for protic unprotected molecules, the use of complexing protective groups can be an interesting way of obtaining the desired grafting.

Acknowledgements The authors wish to thank Ioltech, Inc., French Ministry of Education and Research, and DGA for funding.

References

1. Butts M, Cella J, Wood CD, Gillette G, Kerboua R, Leman J, Lewis L, Rubinsztajn S, Schattenmann F, Stein J, Wicht D, Rajaraman S, Wengrovius J (2002) Kirk-Othmer encyclopedia of chemical technology. John Wiley and Sons, Hoboken (NJ) (2006) 22: 547
2. Marciniec B (1992) Comprehensive handbook on hydrosilylation. Pergamon Press, Oxford
3. Marciniec B (2002) Silicon Chemistry 1: 155
4. Brook MA (2006) Biomaterials 27: 3274
5. Chalk AJ, Harrod JF (1965) Journal of the American Chemical Society 87: 16
6. Coqueret X, Wegner G (1991) Organometallics 10: 3139
7. Lewis LN, Stein J, Gao Y, Colborn RE, Hutchins G (1997) Platinum Metals review 41: 66
8. Stein J, Lewis LN, Smith KA, Lettko KX (1991) Journal of Inorganic Organometallic Polymers 1: 325
9. Mukbaniani O, Zaikov G, Pirckheliani N, Tatrishvili T, Meladze S, Pachulia Z, Labartkava M (2007) Journal of Applied Polymer Science 103: 3243
10. Marty J-D, Tizra M, Mauzac M, Rico-Lattes I, Lattes A (1999) Macromolecules 32: 8674
11. Mauzac M, Hardouin F, Richard H, Achard M-F, Sigaud G, Gasparoux H (1986) European Polymer Journal 22: 137
12. Leroux N, Mauzac M, Noirez L, Hardouin F (1994) Liquid Crystals 16: 421
13. Mauzac M, Hardouin F, Richard H, Achard M-F, Sigaud G, Gasparaoux H (1986) European Polymer Journal 22: 137
14. Apfel MA, Finkelmann H, Janini GM, Laub RJ, Lühmann B-H, Price A, Roberts WL, Shaw TJ, Smith CA (1985) Analytical Chemistry 57: 651
15. Gray GW (1989) In: McArdle CB (ed) Side chain liquid crystal polymers. Blackie, London p. 106
16. Cancouët P, Pernin S, Hélary G, Sauvet G (2000) Journal of Polymer Science, Part A: Polymer Chemistry 38: 837
17. Binet C, Bourrier D, Dilhan M, Estève D, Ferrère S, Garrigue J-C, Granier H, Lattes A, Gué A-M, Mauzac M, Mingotaud A-F (2006) Talanta 69: 757
18. Boutevin B, Guida-Pietrasanta F, Ratsimihety A (2000) In: Jones RG, Ando W, Chojnowski J (eds) Silicon containing polymers. Kluwer Academic Publishers, Dordrecht p. 79
19. Belorgey G, Sauvet G (2000) In: Jones RG, Ando W, Chojnowski J (eds) Silicon containing polymers. Kluwer Academic Publishers, Dordrecht p 43
20. Ma Q, Feng S (2006) Carbohydrate Polymers 65: 321
21. Burgemeister D, Farrell T, Schmidt C (2007) Macromolecular Chemistry and Physics 207: 396
22. Coqueret X, Hajaiej A, Lablache-Combier A, Loucheux C, Mercier R, Pouliquen L, Randrianarisoa-Ramanantsoa L (1990) Pure and Applied Chemistry 62: 1603

23. Yu JM, Teyssié D, Boileau S (1993) Journal of Polymer Science, Part A: Polymer Chemistry 31: 2373
24. von Braunmühl V, Stadler R (1998) Polymer 39: 1617
25. Matisons JG, Provatas A (2000) Silicones and silicone-modified materials, vol 729. ACS SYmposium Series p. 128
26. Ging C, Fréchet JMJ (2000) Journal of Polymer Science, Part A: Polymer Chemistry 38: 2970
27. Suh Y-W, Kung MC, Wang Y, Kung HH (2006) Journal of the American Chemical Society 128: 2776
28. Marty J-D, Mauzac M, Fournier C, Rico-Lattes I, Lattes A (2002) Liquid Crystals 29: 529
29. Marty J-D, Mauzac M (2005) Advances in Polymer Science 172: 1
30. Binet C, Ferrère S, Lattes A, Laurent E, Marty J-D, Mauzac M, Mingotaud A-F (2007) Analytica Chimica Acta 591(1): 1
31. Weyland M, Ferrère S, Lattes A, Mingotaud A-F, Mauzac M (2007) accepted by Liquid Crystals (2008) 35(2): 215
32. Palaprat G, Marty J-D, Routaboul C, Lattes A, Mingotaud A-F, Mauzac M (2006) Journal of Physical Chemistry A 110: 12887
33. Boileau S, Bouteiller L, Kowalewska A (2003) Polymer 44: 6449
34. Rutnakornpituk M, Ngamdee P (2006) Polymer 47: 7909
35. Steffanut P, Osborn JA, De Cian A, Fisher J (1998) Chemistry European Journal 4: 10 2007, and references therein

High Refraction Index Polysiloxanes via Organometallic Routes – An Overview

Włodzimierz A. Stańczyk, Anna Czech, Wojciech Duczmal, Tomasz Ganicz, Małgorzata Noskowska, and Anna Szeląg

Abstract Processes based on four organometallic routes leading to phenylethenyl substituted siloxanes are reviewed. Such materials, exhibiting high refraction indices, can be made by hydrosilylation of phenylacetylenes with Si-H moieties containing siloxane oligomers and polymers. On the other hand reactions between vinyl containing siloxanes with haloaryls or styrenes lead to the same systems via Heck or metathetic and silylative coupling. The effectiveness of the catalytic processes is analyzed from the point of view of side reactions, required catalyst concentration and reaction conditions. At the current state of knowledge the Heck coupling offers synthesis of the desired phenylethenyl substituted silicone fluids using technologically most attractive pathway.

1 Introduction

In recent years one observes a growing industrial demand for organosilicon materials having properties, which can not be found in conventional polymers. These also include silicone fluids, characterized by high refraction indices, such as ~1.50, utilized extensively in personal care applications. An important class of such systems are siloxanes having phenylethenyl type substituents along polymer backbone (Fig. 1).

Current progress in catalytic studies afforded a number of organometallic pathways leading to such structures.

W.A. Stańczyk
Centre of Molecular and Macromolecular Studies, Polish Academy of Sciences, Sienkiewicza 112, 90-363 Łódź, Poland
e-mail: was@bilbo.cbmm.lodz.pl

F. Ganachaud et al. (eds.), *Silicon Based Polymers*,

Fig. 1 Siloxane monomeric unit allowing for high RI

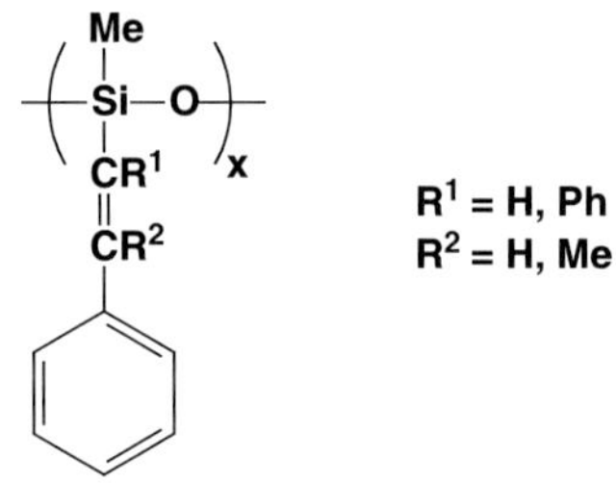

2 Hydrosilylation

First synthetic approach to such siloxane oligomers and polymers was described by L.N. Lewis et.al., [1] applying a hydrosilylation route [2] and based on phenylacetylene, diphenylacetylene and methyl(hydrido)siloxanes as substrates (Fig. 2).

Although, several model studies on hydrosilylation of alkynes using heterogeneous catalytic systems (Ru/MgO, Pt/SiO_2, Pt/MgO) have been recently reported [3–5], the homogeneous catalysts seem to be still the most efficient ones. The most efficient catalyst, even in reactions involving relatively high molecular weight silicon fluids, proved to be Karstedt catalyst used in the ratio of 40–100 ppm to phenylacetylene [1]. However, in the past the use of rhodium complexes in reaction with low molecular weight silanes (e.g. R_3SiH, R_2SiH_2) was also described [6–8].

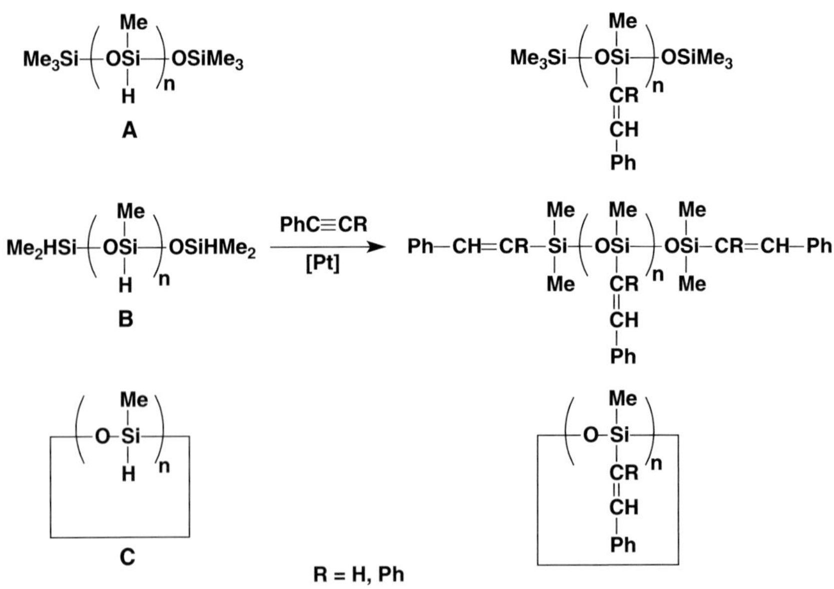

Fig. 2 Hydrosilylation of phenylacetylenes

Table 1 Phenylethenylsiloxanes prepared by hydrosilylation

Siloxane	n	Acetylene	1-Hexene mol %	RI
A	10	PhC≡CH	15	1.546
A	80	PhC≡CH	–	1.572
B	15	PhC≡CH	–	1.574
C	4	PhC≡CH	–	1.574
A	4	PhC≡CPh	20	1.561
C	4	PhC≡CPh	50	1.527

The amount of catalyst in such cases is rather high 1000–5000 ppm and selectivity towards anti-Markovnikov addition is lower (80–90%), compared to hydrosilylation in the presence of platinum based catalyst. The synthesis of phenylethenyl substituted siloxanes is of commercial importance, driven by potential application in personal care products. Such materials should be in the form of fluids and thus in order to preserve this requirement two approaches have been exploited. One of them involved substitution of less than 100% phenylethenyl moieties, the other made use of 1-hexene as a co-reactant, leading to decreased crystallinity of the final materials. Depending on the structure of (methylhydrido)siloxanes and reaction conditions the resulting silicon fluids exhibited refraction indices ranging from 1.527 to 1.574 (Table 1).

The RI values obtained for such phenylethynyl substituted siloxanes are higher then that reported for traditional aromatic-based systems [9] or the phenol modified ones (1.50–1.53) [10]. The synthesis of high refractive index (methyl)(diphenyle thenyl)-dichlorosilane via hydrosilylation was also described [1]. Such monomer was later hydrolyzed and condensed into silicone fluid. Similar process was also presented, applying silylative coupling process in the synthesis of an analogous (methyl)(phenylethenyl)diethoxysilane [11], so the two reactions shall be discussed in the following section.

3 Metathetic and Silylative Coupling

Both metathesis and silylative coupling processes can provide theoretically efficient synthetic routes leading to phenylethenyl substituted silicones. Although they both lead to the same final product, the catalyst structure and mechanisms are significantly different. Metathesis involves C=C bond cleavage (Fig. 3a), while silylative coupling occurs *via* =C–Si and =C–H bond breaking (Fig. 3b), as proved in earlier work concerning low molecular weight vinylsilanes and deuterium labelled styrene [12, 13]. In Fig. 3, both these pathways are presented for reactions of styrene with vinylsiloxanes, leading to the desired phenylethynyl substituted polymer systems.

Comparison of effectiveness of the two processes, carried out in the presence of metallacarbene (metathesis) and non-metallacarbene catalysts (silylative coupling) is shown in Table 2.

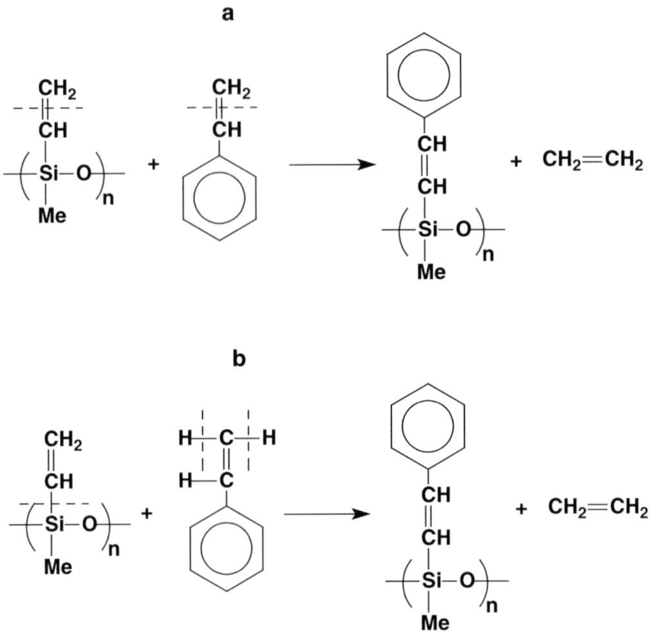

Fig. 3 Metathesis **(a)** and silylative coupling **(b)**

A relatively small extent of coupling has been observed *via* the metathetic route (max. 45%), while in all the studied cases significant amount of polystyrene was being produced (entries 1–4). Silylative coupling of styrene and poly-(dimethylsiloxane-*co*-methylvinylsiloxane) [11] or tetramethyltetravinylcyclotetrasiloxane [11, 14] offers much higher conversion (70–100%). Formation of the side product – polystyrene can be limited or totally avoided by the use of free radical scavengers. The undesired, side process of styrene free radical polymerization can be also eliminated once the styrene as a reactant is substituted by α-methylstyrene. RI of the silicone fluids obtained by organometallic coupling methods falls into the range 1.51–1.59, depending on the extent of phenylethenyl groups substitution, which is typically smaller for cyclosiloxane, due to steric effects. An alternative synthetic method has been also described, based on hydrolytic condensation of methylphenylethenyl silane (Me)(PhCH=CR)SiR^{1}_{2} (R=Ph, R^1=Cl [1]; R=H, R′=OEt [11, 15] (Fig. 4).

In both cases, using dichloro- and (diethoxy)(methyl)-(phenylethenyl)-type silanes an oligomeric silicone fluid was formed, having RI in the range of 1.50 – 1.58.

4 Heck Reaction

A fundamental method of C-C bond formation, which can be also applied in synthesis of phenylethenyl substituted siloxanes is the Heck coupling process [16, 17] (Fig. 5).

Table 2 Comparison of effectivness of metathesis and silylative coupling catalysts

Entry	Siloxane[a]	Styrene[b] (excess, %)	Catalyst[c] (wt %)	Inhibitor[d] (wt %)	Solvent	Temp. (°C)	React. time (h)	Conversion (mol %)	Polystyrene (mol %)	R.I.
1	I	St (50)	A (2)	–	CH_2Cl_2	40	6	45	45	–
2	I	St (50)[e]	A (2)	–	CH_2Cl_2	40	6	8	18	–
3	I	St (50)	A (2)	QUIN(0.2)	toluene	110	12	20	28	–
4	I	St(100)	B (1.5)	–	CH_2Cl_2	40	96	36	7	–
5	I	St (50)	C (2)	–	n-decane	110	48	58	19	1.508
6	I	St(100)	C (2)	BHT(0.5)	toluene	100	92	86	0	1.518
7	I	St(100)	C (2)	QUIN(0.5)	toluene	100	102	100	0	1.532
8	I	St (50)	C (2)	CAT (0.5)	toluene	100	144	72	0	1.530
9	II	St(100)	C (2)	QUIN(0.5)	toluene	100	48	90	18	1.567
10	II	St(100)	C (2)	BHT (0.5)	toluene	100	96	68	21	1.549
11	II	MeSt(100)	C (2)	–	toluene	100	96	39	0	1.583
12	I	MeSt(100)	C (2)	–	toluene	100	96	70	0	1.581

[a] I – poly(dimethylsiloxane-*co*-methylvinylsiloxane) (M_n = 3200, -MeViSiO- = 65%); II – tetramethyltetravinylcyclotetrasiloxane.
[b] St – styrene; MeSt – α-methylstyrene.
[c] A – benzylidenebis(tricyclohexylphosphine)dichlororuthenium (Grubbs catalyst); B – diisopropylphenylimidonephenylidene-(racemic BIPHEN)molybdenum (Schrock catalyst); C – carbonylchlorohydridotris(triphenylphosphine)-ruthenium.
[d] QUIN – hydroquinone; BHT – di-*tert*-butylmethylphenol; CAT – *tert*-butylcatechol.
[e] Step-wise addition of styrene (5 h)

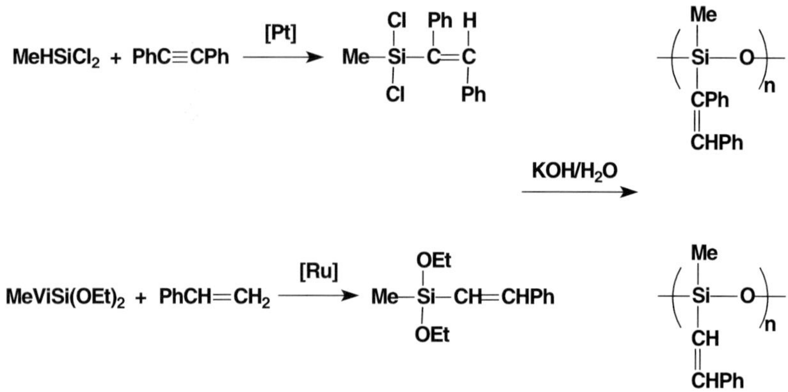

Fig. 4 Hydrolysis and condensation of phenylethenylsilanes

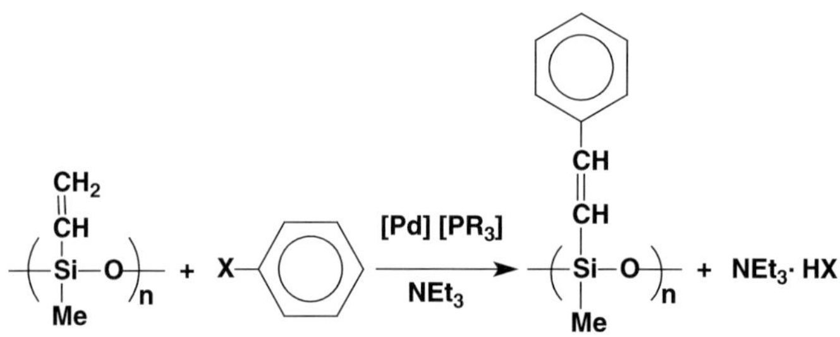

Fig. 5 Heck coupling reaction between haloaryls and vinylsiloxanes

Although such approach has been described earlier [18], the reactions were run in solution, which did not make the process technologically attractive. Currently, a systematic study presented a selection of efficient palladium catalysts, which can be used in the bulk, for the coupling of aryl halides and poly-(dimethylsiloxane-*co*-methylvinylsiloxane) and tetramethyltetravinylcyclo-tetra-siloxane as organosilicon substrates [19]. It has been shown that for both organosilicon substrates the reaction proceeds smoothly under solventless conditions and its rate is in fact higher by a factor of 2, when compared to analogous processes carried out in toluene, at the same temperature. The results obtained for the three most efficient palladium catalysts: $Pd_2(dba)_3$, $PdCl_2(cod)$ and $Pd(OAc)_2$ are shown in Table 3 [19].

The RI values for siloxane copolymers are lower due to the "dilution" of the reactive units in thepolymer with -Me_2SiO- monomeric units. On the other hand the conversion is lower for coupling of aryl halides with tetramethyltetravinylcyclo-tetrasiloxane but the correponding R.I. values are much higher (1.54–1.59). It also appears that substitution of methyl groups into benzene ring increases these values, enhancing the electron density in the phenylethenyl conjugated system. Chlorobenzene can be also applied as a substrate, in the presence of CuI, however, reaction time has to be prolonged.

Table 3 Phenylethenyl-substituted slicones via Heck coupling reaction

Entry	Siloxane[a]	Aryl halide ArX	Catalyst	[Pd][b]	Temp. °C	React. Time h	Conversion %	R.I.
1	I	C_6H_5Br	$[Pd_2(dba)_3]$[c]	$1.4\ 10^{-3}$	100	12	73	1.477
2	I	C_6H_5Br	$[Pd_2(dba)_3]$	$5.0\ 10^{-3}$	100	12	83	1.504
3	II	C_6H_5Br	$[PdCl_2(cod)]$[d]	$5.0\ 10^{-4}$	80	96	59	1.541
4	II	C_6H_5Br	$[PdCl_2(cod)]$	$5.0\ 10^{-4}$	100	12	95	1.582
5	II	*p*-MeC_6H_5Br	$[PdCl_2(cod)]$	$1.4\ 10^{-3}$	100	24	78	1.565
6	II	*m*-MeC_6H_5Br	$[PdCl_2(cod)]$	$1.4\ 10^{-3}$	80	24	47	1.581
7	II	C_6H_5Cl[f]	$[PdCl_2(cod)]$	$1.4\ 10^{-3}$	100	96	56	1.542
8	II	*p*-MeC_6H_5Br	$Pd(OAc)_2$[e]	$1.4\ 10^{-3}$	100	24	53	1.525
9	II	*m*-MeC_6H_5Br	$Pd(OAc)_2$	$1.4\ 10^{-3}$	100	24	53	1.591

[a] I – poly(dimethylsiloxane-*co*-methylvinylsiloxane) (M_n = 3200, -MeViSiO- = 65%); II – tetramethyltetravinylcyclotetrasiloxane.
[b] [-MeViSiO-]:[ArX]:[Et_3N] = 1:1:1; [Pd]:[P(*o*-Tol)$_3$]:[-MeViSiO-] = 1:2:715–2000;
[c] $[Pd_2(dba)_3]$ – tris(dibezylideneacetone)dipalladium(0);
[d] $[PdCl_2(cod)$ – (1,5-cyclooctadiene)dichloropalladium(II);
[e] Pd(OAc) – palladium(II) acetate;
[f] In the presence of Cu

5 Conclusions

The four organometallic routes – hydrosilylation, metathesis, silylative coupling and Heck reaction, described above, lead to the important class of silicon fluids, having high refraction indices (up to RI ~ 1.59). Hydrosilylation of Si-H containing siloxanes with phenylacetylenes is an efficient process run at low concentration of platinum catalyst (10^{-5}), being though rather expensive. The other disadvantage seems to be the side β-addition process, leading to crosslinking via subsequent hydrosilylation with Si-H siloxane as well as lowering of the RI values. Metathesis, involving metallacarbene type catalysts for coupling of vinylsiloxanes and styrenes is the least efficient route due to the parallel process of free radical polymerization of olefins, which have to be used in two fold excess. A relatively large concentration of the catalysts ([Ru], [Mo] ~ 10^{-2}) requires also removal of metal complexes from olefin metathesis products (*via* complexing with $P(CH_2OH)_3$) [20]. Similar restrictions concern the silylative coupling process. The required ratio of olefin to vinylsilyl moieties in some reactions (even for low molecular weight siloxane substrates) is claimed to be 6:1 [12]. However the side polymerization process can be eliminated or limited by the use of α-methylstyrene as a substrate or by addition of free radical scavengers to the reaction system. The Heck coupling has the technological advantage of running the process without any solvent at a relatively low concentration of palladium based catalysts (10^{-4}). In the light of current knowledge the effectiveness of the four synthetic approaches is as follows: metathesis<silylative coupling<hydrosilylation<Heck coupling reaction.

Acknowledgements The authors thank GE Silicones (MPM) for financial support. WAS and AS wish also to acknowledge additional financing via Polish Ministry for Science and Information Technology Grant PBZ-KBN-118.T09/2004.

References

1. Lewis LN, Nye SA. (1996) Aryl substituted silicone fluids having high refractive indices and method for making. US Patent 5,539,137
2. Marciniec B. (2002) Catalysis of hydrosilylation of carbon-carbon multiple bonds. Silicon Chem. 1:155–174
3. Jimenez R, Lopez JM, Cervantes J. (2000) Metal supported catalysts obtained by sol-gel in the hydrosilylation of phenylacetylene with R_3SiH organosilanes ($R_3=Ph_3$, Ph_2Me, $PhMe_2$). Can. J. Chem. 78:1491–1495
4. Jimenez R, Martinez-Rosales JM, Cervantes J. (2003) The activity of Pt/SiO_2 catalysts obtained by sol-gel method in the hydrosilylation of alkynes. Can. J. Chem. 81:1370–1375
5. Ramirez-Oliva E, Hernandes A, Martinez-Rosales JM et al. (2006) ARKIVOC (v):126–136
6. Lappert MF, Maskell RK. (1984) Carbene transition metal complexes as hydrosilylation catalysts. J.Organomet.Chem. 264:217–228
7. Bruckmann M, Dieck HT, Klaus J. (1986) Rhodium (I) mono- und diazadienkomplexe, synthese, spektroskopische charakterizirung, oxidative additionsreaktionen und einsatz in der homogenen katalyse zur hydrosililierung. J.Organomet.Chem. 301:209–226

8. Watanabe H, Kitahara T, Motegi T et al. (1977) The stereoselective addition of phenyldimethylsilane to phenylacetylene by rhodium triphenylphosphine complexes. J.Organomet.Chem. 139:215–222
9. Salmone JC, Kunzler JF, Ozark RM et al. (2006) High refractive index aromatic-based siloxane difunctional macromonomers. US Patent 7,009,024
10. Raleigh WJ, Campagna JA, Lucarelli MA. (1996) High refractive index phenol-modified siloxanes. US Patent 5,541,278
11. Ganicz T, Kowalewska A, Stańczyk WA et al. (2005) A novel organometallic route to phenylethenyl-modified polysiloxanes. J. Mater. Chem. 15:611–619
12. Marciniec B. (2007) Catalytic coupling of sp^2- and sp-hybridized carbon-hydrogen bonds with vinylmetalloid compounds. Acc. Chem. Res. 14:527–538
13. Marciniec B, Pietraszuk C. (1997) Silylation of styrene with vinylsilanes catalyzed by $RuCl(SiR_3)(CO)(PPh_3)_2$ and $RuHCl(CO)(PPh_3)_3$. Organometallics 16:4320–4326
14. Itami Y, Marciniec B, Kubicki M. (2003) Functionalization of vinyl-substituted cyclosiloxane and cyclosilazane via ruthenium-catalyzed silylative coupling reaction. Organometallics 22:3717–3722
15. Marciniec B, Walczuk-Gusciora E, Pietraszuk C. (2001) Activation of the vinylic =C–H bond of styrene by a rhodium siloxide complex: the key step in the silylative coupling of styrene with vinylsilanes. Organometallics 20:3423–3428
16. Heck RF. (1975) Vinylic substitution reactions. US Patent 3,922,299
17. Trzeciak AM, Ziól׳kowski JJ. (2005) Structural and mechanistic studies of Pd-catalyzed C–C bond formation: the case of carbonylation and Heck reaction. Coord Chem. Rev. 249:2308–2322
18. DeVries RA, Schmidt GF, Frick HR et al (1993) Process for preparing vinylically unsaturated compounds. US Patent 5,264,646
19. Czech A, Duczmal W, Ganicz T et al. (2007) Phenylethenyl-substituted silicones via Heck coupling reaction. J. Organomet. Chem.: submitted
20. Maynard HD, Grubbs RH. (1999) Purification technique for the removal of ruthenium from olefin metathesis reaction products. Tetrahedron Lett. 40:4137–4140

Grafting β-Cyclodextrins to Silicone, Formulation of Emulsions and Encapsulation of Antifungal Drug

Ahlem Noomen, Alexandra Penciu, Souhaira Hbaieb, Hélène Parrot-Lopez, Noureddine Amdouni, Yves Chevalier, Rafik Kalfat

Abstract Emulsions of silicone polymers having β-cyclodextrin units as lateral chains have been prepared and used for the encapsulation of the antifungal drug griseofulvin. Such technology enables the formulation of active substances that are not soluble in water as dosage forms for topical administration.

Chemical grafting of monoallyl derivative of β-cyclodextrins to poly(methylhydr osiloxane) polymer or to poly(methylhydrosiloxane-*co*-dimethylsiloxane) copolymer through hydrosilylation reaction was investigated. Grafting of β-cyclodextrins to the Si-H units was limited by steric hindrance of the bulky cyclic D-heptaglucopyranoside. Therefore the kinetics of the grafting reaction slowed down dramatically as the conversion increased. The way to achieve high grafting rates was to run the grafting reaction for a very long time. The full conversion of the poly(methylhydrosiloxane) homopolymer could not be reached.

The grafted silicone polymers have been emulsified by the "spontaneous emulsification" method. The emulsification took advantage of the very low solubility of the polymers in water. Stable emulsions with sizes ranging between 200 and 500 nm have been produced.

The emulsions of β-cyclodextrins grafted on silicone could encapsulate the antifungal substance griseofulvin inside the β-cyclodextrin cavity by formation of inclusion complex. The encapsulation rate was limited to the 1:1 stoichiometry of the complex. Supplementary amount of griseofulvin slowly precipitated as crystalline particles in the aqueous phase.

There are several benefits of using a silicone polymer with grafted β-cyclodextrin units. The silicone backbone is biocompatible and adherent to the skin. Grafted β-cyclodextrin units enable the incorporation of griseofulvin in silicone oil. Lastly, the encapsulation of griseofulvin as inclusion complexes during the "spontaneous emulsification" process is quite easy. Let us recall that the formation of an inclusion

---

Y. Chevalier
University of Lyon, Laboratoire d'Automatique et de Génie des Procédés (LAGEP), UMR 5007 CNRS – Université Claude Bernard Lyon 1, 69622 Villeurbanne, France.

F. Ganachaud et al. (eds.), *Silicon Based Polymers*,

complex of griseofulvin and the water-soluble β-cyclodextrin is a very difficult task owing to the low solubility of griseofulvin in water.

Keywords Cyclodextrin · silicone · nanoparticle · spontaneous emulsification · drug delivery

1 Introduction

Most active substances that are currently used in pharmaceutical, cosmetic or agrochemical applications are not soluble in water. This feature is a major trouble regarding their formulation. The simplest technology consists in solubilizing the active substance either in the oil phase of oil-in-water (o/w) emulsions or inside the particles of aqueous polymer suspensions [1]. The formulation problem is becoming even harder when the active substance is also not soluble in the oils and the polymers used in the application. Griseofulvin is a typical case of such difficult problem of formulation. This antifungal molecule can enter the lipidic membrane of cells and disorganize their functional activity [2,3]. It has also an antibacterial activity by the same mechanism. Griseofulvin is insoluble in water as well as in non-polar oils, as are several other molecules whose site of action is located in the membranes (amphotericin B). The polar organic solvents that may solubilize such active substances do not meet the safety constraints of the pharmaceutical applications. Most dosage forms used in pharmaceutical applications are water-based, which limits a lot the choice of the available formulations. Solubilization inside supramolecular assemblies, micelles, microemulsions or liposomes is one possibility which suffers from the low solubilising capacity of surfactants for such substances. The solubilization of griseofulvin in micelles requires quite large concentrations of detergents [4–8] that are the origin of many troubles such as hemolysis in case of parenteral administration or skin irritancy for topical delivery. The same limits are faced with aqueous suspensions of polymer particles [9]; the encapsulation rate of griseofulvin inside polycaprolactone particles is 7 mg/g of polymer [10].

An alternative is the solubilization with the help of cyclodextrins because these are soluble in water and can incorporate organic molecules inside their hydrophobic cavity [11–13]. β-cyclodextrin is the most useful regarding the typical size of molecules to be solubilized. Griseofulvin forms inclusion complexes of 1:1 stoichiometry with β-cyclodextrin [14, 15]. One possible problem is the moderate solubility of β-cyclodextrin in water (18.5 g/L) and the even lower solubility of most inclusion complexes. A more dramatic problem is the preparation of inclusion complexes of water-soluble cyclodextrins and organic molecules that are not soluble in water. The complexation takes place by means of hydrophobic interactions inside the cavity, which require the presence of water as a solvent.

The chemical modification of β-cyclodextrin is often proposed as a way to improve the utilization of inclusion complexes with β-cyclodextrins. Grafting strongly polar groups to β-cyclodextrin increases its solubility in water. Alternatively, attachment of hydrophobic grafts yields amphiphilic cyclodextrins that most often are no more soluble in water. Such amphiphilic cyclodextrins can be emulsified in water as suspensions of nanoparticles. The complexation may be made easier when it is performed in association with the emulsification process. In particular, inclusion complexes with amphiphilic cyclodextrins are easily formed when the emulsification is performed by the "nanoprecipitation" or "spontaneous emulsification" process [16–19]. Therefore, there are obvious improvements using modified cyclodextrins, especially hydrophobically modified because the complexation of slightly hydrosoluble molecules can be performed in an organic solvent during the "spontaneous emulsification" process. In the present work, hydrophobically modified cyclodextrins are prepared by grafting β-cyclodextrins to a silicone polymer backbone. The "hybrid" material combining cyclodextrin and silicone polymer can be dispersed as an o/w emulsion. It has been shown in a previous work that silicone polymers grafted with β-cyclodextrins were successful for the encapsulation of griseofulvin in an o/w emulsion [20]. However, the first attempts disclosed some difficulties regarding the synthesis of the polymers. The grafting rate of β-cyclodextrin remained limited (20%) and the encapsulation rate as a 1:1 inclusion complex was also limited [20].

The particles of cyclodextrin-grafted silicone are also the vehicles for the transport of the active substance to the site of its therapeutic action. In the case of antifungal dosage form for topical administration action, penetration and slow release in the skin is searched for [21–23]. The *stratum corneum* top layer is made of corneocytes separated by a lipidic intercellular medium. Internal layers, epidermis and dermis are essentially aqueous media. The top layer is hydrophobic whereas the internal layers are hydrophilic. The penetration into the skin is controlled by the hydrophobic character of the particles. Hydrophobic materials easily penetrate and possibly accumulate in the *stratum corneum*. A slow release of the drug into the deep hydrophilic layers is possible from this medium. Silicone emulsions are often selected for cosmetic formulations because of the favorable spreading of silicone oils at the skin surface together with their low-irritancy properties.

The low grafting degree achieved in our previous work [20] led to poor encapsulation. Our goal is to overcome this limitation by increasing the grafting degree of β-cyclodextrin to the silicone backbone. This paper reports on the synthesis of a silicone-based polymer bearing β-cyclodextrin as grafted side-chains by hydrosilylation reaction of poly(hydromethylsiloxane) and an allyl derivative of β-cyclodextrin. Higher grafting degree of β-cyclodextrin could be achieved and highly grafted silicone polymers are evaluated with respect to their ability to be emulsified by means of the "spontaneous emulsification" process and their ability for the encapsulation of griseofulvin inside the emulsion droplets. The investigation of the grafting and the preparation of highly grafted functional polymer are described in a first section. The emulsification of the polymers as oil-in-water nanoemulsions by means of the "spontaneous emulsification" method is reported in a second part. The encapsulation of griseofulvin inside the polymer droplets is presented in the third part.

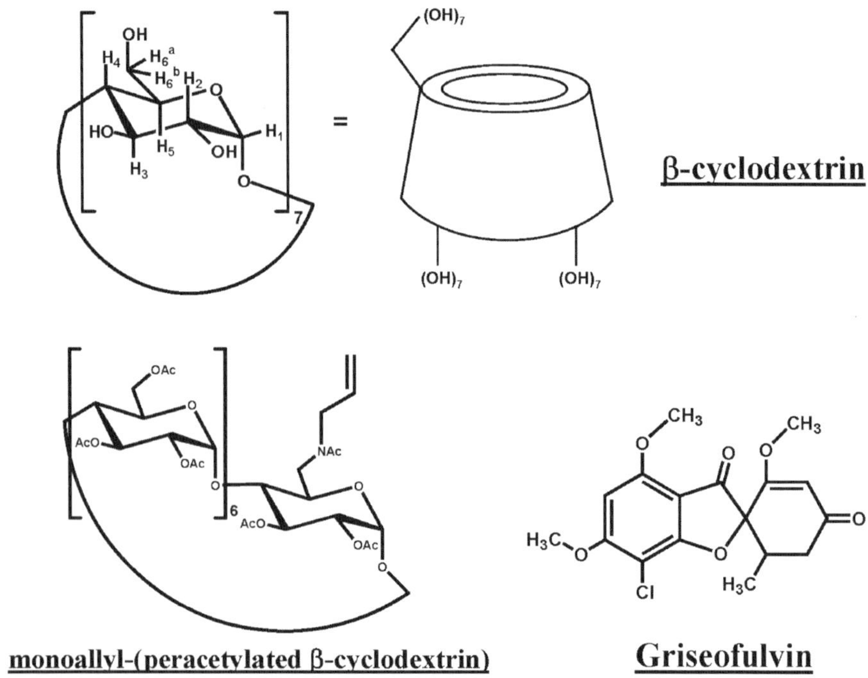

Fig. 1 Structures of native β-cyclodextrin, the monoallyl derivative of β-cyclodextrin and Griseofulvin. Proton numbers given for β-cyclodextrin are used in the assignment of ^{1}H NMR data reported in the "Materials and methods" section

The chemical structures of the starting materials β-cyclodextrin, monoallyl-β-cyclodextrin and griseofulvin are shown in Fig. 1.

2 Materials and Methods

Materials: The β-cyclodextrin as its native form was kindly provided by Roquette (Roquette Frères, Lestrem, France); it was dried under vacuum before use. The Karstedt catalyst, platinum(0)-1,3-divinyl-1,1,3,3-tetramethylsiloxane complex, was supplied by Aldrich as its 2% solution in xylenes. Toluene was distilled and stored under dry conditions. The hydrosilane polymers were chemicals from Gelest purchased from ABCR: the poly(methylhydrosiloxane) (PMHS) homopolymer HMS-991, CAS [63148-57-2], of number average molar mass 1600 g/mol was a liquid of viscosity 15–25 mPa.s; the poly(methylhydrosiloxane-*co*-dimethylsiloxane) copolymer HMS-301, CAS [68037-59-2], of number average molar mass 2000 g/mol contained 25% methylhydrosiloxane units and was a liquid of viscosity 25–35 mPa.s. Silicone oil (polydimethylsiloxane) from Dow Corning was the grade 200® light 20cSt of viscosity 20 mPa.s. Reagent grade 1-pentene from Aldrich was used as received. The Polysorbate 20 emulsifier was purchased from Aldrich under the trade name Tween® 20. Griseofulvin from *Penicillium griseofulvum* (≥ 95% purity) was purchased from Sigma.

The monoallyl derivative of β-cyclodextrin in its peracetylated form, namely mono-(6-N-allylamino)-6-deoxy)-peracetylated-β-cyclodextrin, was synthesized in 3 steps from native β-cyclodextrin. The synthesis of this derivative is reported in ref [20]. The NMR and IR analyses were consistent with the expected chemical structure.

^{1}H NMR ($CDCl_3$) δ(ppm): 2.11 (s, 63H, CH_3CO); 3.07–3.15 (m, 1H, $H_{6'}^a$); 3.37–3.49 (m, 1H, $H_{6'}^b$); 3.71–3.80 (m, 7H, H_4); 3.90–4.05 (m, 2H, CH_2NAc); 4.06–4.65 (m, 19H, H_5, H_6^a, H_6^b); 4.79–4.83 (m,7H, H_2); 4.95–5.12 (m, 2H, CH_2=CH); 5.14–5.20 (m, 7H, H_3); 5.26–5.33 (m, 7H, H_1); 5.55–5.77 (m, 1H, CH_2=C*H*).

^{13}C NMR ($CDCl_3$) δ(ppm): 21.17 (CH_3-C); 41.3 ($C_{6'}$); 62.89 (C_6); 69.4 (CH_2-NAc); 69.96–71.69 (C_2, C_3, C_5); 76.62–77.84 (C_4); 97.12–97.42 (C_1); 124.35 (CH_2=*C*H); 133.17 (CH_2=CH); 149 (C=O amide); 169.6–171.1 (C=O ester).

IR (KBr, ν cm^{-1}): 3460 (N-H); 2960 (C-H); 1740 (C=O, acetyl); 1642 (C=C); 1156 (C-N); 1040 (C-O-C).

The mono-(6-N-allylamino)-6-deoxy)-peracetylated-β-cyclodextrin polymer having 40% grafting rate, PMHS-CD(40), was prepared by hydrosilylation of PMHS with the monoallyl peracetylated β-cyclodextrin and final capping with *n*-pentene. A solution of poly(methylhydrosiloxane) (HMS-991) polymer (0.051 g, 8×10^{-4} mol of Si-H) and monoallyl peracetylated β-cyclodextrin (0.66 g, 3.2×10^{-4} mol) in 30 mL dry toluene was heated at 85°C for 24 h under dry nitrogen. The Karstedt catalyst was added in several portions of 10 μL during the course of the reaction. 10 μL of the catalyst solution (2% of platinum content) corresponded to a concentration of 6.7 ppm in the reaction medium. The disappearance of the monoallyl β-cyclodextrin was monitored by thin layer chromatography. After full conversion of monoallyl β-cyclodextrin, 4 g of pentene were added and the reaction mixture was stirred overnight at 85°C. The reaction completion was confirmed by the disappearance of the IR absorption peak at 2160 cm^{-1} corresponding to the Si-H function of HMS-991. The solvent was evaporated under reduced pressure and the product was purified by column chromatography on silica eluted with chloroform/methanol (90/10) mixture. Evaporation of the solvent gave PMHS-CD(40) as a pale yellow solid with 60% yield.

TLC: R_f: 0.64 (chloroform/methanol 9:1).

^{1}H NMR ($CDCl_3$) δ(ppm): 0.01–0.12 (m, Si-CH_3); 0.38–0.49 (m, Si-CH_2-CH_2); 0.84–0.86 (m, Si-CH_2-CH_2 and CH_2-CH_3 of pentyl); 0.96–0.98 (m, CH_2-CH_2-CH_3 of pentyl); 1.01–1.19 (m, CH_2-CH_3 of pentyl); 1.55–1.58 (m, CH_2-N(Ac)-CD); 1.98–2.05 (m, CH_3CO); 3.05–3.15 (m, $H_{6'}^a$); 3.31–3.43 (m, $H_{6'}^b$); 3.64–3.65 (m, H_4); 4.12–4.40 (m, H_5, H_6^a); 4.47–4.60 (m, H_6^b); 4.73–4.80 (m, H_2); 5.01–5.04 (m, H_3); 5.20–5.26 (m, H_1).

^{13}C NMR ($CDCl_3$) δ(ppm): 1.98 (Si-CH_3); 14.09–19.68 (Si-CH_2); 20.74–27.28 (Si-CH_2-CH_2, CH_2-CH_2-CH_3, CH_2-CH_3, CH_2-CH_3 and CH_3-CO); 46.27 ($C_{6'}$; 62.48 (C_6); 69.56 (CH_2-NAc-CD); 70.39–71.27 (C_2, C_3, C_5); 76.68–77.52 (C_4); 96.71 (C_1); 169.28–170.91 (C=O).

IR (KBr, *ν* cm^{-1}): 3581 (N-H); 2962, 2926 and 2817 (CH_2 and CH_3)); 1746 (C=O acetyl); 1255 and 1282 (Si-C); 1165 (C-N); 1040 (C-O-C); 1099 (Si-O-Si).

PMHS-CD with 10, 20 and 30% grafting degrees were synthesized in the same way using the suitable stoichiometric amount of monoallyl β-cyclodextrin.

Kinetic experiments: Either PMHS or HMS-301 reacted in toluene at 60°C with peracetylated monoallyl β-cyclodextrin in stoichiometric amounts with respect to Si-H. The concentration of β-cyclodextrin and Si-H was 0.032 mol/L. The concentration of Karstedt catalyst was 20 ppm of platinum content. Aliquots of the reaction medium were collected at different times and the solutions were directly analyzed by IR spectroscopy in a liquid cell. The absorbance of the Si-H band at 2160 cm^{-1} was normalized to the absorbance of the C=O stretching band of the ester at 1740 cm^{-1}. The decrease of the Si-H band was plotted as a function of time. The conversion is the complement to 100%.

Methods: ^{1}H and ^{13}C NMR spectra were recorded with a Bruker DRX300 spectrometer working at 300 MHz and 75 MHz Larmor frequencies for ^{1}H and ^{13}C respectively. Solutions in $CDCl_3$ did not contain tetramethylsilane as internal standard because the TMS line was too close to the CH_3-Si lines of the polymer in the ^{1}H NMR spectra. The calibration on a right ppm scale was made using the line of the $CDCl_3$ solvent at 7.28 ppm from the TMS standard. The abbreviations used are: s (singlet), d (doublet), t (triplet) and m (multiplet). Fourier transform infrared spectra were recorded with a Perkin Elmer 1000 spectrometer in transmission mode between wave numbers 400 and 4000 cm^{-1} using pellets made with 0.1 g KBr and 10 mg of each compound. IR spectra of the kinetic investigations were recorded with a Nicolet Protégé 460 spectrometer in transmission mode through a liquid cell EZ-Fill BDH kit FT54-019 with AgBr windows and 0.1 mm Teflon spacer.

Emulsification: Typical preparation of an unloaded emulsion is as follows. The organic phase was made up with 10 mg of polymer dissolved in 2.5 mL of acetone and the aqueous phase contained 5 mg of the Tween® 20 emulsifier in 5 mL water. The organic phase was rapidly poured into the aqueous phase under gentle stirring with a magnetic stir bar; the mixture turned milky as the emulsion formed. The acetone was then evaporated under reduced pressure until the mass of the sample corresponded to the expected mass corresponding to all the ingredients excepted acetone.

The same process was applied with the griseofulvin solubilized in the organic phase for the formulation of emulsions loaded with griseofulvin. Different loadings of griseofulvin could be easily checked.

Characterization methods: Granulometric analyses of the submicronic emulsions were performed by means of dynamic light scattering with a Malvern NanoZS instrument. The suspensions were diluted with water until the count rate was 200 kCounts/s and the autocorrelation function of the back-scattered light was measured with a sample time of 0.5 μs. The data was analyzed for the *z*-average diameter by the method of cumulants and for the size distribution using the CONTIN algorithm. Complementary granulometric analyses were performed with a small angle laser light scattering instrument Coulter LS230. The emulsions were diluted until the transmission was 88%, as recommended by the supplier for optimum operational conditions. The real part of refractive index of the particles was taken as that of silicone oil (1.40) and the imaginary part was zero; the refractive index of water was 1.33.

Optical microscopy pictures of the emulsions were taken with a Leica DML microscope. A drop of the aqueous suspension was observed in transmission mode between glass slide and cover slip at magnifications of the objective ranging from ×10 to ×100.

3 Results and Discussion

3.1 Synthesis of Polymeric Materials Bearing Cyclodextrins

Synthesis (Fig. 2) implies as first step the preparation of a monoallyl derivative of β-cyclodextrin that will subsequently be grafted to poly(methylhydrosiloxane) (PMHS) or a copolymer poly(methylhydrosiloxane-co-dimethylsiloxane) (HMS-301) by means of hydrosilylation reaction. The cyclodextrin derivative should bear a single vinyl group to be grafted to the poly(hydrosiloxane) chain in order to avoid cross-linking during the grafting reaction. The formation of a gel through multiple cross-linking between β-cyclodextrin molecules and polysiloxane chains is to be avoided because it prevents any subsequent emulsification. The hydroxyl groups of β-cyclodextrin are deactivated by acetyl protective groups because they are also reactive towards the hydrosilane function. Such peracetylated monoallyl β-cyclodextrin derivative, namely mono-(6-N-allylamino)-6-deoxy)-peracetylated-β-cyclodextrin has been prepared in good yield [20]. The limiting reaction was the grafting to the polymer backbone by a hydrosilylation reaction catalyzed by the Karstedt catalyst as shown in Fig. 2. A grafting degree of only 20% was reached in our previous work.

The present kinetic investigation aims at understanding the limiting factors of the hydrosilylation reaction with the poly(hydromethylsiloxane). This reaction might be impeded either by the poor reactivity of the allyl derivative of β-cyclodextrin,

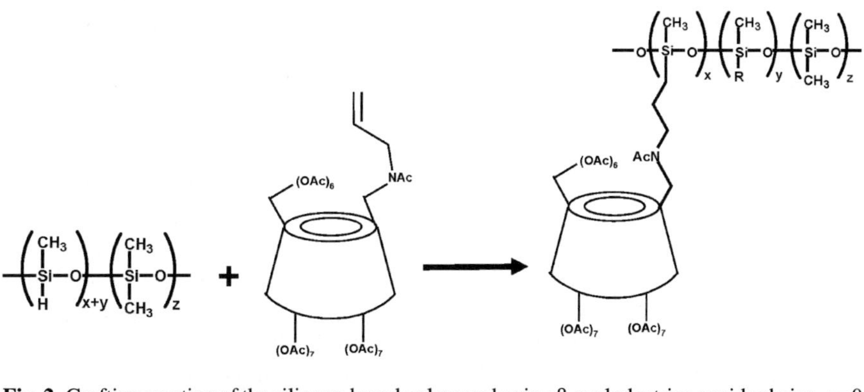

Fig. 2 Grafting reaction of the silicone-based polymers having β-cyclodextrins as side chains. $z=0$ and $x+y=1$ for the PMHS homopolymer; $z=0.75$ and $x+y=0.25$ for the HMS-301 copolymer. R is either H after incomplete grafting of β-cyclodextrin units, or the pentyl substituent after deactivation of the Si-H by reaction with 1-pentene. The final compound is called PMHS-CD(x) for *x* % grafting rate

or a deactivation of the Karstedt catalyst by the cyclodextrin, or the steric hindrance of the grafted β-cyclodextrins that blocks further conversion.

The hydrosilylation reaction with the polymer or the copolymer was performed with the use of Karstedt catalyst (platinum(0)-1,3-divinyl-1,1,3,3-tetramethylsiloxane complex) in toluene at 80 °C under dry and inert atmosphere. Grafting to the full hydrosilane groups of the PMHS homopolymer could not be achieved after 72 h reaction time of a stoichiometric amount of monoallyl β-cyclodextrin with respect to the Si-H. Several reports of such incomplete grafting reactions have been reported in the literature [24–27]. In particular, evidence of limited grafting caused by steric hindrance of the bulky tris(trimethylsilyl)methyl group has clearly been given [24]. Regarding grafting of cyclodextrins similar to ours, Cousin et al. [28] grafted the permethylated monooctenyl β-cyclodextrin to a poly(methylhydrosiloxane-*co*-dimethylsiloxane) copolymer using hexachloroplatinic acid catalyst in THF. 8.5% residual Si-H were analyzed at the end of the grafting reaction for a copolymer having 12.3 mol% Si-H content, corresponding to a grafting degree of 4%; therefore the yield was 30% with respect to the starting Si-H. It is quite common that such reaction does not go to completion, giving partial grafting, even at low Si-H content in the copolymer. Larger grafting degrees can be obtained with substituents less bulky than cyclodextrin such as fluorinated alkyl chains [29] and bromoundecyl group [30]. Full conversion of the Si-H of PMHS was easily achieved with *n*-pentene in the present work. Complete grafting by hydrosilylation has been achieved with copolymers containing low contents of methylhydrosiloxane units, possibly because the reactive Si-H functions were remote from each other [31, 32].

Because of incomplete grafting, the polymers contained residual hydrosilane groups that are reactive in the presence of traces of water. Hydrolysis of Si-H into Si-OH and subsequent condensation of the formed silanols may involve a slow cross-linking of the polymer if it is not stored safe from moisture. A high degree of cross-linking would cause gelation and would impede the emulsification of the polymer in water. Even if the hydrolysis is prevented during storage, the final utilization is an emulsion in water where the hydrolysis and cross-linking will take place. A slow drift of the encapsulation and release properties would result from this uncontrolled cross-linking. Therefore, the residual Si-H functions have been deactivated by reaction with pentene after the end of grafting reaction of cyclodextrin.

In order to understand the limitations and suggest improvements, the kinetics of the grafting reaction of β-cyclodextrin was investigated. Thus, the hydrosilylation reaction was run with either PMHS homopolymer or HMS-301 copolymer and stoichiometric amounts of peracetylated monoallyl β-cyclodextrin in toluene solvent at 60°C. The kinetics was followed by IR analyses of the residual Si-H at 2160 cm^{-1} [33]. Typical spectra are shown in Fig. 3.

As shown in Fig. 4, the Si-H groups were rapidly converted at the beginning of the reaction. The reaction of the PMHS homopolymer slowed down dramatically after 2 h. Full conversion of the Si-H could not be reached after 48 hreaction time. On the contrary, the fast kinetics was followed up to full conversion of the Si-H with the HMS-301 copolymer.

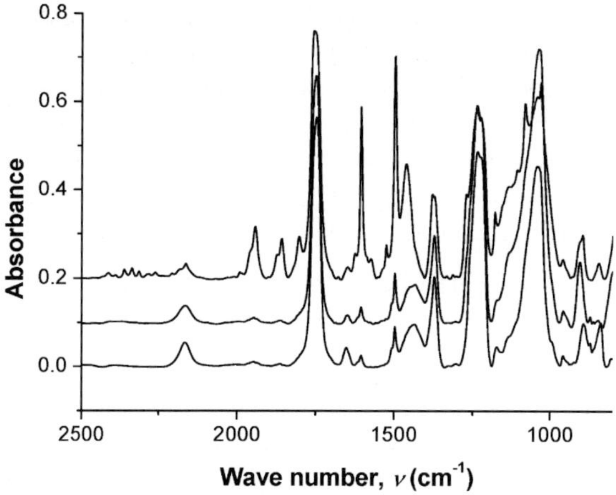

Fig. 3 IR spectra of the reaction medium in the kinetics of the reaction to the PMHS homopolymer. The band of the C=O ester at 1740 cm^{-1} was normalized to a constant absorbance. The absorbance of the Si-H band at 2160 cm^{-1} was used for investigating the kinetics of the reaction. The spectra were vertically shifted by 0.1 absorbance units for clearer reading. Reaction times from bottom to top are: $t=0$; $t=1$ h30; $t=8$ h

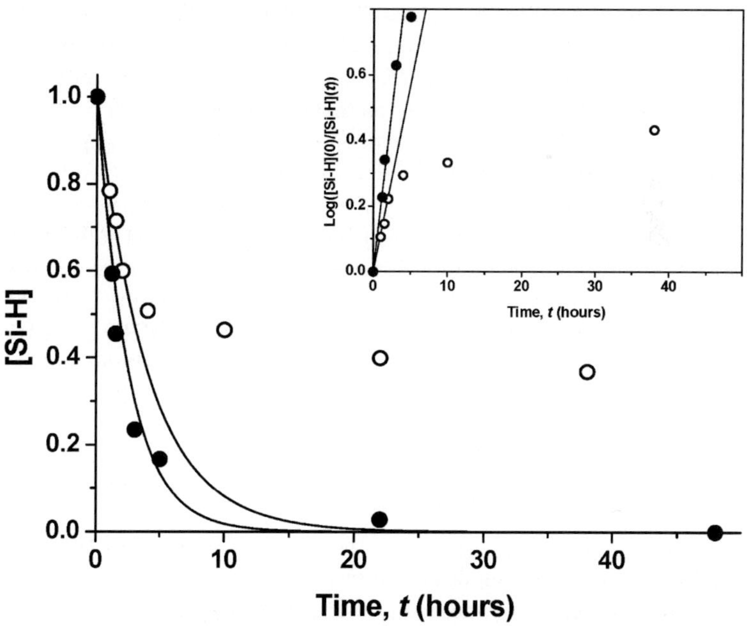

Fig. 4 Kinetics of the conversion of Si-H by the hydrosilylation reaction of monoallyl β-cyclodextrin with PMHS (○) or HMS-301 (●). The solid lines show the first order kinetics fitted to the data at short reaction times (see text). The linearized plot of a first order kinetics, log([Si-H](0)/[Si-H](t)) vs t, is shown in the Inset

It has been shown that the kinetics of hydrosilylation of α-olefins (1-decene) by PMHS was first order with respect to the concentrations of Si-H and Karstedt catalyst [33]. The first order kinetics was observed for the HMS-301 copolymer in the present instance since the plot of

$$\log\left(\frac{[\mathrm{Si\text{-}H}](0)}{[\mathrm{Si\text{-}H}](t)}\right)$$

against time, t, was linear. The first order kinetics was only followed at the very beginning of the reaction of the PMHS homopolymer. The plot of

$$\log\left(\frac{[\mathrm{Si\text{-}H}](0)}{[\mathrm{Si\text{-}H}](t)}\right)$$

against time departed from the logarithmic law of the first order kinetics after 2 h reaction time. The conversion at which the data left the first order kinetics was 50%.

The origin of such slowing down occurring at intermediate conversion was obviously the steric hindrance of the grafted bulky substituents. The distance between two adjacent Si-H groups in the PMHS homopolymer is 3.78 Å [34], whereas the largest diameter of the β-cyclodextrin molecule is 15.3 Å [12]. These figures demonstrate the obvious steric hindrance effect of the β-cyclodextrin substituent. The conversion at which the kinetics departed from the first order kinetics was 50% for the homopolymer. Therefore full conversion of the HMS-301 copolymer could be reached because the Si-H content was 25% only. The mean distance separating Si-H groups is 4 times larger with the poly(methylhydrosiloxane-*co*-dimethylsiloxane) copolymer containing 25% of methylhydrosiloxane units. The comparison of the grafting kinetics of the homopolymer and copolymer strongly suggested that steric hindrance was the origin of the limited grafting to the homopolymer.

Grafting rates up to 40% could be reached through the reaction at full conversion of the corresponding amounts of β-cyclodextrin and the PMHS homopolymer. This was achieved according to the synthesis method described in the "Materials and methods" section. The PMHS-CD(20), PMHS-CD(30) and PMHS-CD(40) derivatives were prepared and the residual Si-H groups were capped with *n*-pentene. They were evaluated for their emulsification and loading with griseofulvin.

3.2 Emulsification of Highly Grafted Polymers

Functional polymers have been dispersed in water by the "spontaneous emulsification" (or "nanoprecipitation") method [35, 36]. Such method consists in mixing a polymer solution in acetone to a large amount of water. Complete dissolution of acetone in water takes place readily and leaves the polymer as a supersaturated solution in water. The polymer forms a colloidal suspension upon precipitation. Thereafter, acetone is evaporated under reduced pressure. The resulting emulsion is made of nanometric polymer droplets in the range of 100 nm to 1 μm in the case of low initial polymer concentrations. Many polymers and oils can be emulsified using this method [37]. The emulsion is stable when a stabilizing agent has been added to the formulation, either in the aqueous phase, or in the organic phase.

The adaptation of this method to the emulsification of silicone grafted with cyclodextrins is straightforward since this polymer is soluble in acetone and not soluble in water. The low molar mass of the macromolecules is also favorable [37]. Griseofulvin is also soluble in acetone and can be co-emulsified with the polymer. Encapsulation of griseofulvin inside the cavity of cyclodextrins takes place during emulsification. The stabilizing agent was the mild and low-irritant nonionic emulsifier Polysorbate 20 (Tween® 20, PEG(20) sorbitan monolaurate) introduced in the aqueous phase. Its high HLB = 16.7 is well-suited for the formulation of an oil-in-water emulsion according to the Bancroft rule [38]. Typical recipe is given in Table 1.

The PMHS-CD polymers of substitution degree 20, 30 and 40% were successfully emulsified. Stable milky emulsions were obtained in the presence of emulsifier (Polysorbate 20); no organic material was separated as sediment or supernatant. Emulsions of silicone oil were also prepared in the same way for comparison. Submicrometric emulsions were produced in every case. The z-average diameters of the blank particles (free of griseofulvin) measured by dynamic light scattering are given in Table 2. The granulometric distributions were monomodal and did not show the presence of any particle larger than 1 μm (Fig. 5). Optical microscopy and complementary measurements by small angle laser light scattering did not show the presence of large particles.

Table 1 Chemical composition of both organic and aqueous phases in the final emulsion

	Compound	Mass/Volume	Concentration in the phase	Concentration in the final emulsion
Organic phase	Acetone	2.5 mL		
	PMHS-CD	10 mg	4 g/L	0.2 wt%
	Griseofulvin	0–6 mg	0–2.4 g/L	0–0.12%
Aqueous phase	Water	5 mL		
	Polysorbate 20	5 mg	1 g/L	0.1 wt%

Table 2 Granulometric analyses of the emulsions made of the different β-cyclodextrin-grafted silicones

Polymer	Grafting rate (%)	z-average diameter (nm)		
		with Polysorbate 20	without Polysorbate 20	
			at once	after 15 days storage
Silicone oil DC 200	0	300	unstable	unstable
PMHS-CD(20)	20	140	175	1720
PMHS-CD(30)	30	146	275	2050
PMHS-CD(40)	40	169	424	2600

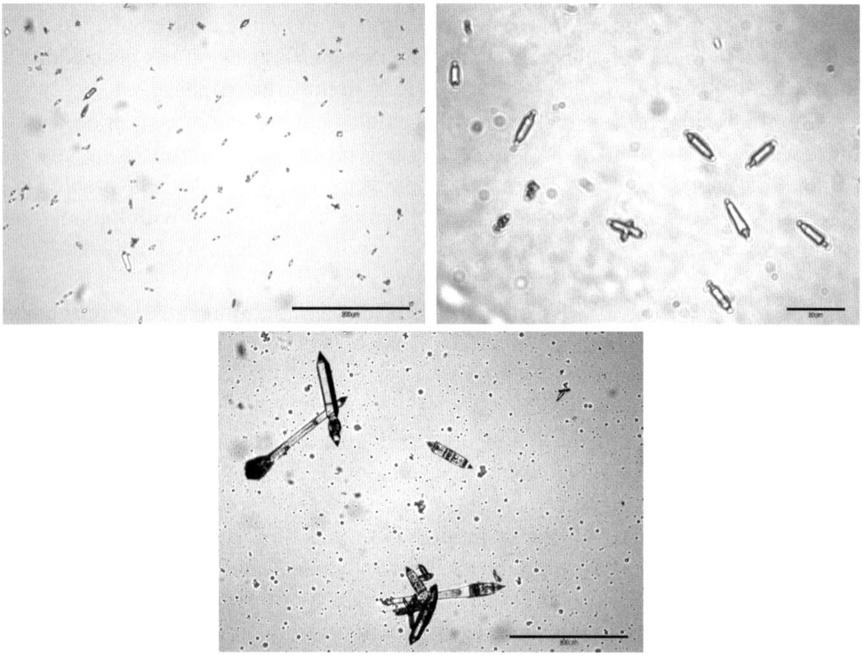

Fig. 5 Microscopy pictures of griseofulvin crystals precipitated in water and out from a silicone emulsion. Top: Griseofulvin crystals precipitated by the spontaneous emulsification method; 10 mg griseofulvin in 2.5 mL acetone poured into 5 mL of 0.1% Polysorbate 20 aqueous solution (*left*: ×20, bar = 200 μm; *right*: ×100, bar = 20 μm). *Bottom*: 10 mg silicone oil and 64 mg griseofulvin in 2.5 mL acetone poured into 5 mL of 0.1% Polysorbate 20 aqueous solution (×20, bar = 200 μm)

Silicone emulsions show several features that depart from emulsions made of classical oils (silicon-free oils) [39]. Therefore, silicone-based emulsifiers are often used for stabilization of silicone emulsions. The classical Polysorbate 20 emulsifier was successfully used in the present case, possibly thanks to implementation of the "spontaneous emulsification" process. The emulsifier did not significantly influence the size of the emulsion droplets. Table 2 shows that the z-average diameters were similar to emulsifier-free formulations. However, the presence of emulsifier was necessary for ensuring the colloidal stability of emulsions over long term storage. Diameters did not significantly change after 3 months storage. On the contrary, the diameter grew quite fast in absence of emulsifier (Table 2 reports nearly 10-fold increase of the droplet diameter upon 15 days storage), emulsions underwent massive creaming since the first day and finally coalesced into macroscopic lumps.

The presence of β-cyclodextrin grafts allows the formation of smaller emulsion droplets in comparison to systems formed using silicone oil. Thus, the diameters of PMHS-CD droplets were smaller than the silicone oil droplets by a factor of 2. The presence of substituents containing ester groups (20 ester groups per graft) makes the PMHS-CD polymers more polar than silicone oil. The backbone of PMHS-CD is also shorter than silicone oil. It is therefore presumed that the solubility of the PMHS-CD in water is higher, so that the supersaturation during the spontaneous emulsification is less. Indeed, it has been shown that smaller supersaturation led to smaller emulsion droplets [37, 40].

3.3 High Loading of Griseofulvin

The solubilization capacity of the PMHS-CD suspension with respect to the antifungal molecule griseofulvin was assessed from optical microscopy observations. The loading of griseofulvin inside emulsions of PMHS-CD(40) (40 mol% grafting degree) is reported as an example. Optical microscopy is quite efficient in determining the loading capacity of the particles [20]. Encapsulation of griseofulvin inside the hydrophobic cavity of β-cyclodextrin prevents its crystallization and precipitation. On the contrary, griseofulvin molecules that are simply co-emulsified with the PMHS-CD polymer and not complexed, undergo slow crystallization. Crystals of griseofulvin can easily be detected with the optical microscope as anisotropic-shaped particles (Fig. 5) that display the characteristic birefringence of crystals when observed between crossed polars. Silicone oil emulsions did not allow the encapsulation of griseofulvin that is not soluble in silicone; griseofulvin crystals were observed under the microscope (Fig. 5).

The encapsulation of increasing amounts of griseofulvin inside PMHS-CD particles was followed by optical microscopy. Griseofulvin was solubilized in the organic phase (acetone) and co-emulsified with PMHS-CD by the spontaneous emulsification method. This method allowed the easy incorporation of griseofulvin as an inclusion complex into the β-cyclodextrin cavity. Loading griseofulvin after the emulsification would be very difficult because the very low solubility of griseofulvin in water (34.3 μM [41]) dramatically slows down the loading process. The same is true for the preparation of inclusion complex of griseofulvin with the water-soluble β-cyclodextrin [42]. Since griseofulvin complexation by cyclodextrin is essentially ensured by hydrophobic interactions inside the cavity, the presence of water is compulsory. Although cyclodextrin and griseofulvin were present in the starting acetone solution used as organic phase, complexation does not take place before this solution is mixed with water. The formation of inclusion complex probably takes place during the emulsification when both griseofulvin and cyclodextrin simultaneously experience the aqueous environment. The spontaneous emulsification process appears a suitable way for an easy preparation of the inclusion complex.

Griseofulvin and PMHS-CD(40) were solubilized together in acetone and subsequently emulsified in water. The emulsions were observed by optical microscopy. No crystal could be observed when the griseofulvin amount in acetone was less than the 1:1 stoichiometric ratio with respect to cyclodextrin. On the contrary, coexistence of needle-shaped crystals and emulsion droplets was observed for larger amounts of griseofulvin, in excess with respect to the cyclodextrin content of PMHS-CD(40).

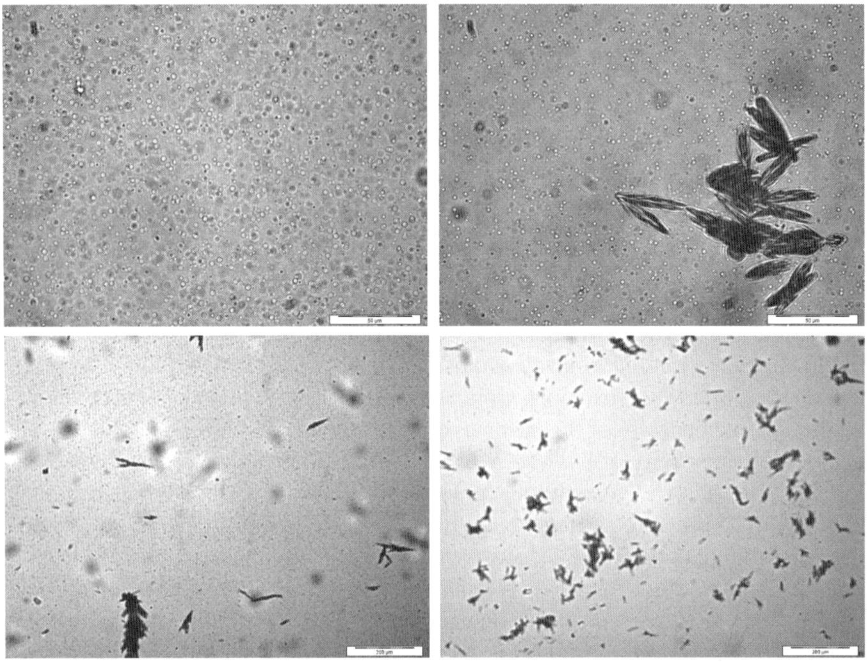

Fig. 6 Microscopy pictures of PMHS-CD(40) emulsion and precipitated griseofulvin crystals for various griseofulvin:β-cyclodextrin stoichiometries. *Top left*: 1:1 stoichiometry where no crystals can be seen (×50, bar=50 μm). *Top right*: 2:1 stoichiometry with few crystals coexisting with the emulsion droplets (×50, bar=50 μm). *Bottom left* and *right*: 5:1 and 10:1 stoichiometries showing larger amounts of precipitated griseofulvin (×10, bar=200 μm)

Figure 6 shows the optical microscopy observation of a PMHS-CD(40) emulsion containing a 1:1 stoichiometric amount of griseofulvin (picture on top-left) that does not display any crystal. Spherical droplets of emulsion are observed, their size being equal to the resolution of the microscope. These features are kept upon long period of storage (3 months). In particular, there is no slow separation of griseofulvin crystals. An emulsion containing a 2:1 stoichiometric amount of griseofulvin with respect to cyclodextrin slowly released small crystals of griseofulvin that can be seen with the microscope within few hours (top-right of Fig. 6). Within weeks of storage, a slow ripening process converted the crystals of irregular shape into needle-shaped crystals of several 10 μm length. The final crystals were identical to those produced by means of direct precipitation and ageing. For larger amounts of griseofulvin, crystals separated and grew faster. In these cases, the amount of precipitated griseofulvin was also larger (bottom of Fig. 6). Similar results were obtained with PMHS-CD(10), PMHS-CD(20) and PMHS-CD(30); namely the less substituted polysiloxanes allowed lower dissolution of griseofulvin.

The same type of observations could be made from granulometric analyses carried out with the help of small angle laser light scattering. This technique allows the observation of both the sub-micronic PMHS-CD particles and the much larger griseofulvin crystals. Coexistence of the two would result in a bimodal distribution. The interpretation of light scattering data in terms of distribution of particle diameters requires an "optical model" of the particles, that is, the refractive index of the material inside the particles. The refractive indices of PMHS-CD(40) loaded with griseofulvin and crystals of pure griseofulvin are not known. According to the design of the instrument software, the optical model is necessarily unique for the full size distribution, whereas loaded PMHS-CD(40) particles and griseofulvin crystals have obviously different refractive indices. In the present case, the refractive index of silicone oil (1.40) was used arbitrarily. Moreover, the griseofulvin crystals are needle-shaped instead of spherical as considered in the model. Therefore the size distributions that can be extracted from the recorded scattered intensity are not accurate and can be used for a qualitative discussion only.

Accordingly, the "diameter" of the griseofulvin crystals was 2 μm and the diameter of the blank PMHS-CD(40) emulsion was 500 nm (Fig. 7). This later diameter was larger than the one measured by dynamic light scattering because the optical model was not accurate. Light scattering showed that the size distribution did not contain particles larger than 1 μm for griseofulvin loads less than the 1:1 stoichiometry of the inclusion complex. On the contrary, a population of particles larger than 1 μm was detected for stoichiometries larger than 1:1. The size distribution of the population of large particles centered at 2 μm was identical to that of pure griseofulvin crystals precipitated by means of the spontaneous emulsification process. Therefore, small angle light scattering experiments reached the same conclusion as optical

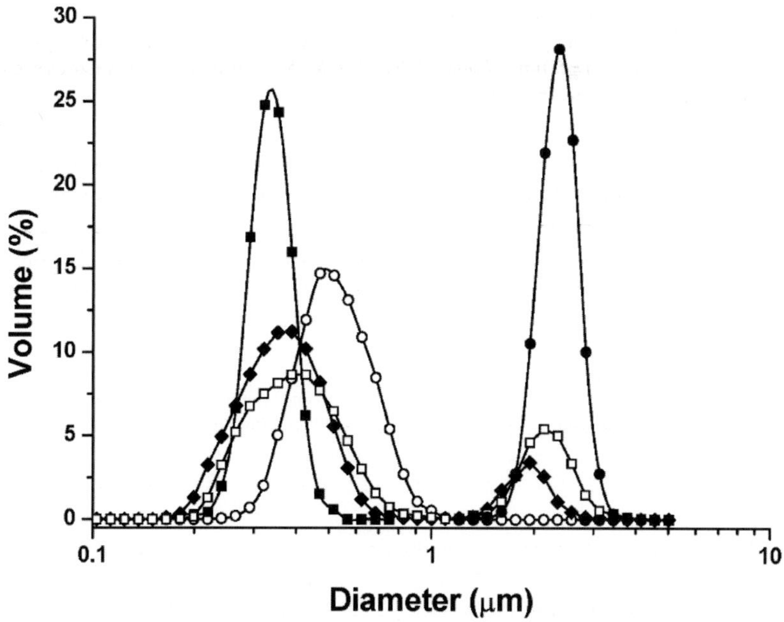

Fig. 7 Granulometric analyses of griseofulvin loaded particles of PMHS-CD(40) drawn from small angle light scattering measurements performed at different loading stoichiometries (griseofulvin: β-cyclodextrin). ○ : 0:1 (blank particles); ■:1:1; ◆ : 2:1; □: 5:1; ●: pure griseofulvin

microscopy. However, the conclusions remained qualitative because of the inadequate choice of the optical model. The correct interpretation of the data including accurate values of refractive indices and a correct description of the crystal shape would be a daunting task.

It appeared that encapsulation of griseofulvin inside emulsion droplets could be achieved for stoichiometries lower than 1:1. High loading of griseofulvin clearly requires the high β-cyclodextrin grafting rates. The presence of the β-cyclodextrin attached on the polymer has a beneficial effect for the encapsulation of griseofulvin.

4 Conclusion

Silicone-based polymers bearing β-cyclodextrin as side chains have been prepared for the encapsulation of hydrophobic drugs that fit the cyclodextrin cavity. The example of the antifungal agent griseofulvin was detailed.

Synthesis in several steps could be performed. The limiting step of the synthesis scheme was the grafting to the polymer backbone. Kinetic investigation of the grafting reaction showed that the limitation came from the size of grafted β-cyclodextrin units by steric hindrance effect of such bulky substituents above 50% conversion. Grafting rate as high as 40 mol% was reached with full conversion of the monoallyl derivative of β-cyclodextrin. This appeared quite a high functionalization rate when it was expressed as weight fraction of substituent: indeed 40 mol% corresponds to 90 wt% of peracetylated β-cyclodextrin.

Such polymers are not soluble in water and were successfully emulsified by means of the spontaneous emulsification method, yielding aqueous emulsions with nanometric sizes (mean diameters from 100 to 500 nm).

Submicronic particles (nanoparticles) produced in that way allowed the encapsulation of the griseofulvin antifungal substance up to the 1:1 stoichiometry with respect to the β-cyclodextrin. A simple oil-in-water emulsion could not achieve such encapsulation because β-cyclodextrin was not soluble in common oils; this was shown with silicone oil in the present study. The spontaneous emulsification process used for the preparation of the particles also allowed the easy encapsulation of the sparingly soluble griseofulvin.

Acknowledgments This work was supported by a grant from French-Tunisian cooperation project (CMCU n°04S1207). The supply of β-cyclodextrin by Roquette Frères (Lestrem) is gratefully acknowledged.

References

1. S. Benita, Ed., *Microencapsulation. Methods and industrial applications*, Marcel Dekker, New York (1996)
2. E. Finkelstein, B. Amichai, M.H. Grunwald, *Int. J. Antimicrobial Agents* 6 (1996) 189–194.
3. I. Smirnova, M. Türk, R. Wischumerski, M.A. Wahl, *Eng. Life Sci.* 5 (2005) 277–280.
4. P.H. Elworthy, F.J. Lipscomb, *J. Pharm. Pharmacol.* 20 (1968) 817–824.

5. A.A. Kassem, N.M. Mursi, *Bull. Faculty Pharm. Cairo Univ.* 9 (1970) 11–23.
6. P.H. Elworthy, M.S. Patel, *J. Pharm. Pharmacol.* 34 (1982) 543–546.
7. S. Tolle, T. Zuberi, W. Warisnoicharoen, M.J. Lawrence, *J. Pharm. Sci.* 89 (2000) 798–806.
8. N.R. Calafato, G. Picó, *Colloids Surfaces B* 47 (2006) 198–204.
9. D. Moinard-Checot, Y. Chevalier, S. Briançon, H. Fessi, S. Guinebretière, *J. Nanosci. Nanotechnol.* 6 (2006) 2664–2681.
10. Z. Zili, S. Sfar, H. Fessi, *Int. J. Pharm.* 294 (2005) 261–267.
11. J. Szejtli, *Cyclodextrin technology*, Kluwer Academic, Dordrecht (1988).
12. D. Duchêne, Ed., *Cyclodextrins and their industrial uses*; Éditions de la Santé, Paris (1987).
13. G. Wenz, *Angew. Chem., Int. Ed. Engl.* 33 (1994) 803–822.
14. M.D. Veiga, P.J. Diaz, F. Ahsan, *J. Pharm. Sci.* 87 (1998) 891–900.
15. M.D. Dhanaraju, K. Senthil Kumaran, T. Baskaran, M. Sree Rama Moorthy, *Drug Develop. Ind. Pharm.* 24 (1998) 583–587.
16. M. Skiba, C. Morvan, D. Duchêne, F. Puisieux, D. Wouessidjewe, *Int. J. Pharm.* 126 (1995) 275–279.
17. M. Skiba, D. Duchêne, F. Puisieux, D. Wouessidjewe, *Int. J. Pharm.* 129 (1996) 113–121.
18. M. Skiba, F. Nemati, F. Puisieux, D. Duchêne, D. Wouessidjewe, *Int. J. Pharm.* 145 (1996) 241–245.
19. E. Lemos-Senna, D. Wouessidjewe, S. Lesieur, F. Puisieux, G. Couarraze, D. Duchêne, *Pharm. Dev. Technol.* 3 (1998) 85–94.
20. A. Noomen, S. Hbaieb, H. Parrot-Lopez, R. Kalfat, H. Fessi, N. Amdouni, Y. Chevalier; *Mater. Sci. Eng. C* 28 (2008) 705–715.
21. W.L. Epstein, V.P. Shah, S. Riegelman, *Arch. Dermatol.* 106 (1972) 344–348.
22. W.L. Epstein, V.P. Shah, H.E. Jones, S. Riegelman, *Arch. Dermatol.* 111 (1975) 268–272.
23. R. Aly, C.I. Bayles, R.A. Oakes, D.J. Bibel, H.I Maibach, *Clin. Experim. Dermatol.* 19 (1994) 43–46.
24. A. Kowalewska, W.A. Stańczyk, S. Boileau, L. Lestel, J.D. Smith, *Polymer* 40 (1999) 813–818.
25. J. Pfeiffer, V. Schurig, *J. Chromatogr A* 840 (1999) 145–150.
26. A. Ruderisch, J. Pfeiffer, V. Schurig, *Tetrahedron Asymmetry* 12 (2001) 2025–2030.
27. V. Schurig, S. Mayer, *J. Biochem. Biophys. Methods* 48 (2001) 117–141.
28. H. Cousin, O. Trapp, V. Peulon-Agasse, X. Pannecoucke, L. Banspach, G. Trapp, Z. Jiang, J.C. Combret, V. Schurig, *Eur. J. Org. Chem.* (2003) 3273–3287.
29. E. Beyou, P. Babin, B. Bennetau, J. Dunogues, D. Teyssie, S. Boileau, *J. Polym. Sci. A* 32 (1994) 1673–1681.
30. H. Touzi, Y. Chevalier, R. Kalfat, H. Ben Ouada, H. Zarrouk, J.-P. Chapel, N. Jaffrezic-Renault, *Mater. Sci. Eng. C* 26 (2006) 462–471.
31. G. Yi, J.S. Bradshaw, B.E. Rossiter, A. Malik, W. Li, M.L. Lee, *J. Org. Chem.* 58 (1993) 4844–4850.
32. G. Yi, J.S. Bradshaw, B.E. Rossiter, A. Malik, W. Li, H. Yun, M.L. Lee, *J. Chromatogr. A* 673 (1994) 219–230.
33. V.V. Antić, M.P. Antić, M.N. Govedarica, P.R. Dvornić, *J. Polym. Sci. A: Polym. Chem.* 45 (2007) 2246–2258.
34. P.J. Flory, *Statistical mechanics of chain molecules*, Wiley, New York (1969) p.175.
35. I. Montasser, S. Briançon, J. Lieto, H. Fessi, *J. Pharm. Belg.* 55 (2000) 155–167.
36. I. Montasser, H. Fessi, S. Briançon, J. Lieto, Brevets FR 2806005 (2000); US 2003059473 (2001); WO 0168235 (2001).
37. F. Ganachaud, J.L. Katz, *Chem. Phys. Chem.* 6 (2005) 209–216.
38. K. Shinoda, H. Kuneida, in *Encyclodepia of emulsion technology*, P. Becher, Ed., Marcel Dekker, New York (1983) Chap. 5, pp. 337–367.
39. P. Somasundaran, S.C. Mehta, P. Purohit, *Adv. Colloid Interface Sci.* 128–130 (2006) 103–109.
40. S.A. Vitale, J.L. Katz, *Langmuir* 19 (2003) 4105–4110.
41. V.M. Rao, M. Lin, C.K. Larive, M.Z. Southard, *J. Pharm. Sci.* 86 (1997) 1132–1137.
42. L. Szente, *Preparation of cyclodextrin complexes*. In *Comprehensive Supramolecular Chemistry*, Elsevier, Oxford (1996) 3, 243–252.

Glycosilicones

Juliette Fitremann, Waêl Moukarzel, and Monique Mauzac

Abstract This article reviews the methods described to date for the preparation of polysiloxanes with well-defined structures containing sugar groups either as side-chain groups, end-groups or included in the main chain.

Keywords Carbohydrate · sugar · glyco · polysiloxane · glycopolysiloxane · hydrosilylation · polymer · glycopolymer

1 Introduction

Synthetic polymers grafted with sugar (or carbohydrate) groups are receiving growing interest. They are often named "glycopolymers", by analogy with the term "glycoconjugates", used for the description of the biological molecules in which oligosaccharides are covalently linked to other non-carbohydrate biological molecules, such as proteins (glycoproteins), lipids (glycolipids) or others [1].

The introduction of sugar residues in synthetic polymers can be made in order to mimic these glycoconjugates, prospecting for interesting biological properties for these glycopolymers. Especially, the grafting of well-defined complex oligosaccharides is the object of many works. For polymers more devoted to industrial applications, the introduction of less complex sugar groups in polymers affords new properties, especially amphiphilicity, hydrophilicity, solubility in water, super-absorbent properties, biocompatibility, biodegradability and better environmental and toxicological profiles. They can be seen as an alternative to polyethyleneoxide (POE or PEG) grafting for

J. Fitremann
Laboratoire des Interactions Moléculaires et Réactivité Chimique et Photochimique,
Université de Toulouse, UMR 5623 CNRS – Université Paul Sabatier,
Bâtiment 2R1, 118 Route de Narbonne, F-31062 Toulouse Cedex 9, France.
e-mail: fitremann@chimie.ups-tlse.fr

F. Ganachaud et al. (eds.), *Silicon Based Polymers*,

enhancing hydrophilicity. Both are hydrophilic, non ionic, and with good biocompatibility, and clearly POE grafting, coming from petrochemicals, is much more developed to date. One other reason for their preferential use lies also in easier synthesis and analysis when using POE instead of sugars. This is due partly to the high functionality of sugars compared with POE. The multiplicity of hydroxyl groups in sugars not only makes the analysis more complex, but also makes the selectivity of the reactions with sugars more difficult to control. This point is crucial in the case of polymers, since cross-linked materials instead of well-defined monodimensional polymers would result from the simultaneous reactions of sugar hydroxyles. However, if one wants to use sugars in applications and high scale productions, syntheses must use directly unprotected sugars, instead of long and time-consuming protective-deprotective strategies. Then it remains only very few methods to link unprotected sugars selectively to polymers in order to avoid cross-linking, by taking advantage of some differential reactivities in sugars and some of their derivatives.

More especially in the case of polysiloxanes, the introduction of sugar residues has been attempted according to various synthetic pathways that will be described more in detail in this review. From a general point of view, the main (and tremendous) problem with polysiloxanes, compared with the other synthetic polymers on which sugars have been grafted (methacrylic, polyethylene, polystyrene…), is the high fragility of the polysiloxane backbone to acidic or basic conditions, especially in polar solvents. This drawback lets very little room for making the grafting without chain scission. The second problem is to overcome the incompatible solubilities of sugars (hydrophilic, soluble in water and few polar -aprotic- solvents) and polysiloxanes (hydrophobic) and to get them to react together.

In this article, we reviewed the different ways described to date for the synthesis of well-defined "glycosilicones" or "glycopolysiloxanes". They have been called also "carbohydrate (-grafted, -functional, -modified) polysiloxanes", "sugar (-grafted…etc) polysiloxanes", or "sweet polysiloxanes". We focussed on the advantages and drawbacks of each method.

From the point of view of properties and applications, there are still quite little data about them. A part is devoted to surface-active properties, when surfactant-like structures are concerned, while very few data are available concerning glycopolysiloxanes as "true" monodimensional polymers. The very high flexibility of the polysiloxane backbone should provide them with unique mechanical properties, compared with other glycopolymers, such as acrylates glycopolymers. Probably more versatile and straightforward syntheses will help the development of these glycosilicones, especially as polymers for biological and medical applications.

In this review, we addressed the synthesis of well-defined monodimensional sugar-grafted polysiloxanes. For this reason, we excluded from this paper the synthesis of silicones (poly)-glycosides as surfactants, obtained by glycosylation of hydroxyl-terminated polysiloxanes and the preparation of polysaccharide-silicone adducts. Both are essentially described in patents, from which straighforward data concerning the structure of the products and the outcome of the reactions are difficult to extract. For silicone glycosides, some descriptions can be found, only in patents, for example in US 5,428,142 [2]. They are usually prepared by a Fischer glycosylation in acidic medium, leading to both carbohydrate oligomerization and siloxane equilibration providing not well-defined materials.

2 Glycosilicones by Using Aminopolysiloxanes and Sugar Lactones

The simplest method for grafting directly and selectively unprotected sugar units on polysiloxanes consists in performing the reaction between an aminopropyl-polysiloxane or an other aminated polysiloxane with a sugar lactone, such as gluconolactone (Fig. 1). This strategy bypasses the selectivity issue by using a sugar with a "differential" electrophilic reactive group, the lactone function. On the other hand the amino groups grafted on the polysiloxane display a strong nucleophilicity much higher compared with hydroxyl functions and leading to a non-hydrolysable amide bond.

This reaction is quite straightforward despite the fact that gluconolactone is not soluble in polysiloxanes solvents (e.g. toluene, THF), and conversely, polysiloxanes are not soluble in gluconolactone solvents such as DMF, DMSO, water, alcohols. The reaction was usually carried out at temperature between 40–90°C, for several hours, but has also been described to occur at room temperature. These conditions do not cleave the polysiloxane backbone, the amines being quite quickly consumed by the reaction with sugar lactone. The solubility issue can be solved either by carrying out the reaction without solvent, with a dispersion of gluconolactone in the aminopoly-siloxane, or with a slight amount of solvent for solubilizing the gluconolactone, such as DMF or isopropanol, avoiding a high increase of the reaction medium viscosity. The formation of glycopolysiloxanes ensures progressive emulsification of the mixture, where the gluconolactone -solution- is getting progressively miscible and then, is fully consumed [3].

Examples of this reaction are found in patents [4]. It has also been used by Wagner et al. [8], Stadler et al. [29]. According to Stadler et al., with unprotected monosaccharides lactones, the reaction proceeded easily. However, when larger saccharides were used, such as an oligosaccharide terminated by an aldonolactone, the miscibility problem could not be solved. In this case, the reaction could not be

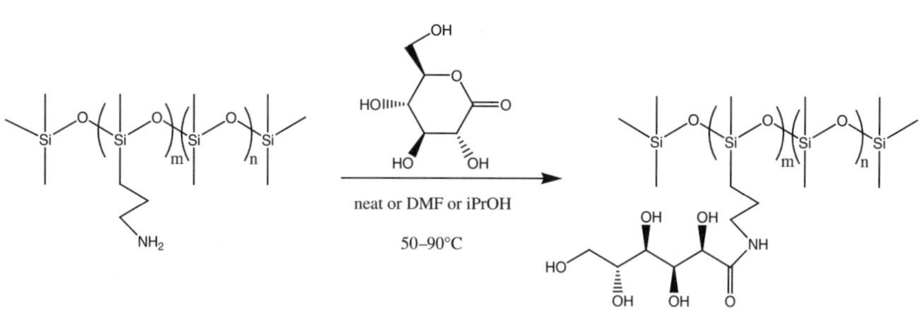

Fig. 1 Grafting of a sugar lactone on an aminopropyl-polysiloxane

made directly on the unprotected sugar lactone, and protective groups have been used (see Section 5.1.3). It has been applied also to sugar-functionalization of sol-gel siloxanes networks [5].

The main drawback of this method is that aminopolysiloxanes are not easily available in all desired grafting ratios and molecular weights. Only a little range of MW (up to 8000) and grafting ratios (1–7%) are available commercially. This is due to limitations when polycondensation reactions are carried out with aminopropylsiloxane units. Methods for obtaining aminopolysiloxanes by hydrosilylation, in their NH_2 form suitable for nucleophilic reactions, are not available either.

3 Glycosilicones by Using Epoxy-Polysiloxanes

3.1 To overcome the problem of the availability of polysiloxanes grafted by nucleophilic amino groups, Wagner et al. developed a method in which epoxy-polysiloxanes are used as intermediates for obtaining amino-grafted polysiloxanes, by reaction with diamines. The resulting amino-polysiloxanes were reacted with sugar lactones to give carbohydrate-silicon surfactants [6–8] (Fig. 2). By this method, aminopolysiloxanes with almost any desired composition and with well-defined structures can be obtained. Actually, epoxy-grafted polysiloxanes are very useful electrophilic polysiloxanes that can react with a variety of nucleophiles.

In a first step the epoxy-functionalized siloxanes were prepared by hydrosilylation, with allyl glycidyl ether and Lamoreaux catalyst (Pt catalyst resulting from $H_2PtCl_6.H_2O$ in octanol), at 130–145°C for 4–10h. The reaction is quantitative,

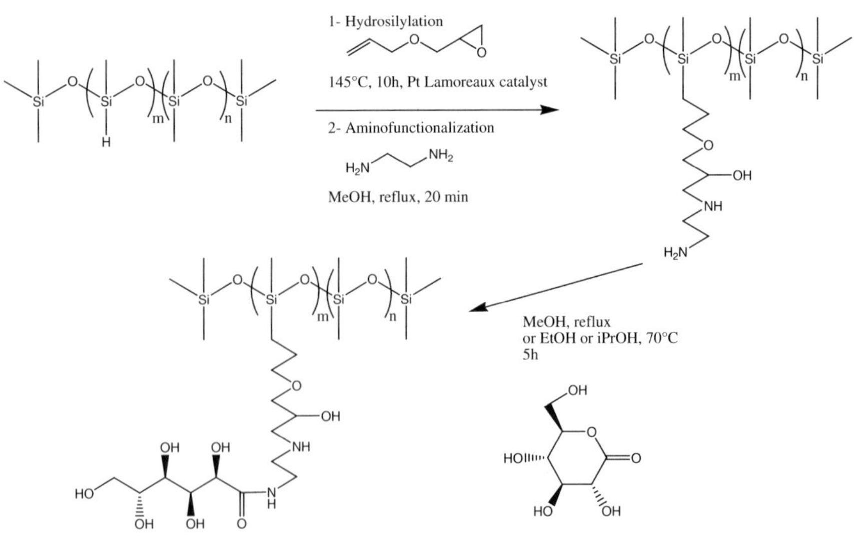

Fig. 2 Grafting of a sugar lactone on an amino-polysiloxane obtained from an epoxy-polysiloxane

occurring easily without side-reactions and preserving the epoxide groups. In a second step, an excess of diamine (1–5 eq.) allowed the nucleophilic epoxide ring opening. The reaction was carried out in refluxing methanol for 20–90 min. Thanks to this excess, the formation of cross-linked by-products that would result from the reaction of the diamine with two epoxide functions can be avoided. Thus, one amino-function ensures the grafting to the epoxide, while the other one remains free. When ethylenediamine and triethylenediamine were used, a quantitative conversion was observed, giving free primary amine silicone derivatives. When 1,2-propylenediamine was used, 8% of "bis"-products was observed, leading to a partially cross-linked elastomer.

In a final step, different mono- and disaccharide lactones (e.g. δ-gluconolactone, glucopyranosyl arabinonic acid lactone etc.) reacted with the synthesized aminopolysiloxanes. The reaction lasted 5–6 h, either in refluxing methanol or in ethanol or isopropanol at 70 °C. The reaction gave the expected sugar silicone derivatives quantitatively as yellow or white powders.

A variation of the method consisted in the preparation of sugar-grafted amines as described in Fig. 3, that will react further with the epoxy-polysiloxanes. However the grafting of the remaining secondary amine on the epoxy-polysiloxane resulted partly in cross-linking, because of competing reactions of the sugar hydroxyl groups. The same method proceeded successfully when the secondary amine was a more nucleophilic cyclic amine derived from piperazine, used in slight molar excess to the epoxides (1:1.05) to avoid cross-linking.

While this method has been applied mainly to short siloxane units, for making surfactants, it has been applied also with success to comb-like polysiloxanes up to DP ≈ 100.

3.2 An other strategy, described in some patents, takes also advantage of epoxy-functional polysiloxanes as electrophilic groups in order to graft sugars directly, without protective groups [9]. When amino-sugars or their derivatives are used, such as N-methylglucamine, it can be expected that the higher nucleophilicity of

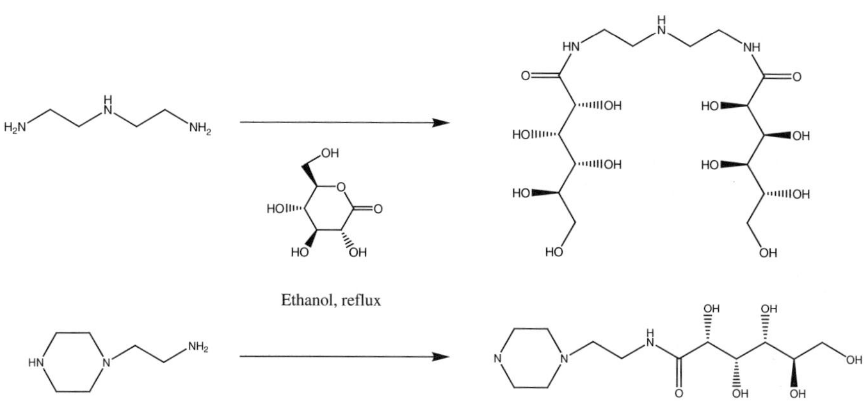

Fig. 3 Preparation of sugar-grafted secondary amines for addition on epoxy-polysiloxanes

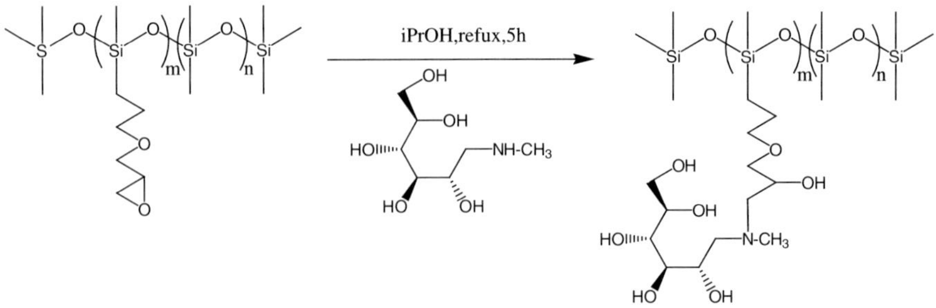

Fig. 4 Direct addition of N-methylglucamine on an epoxy-polysiloxane

the amine compared to hydroxyl groups would lead to selective grafting through an amine bond (Fig. 4). However, as it was observed by Wagner et al., hydroxyl groups of sugar can also react with epoxide in the same conditions to give ether bonds, which should induce cross-linking. This issue has not been pointed out in this example.

Actually, other patents described the use of epoxy-polysiloxanes as electrophilic groups for grafting any kind of sugars, through the formation of ether linkages [10]. If this reaction is used in bulk, a lack of regioselectivity of the grafting and cross-linking should be expected because of the similar reactivity of the multiple OH groups of sugars towards the epoxy-polysiloxanes.

4 Attempts for Grafting Unprotected Sugars on Polysiloxanes by Hydrosilylation

Different authors tried to perform directly the hydrosilylation of unprotected sugars. It would avoid time-consuming protection-deprotection steps. It would avoid also the risk of equilibration of the silicone backbone occuring under the acid or basic deprotection conditions. Usually the sugar hydroxyl groups react with hydrosilanes to form silylethers, namely Si-O-C bonds in the presence of a hydrosilylation catalyst. This dehydrocondensation side reaction has to be controlled in order to avoid cross-linking, by using selected solvents and catalysts.

4.1 In a model reaction, Stadler et al. have successfully hydrosilylated 1-allyloxy-2, 3-propanediol using THF or toluene as solvents and dichloro-dicyclopentadienyl-platinum(II) as catalyst, at 70°C (Fig. 5). The reaction was quantitative and no side-reactions, either dehydrocondensation or other reactions, were observed [12].

An attempt to perform the same reaction with 6-O-allylgalactopyranose without protection of the hydroxyl groups failed due to the large difference between the miscibility of sugars and polysiloxane. They could not find a suitable solvent chemically inert able to solubilise both reagents. While THF/methanol mixture dissolves both of the components, in this case, even traces of methanol in the reaction media react with the polymer hydrosilane functions to form ether bonds.

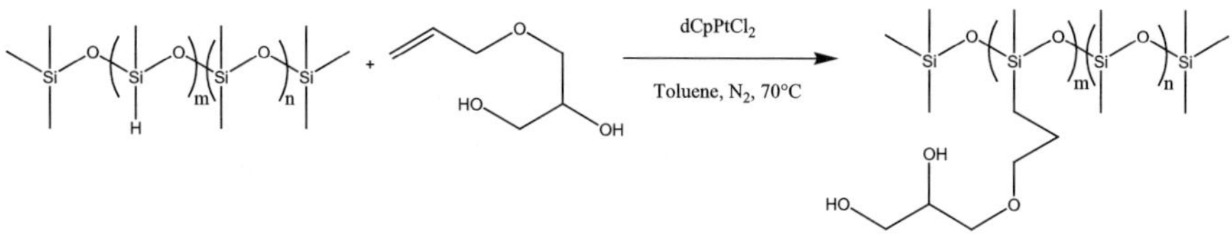

Fig. 5 Hydrosilylation without protective groups: example of 1-allyloxy-2-3-propanediol

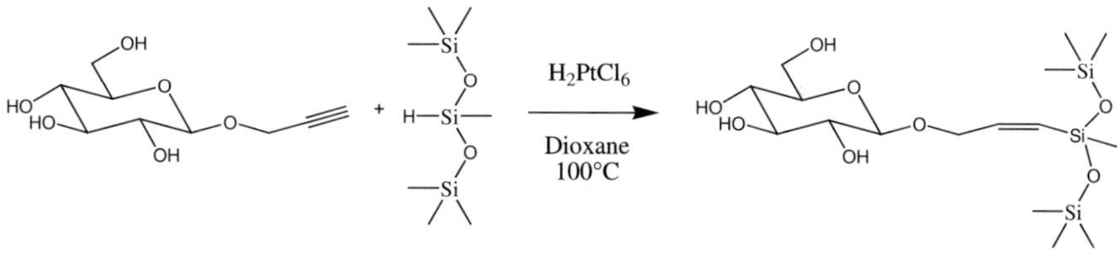

Fig. 6 Hydrosilylation without protective groups: example of propargyglucoside

4.2 To obtain non-ionic siloxane surfactants, Wersig et al. [11] performed the hydrosilylation of a non-protected oligoethoxylated but-2-yne dialcohol. Dioxane was applied as solvent, hexachloroplatinic acid as the reaction's catalyst at 100°C. The difference between the activation energies of dehydrocondensation and the triple bond hydrosilylation reactions can be increased by the use of an appropriate solvent, such as dioxane. The selectivity toward hydrosilylation has also been explained by the "protective" association of hydroxyl groups in coiled POE chains. Thus, in this case the hydrosilylation reaction has been carried out successfully without any protection.

4.3 Wagner et al. [6] used the same method for the hydrosilylation of an unprotected propargyl glucoside (Fig. 6). Dioxane was used as solvent and hexachloroplatinic acid in isopropanol as catalyst at 100°C. However, the weak solubility of the unprotected propargyl glycosides (from mono-, di- or oligosaccharides) and their steric hindrance required long reaction times (20–38 h). Additionally, the multifunctional carbohydrate had an inhibiting effect on the reaction so that the catalyst has to be used in a 10–100-fold excess compared with the usual amounts.

4.4 At last, Hamaide et al. described also shortly their attempts for performing directly the hydrosilylation without protective groups. In addition to incompatible solubilities, they observed up to 50% O-silylation, instead of C-silylation with Karstedt's catalyst (Platinium divinyltetramethyldisiloxane complex) [21].

5 Grafting of Allyl-Functionalized Carbohydrates with Protective groups by Hydrosilylation

As seen before, direct hydrosilylation cannot be performed with satisfactory results with unprotected double-bond-functionalized sugars. Thus, the introduction of protective groups on sugars is necessary if hydrosilylation is selected as the coupling method. It affords two additional advantages. The first one is to provide only one grafting point on the sugar residue, ensuring high regioselectivity and no risk of cross-linking (unless partial deprotection during hydrosilylation would occur). A second one is to provide a sugar derivative soluble in organic, apolar solvents also suitable for polysiloxanes. The reactions thus occur in homogeneous conditions. After hydrosilylation, these groups have to be removed without chain scission of the polysiloxane backbone. We will see that this point is far to be obvious, since virtually any condition necessary for the cleavage of C-O-C(X) or C-O-Si(X) bonds will also cleave the Si-O-Si bonds of the polysiloxane. In the case of sugars, more especially, it can be pointed out too that the deprotection of all the OH groups can require quite long reaction times to go to completion, which can explain probably some different successes with a same reaction, depending more precisely on the structure of the grafted polysiloxane or the protected alcohol. Comparing the two main protective groups used, acetyl and trimethylsilyl, from a general point on view, it is quite clear that trimethylsilyl would be preferable, since C-O-Si bonds can be cleaved in milder conditions compared with C-OCOR bonds. It should be possible to take advantage of the slight differential reactivity between C-O-Si and Si-O-Si for controlling the chemoselectivity of the cleavage.

Because of these critical points, we tried to extract as precisely as possible the successes and failures obtained by this strategy, by the different authors. Despite many complete and careful studies, no clear-cut successes were obtained. We found also that measurements allowing to check the status of the polysiloxane chain after the reactions, especially size exclusion chromatography (SEC), remained quite scarcely described. While the hydrosilylation step gave generally good results, the deprotection step remained critical.

5.1 Stadler et al. has synthesized several sugar-grafted polysiloxanes obtained by multistep procedures, including the preparation of the protected allyl-functionalized-sugars, the hydrosilylation with a statistical poly(dimethyl-co-hydrogenomethyl)-siloxane and finally the removal of the carbohydrate protective groups.

5.1.1 In a first strategy [12], monosaccharides (glucose and galactose) were protected by isopropylidene groups and subsequently treated with sodium hydride/allyl bromide to give the corresponding protected mono-O-allyl-sugars. Subsequently hydrosilylation proceeded in THF or toluene at 70°C using dichloro-dicyclopentadienyl-platinum (II) as a neutral catalyst. A mild acidic mixture of THF and aqueous formic acid was used for the deprotection of the isopropylidenes. They found that, because of the extremely hydrophobic silicone backbone the acidic solution could not cleave easily the protective groups. Thus very long reaction times were necessary and the silicone backbone was exposed to the risk of equilibration or cleavage, giving no reproducible results. To overcome this problem, trimethylsilyl protective groups were substituted to isopropylidene groups and were cleaved using a THF/methanol mixture for 48 h at room temperature, leading to the desired sugar-grafted polysiloxanes (Fig. 7). The glass transition temperatures (Tg) of the latter were determined by DSC, showing a large upward shift of Tg for polymers with a high content of unprotected sugars. The polymers were described as being "highly viscous and rubbery" for high substitution degree.

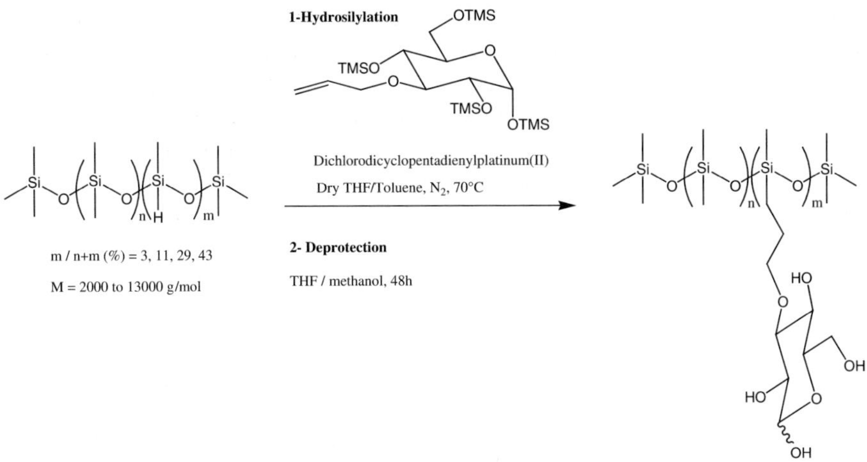

Fig. 7 Hydrosilylation with trimethylsilyl protected allyl-grafted sugar followed by deprotection

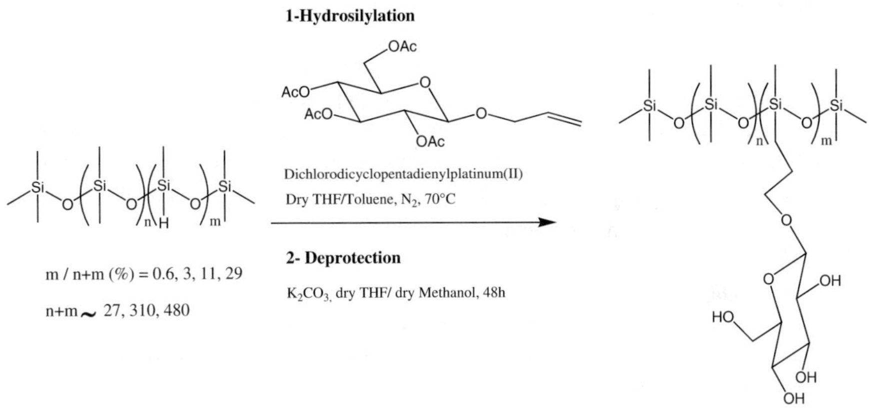

Fig. 8 Hydrosilylation with acetyl protected allyl-grafted sugar followed by deprotection

5.1.2 In another study of the same group, mono, di- and oligosaccharides (heptasaccharides) were protected by O-acetylation and allyl alcohol was introduced in anomeric position [13]. These sugar derivatives were grafted on statistical poly-(hydromethyl-co-dimethyl)-siloxanes (with 0.6–29% of Si-H) by hydrosilylation with dicyclopentadienylplatinum-(II) dichloride (70°C, 15–120 min). The deprotection was achieved by reacting the grafted polymer with traces of K_2CO_3 in a 1/1 vol. mixture of dry THF/dry methanol, for 48 h at room temperature (Fig. 8). After completion, the reaction was quenched with an acidic ion exchange resin. The reaction was monitored by IR spectroscopy and NMR.

Further physicochemical characterizations were made on both the protected and deprotected carbohydrate-grafted polysiloxanes in a later publication [14]. Size Exclusion Chromatography (SEC), viscosimetric and static light scattering measurements on the acetyl-protected sugar-grafted polysiloxanes showed that hydrosilylation did not affect the polymer integrity and did not lead to cross-linking. After deprotection, static light scattering experiments were performed directly on solutions in toluene of the mono- or disaccharide-grafted polysiloxanes with a grafting rate of 0,6 or 3%, in order to highlight the self-aggregating behavior of these amphiphilic polysiloxanes. With glucose and cellobiose, polymers with 3% grafting remained soluble in toluene. Apparent molecular weights calculated from static light scattering showed values higher than the expected molecular weights. It was explained by the formation of clusters comprising from 1 to 14 polymer chains. The formation of these clusters appeared to be very dependent on the grafted saccharide (mono- di- or oligo-saccharides and stereochemistry of the 4-OH) and the grafting ratio. Dynamic light scattering and viscosity measurements were also performed on some of these deprotected polymers, pointing out again the formation of aggregates.

Attempts from Wagner et al. to implement the same strategy [6], including hydrosilylation of an acetyl-protected-propargyl-carbohydrate followed by deprotection failed. They did not success in the deprotection of the OH groups without cleavage of the siloxane bonds.

5.1.3 Another study from Stadler et al. reported the statistical side-chain grafting of polydimethylsiloxanes with aldonamides derived from mono-, di- or oligo-saccharides (heptaoses) [15]. A first method consisted in reacting amino-polysiloxanes with peracetylated lactones. A second method used peracetylated N-allyl-aldonamides in a hydrosilylation reaction with poly(dimethyl-co-hydromethyl)siloxane. A third method used peracetylated-1-O-allyl-glucosamine, grafted again by hydrosilylation. All syntheses are followed by the deprotection step using either methanol/KCN or methanol/K_2CO_3 (Fig. 9).

Concerning the first method, the availability of primary amine functionalized polysiloxanes (e.g. poly(aminopropylmethyl-co-dimethyl)siloxane) is limited. Whereas the second method allows the synthesis of polysiloxanes with varying amounts of grafts and a wide range of molecular weights via synthesized statistical poly(dimethyl-co-hydromethyl)siloxane.

Because of the amide bonds included in the structure of the allylated sugars derivatives, classical hydrosilylation catalysts failed, including dicyclo-pentadienyl-platinum(II) dichloride. Thus, in this study, a very interesting comparison of catalysts for the hydrosilylation of amide-containing derivatives was carried out. More especially, bis(dialkylsulfido)platinum(II) salts e.g. $(Et_2S)_2PtCl_2$, $(Bn_2S)_2PtCl_2$ have been shown to be efficient catalysts well compatible with the amide bond and the acetyl protective groups and led to quantitative conversions of Si-H bonds into Si-C bonds. After deprotection, preliminary solubility studies of the resulting sugar-grafted polysiloxanes were performed, but without more physicochemical studies.

5.2 The hydrosilylation of acetyl-protected allylglucoside with a Si-H-ended telechelic polysiloxane, followed by deprotection has also been described by Nagase et al. for the preparation of “glyco-oligodimethylsiloxanes” as transdermal penetration enhancers (Fig. 10) [16–18]. The status of the siloxane chain after deprotection has been checked by size exclusion chromatography (SEC). When they used MeOH/K_2CO_3 for deprotection, they observed by SEC the formation of two peaks

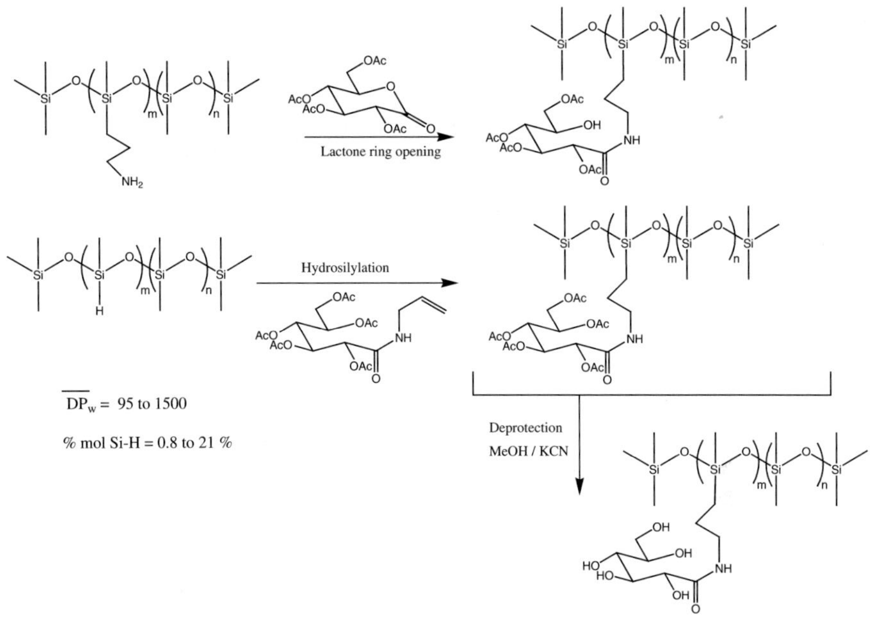

Fig. 9 Grafting of acetyl protected sugars by lactone ring opening or hydrosilylation, followed by deprotection

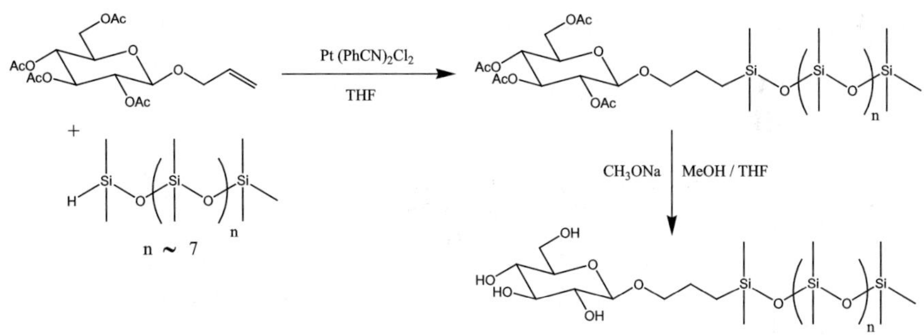

Fig. 10 Grafting of an acetyl protected sugar by hydrosilylation on a Si-H-ended oligosiloxane

at lower molecular weights. However, with 1N sodium methylate in MeOH/THF, as deprotecting agent, at 0°C for 1.5 h, these peaks did not appeared, thus providing the sugar-ended polysiloxane without chain scission. It could be pointed out that, in this case, only one sugar group per chain has to be deprotected. Perhaps it allows a full deprotection of the hydroxyle groups in a short enough reaction time, before the chain scission occurred.

5.3 The protection-deprotection strategy was also applied by Gruber et al. to the synthesis of sugar-grafted dialkoxysilanes, trialkoxysilanes or cyclotetrasiloxanes as building blocks that can be further involved in polycondensation or equilibration reactions to make polysiloxanes or cross-linked materials [19]. In a first part

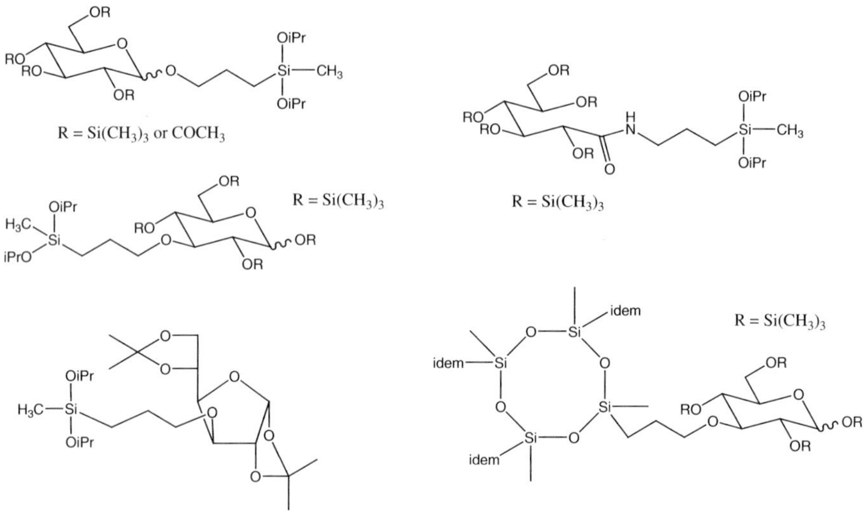

Fig. 11 Preparation of alkoxysilanes functionalized with protected sugars for using in polycondensation reactions

of their work, the preparation of these building blocks by hydrosilylation of various protected allyl-functional sugars has been detailed (Fig. 11), pointing out that it was necessary to adjust the reaction conditions, especially the catalyst. The allyl-functional protected sugars were prepared either from glycosylation (with allylic alcohol), sugar lactone opening (with allylamine) or by nucleophilic substitution. They have been protected with acetyl, trimethylsilyl or dimethylacetal groups. Preliminary equilibration or polycondensation reactions with TMS-protected derivatives were carried out, followed by deprotection in methanol/water, but the resulting polymers or cross-linked materials were not characterized at this stage. However, these building blocks are quite bulky, what could hinder the polycondensation.

5.4 The same kind of protected allyl-functional sugars were used by Thiem et al. for the preparation of sugar-ended polysiloxanes, by hydrosilylation with Si-H-ended telechelic polysiloxanes of $M_n = 590\,g/mol$ (oligosiloxanes) [20]. Both acetyl, isopropylidene acetal and trimethylsilyl protective groups are involved (Fig. 12). For hydrosilylation of 1-allyloxy-(2,3,4,5)-tetraacetyl-glucopyranose and (1,2-5,6) di-isopropylidene-3-allyl-glucofuranose, contradictory results were obtained

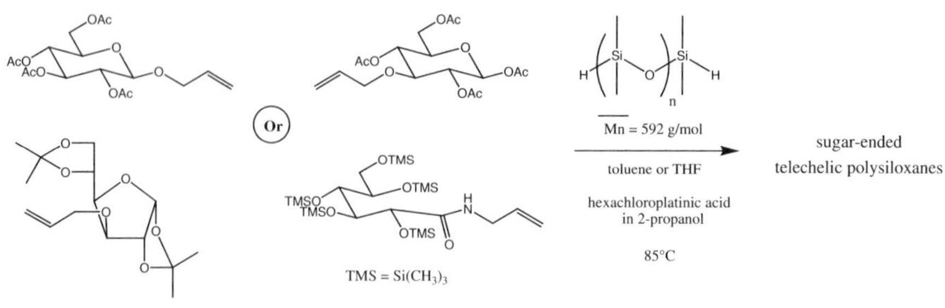

Fig. 12 Preparation of sugar-ended telechelic oligosiloxanes by hydrosilylation with protected sugars

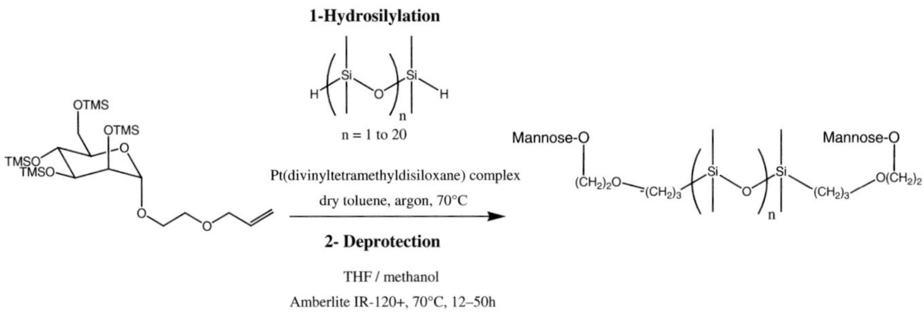

Fig. 13 Preparation of sugar-ended telechelic oligosiloxanes by hydrosilylation with trimethylsilyl protected sugars

by Gruber et al. (previous Section 5.3), and by Thiem et al. (this section). The latter reported no impairment of hydrosilylation of these compounds with hexachloroplatinic acid (Speiers' catalyst), while Gruber et al. get low yields with this catalyst, preferring Karstedt's catalyst (Platinium divinyl-tetramethyl-disiloxane complex).

Acetyl protective groups were removed by sodium methylate in methanol for several days. Isopropylidene acetals were cleaved by acetylbromide in methanol and trimethylsilyl protective groups were cleaved by simple heating in methanol for 45 min. The resulting deprotected sugar-ended polysiloxanes were analysed by NMR and MALDI-TOF. Smaller fragments were observed in the two first method compared with the third one.

5.5 Hamaide et al. described the preparation of sugar-grafted polysiloxanes (telechelic and comb-like), by hydrosilylation of trimethylsilyl-protected allyloxyethylglycosides (from mannose, glucose, galactose), followed by deprotection (Fig. 13) [21, 22]. First, when simple reflux in MeOH/THF for 48 h was performed, nearly no deprotection was observed. Then the following conditions were selected: THF/MeOH, with Amberlite IR-120+, 70°C for 12–50 h depending on the structure of the polysiloxane. The integrity of the siloxane chain for telechelic polymers was kept by this method. It was checked by comparing NMR spectra before and after deprotection, since these polysiloxanes were short enough to get significant integration ratio for saccharide and polysiloxane signals. Attempts to characterize the deprotected polysiloxanes by MALDI-TOF failed.

5.6 Fleury has described a similar method, in which N-acetyl-N-allylglycosylamines protected either with acetyl or with trimethylsilyl groups are grafted by hydrosilylation [23]. Deprotection of trimethylsilyl groups was made by the same method as in Section 5.5.

6 Grafting of Propargyl-Functionalized Sugars by "Click-Chemistry"

A completely different method for grafting unprotected sugars on oligosiloxanes has been described recently by Fleury et al. in order to prepare surface-active silicone-carbohydrate derivatives [24, 25]. The bonding was achieved by "click chemistry"

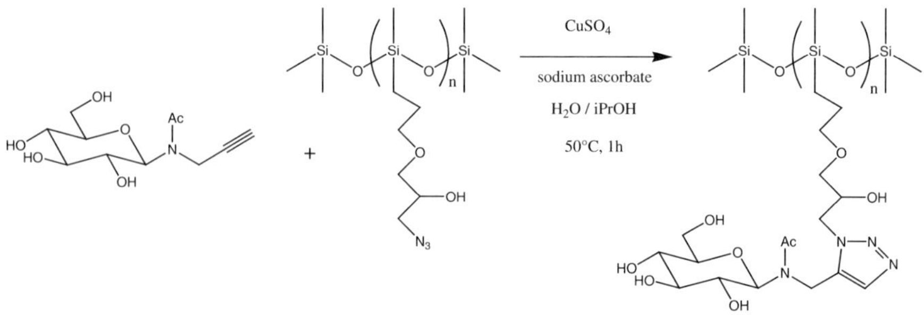

Fig. 14 Grafting of unprotected sugars on azido-polysiloxanes by "click chemistry"

(Huisgen reaction), corresponding to a 1,3-dipolar cycloaddition involving an azido group N_3 and a C-C (or C-N) triple bond. The reaction leads to the formation of a 1,2,3-triazole (or tetrazole) heterocycle (Fig. 14). Azido-functional oligosiloxanes (trisiloxane and telechelic dodecasiloxane) were prepared by the opening of an epoxy-polysiloxane by sodium azide in a mixture of iPrOH/H_2O/AcOH (pH = 6), for 4 h at 50°C. On the other hand, N-acetyl-N-propargyl-glycosylamines are prepared directly without protection-deprotection steps, by reaction with propargylamine followed by N-acetylation. The cycloaddition is achieved in water/iPrOH with sodium ascorbate and copper sulfate, at 50°C for 1 h. The azido-functional trisiloxanes were characterized by mass spectroscopy, IR, and ^{1}H and ^{13}C NMR. The glyco-trisiloxanes were first purified on silica gel and characterized by IR, ^{1}H and ^{13}C NMR. With this method, the sugar residue was thus directly introduced on the trisiloxane without protective groups.

7 Grafting of Carbohydrates by Acetalation

7.1 Grafting of sugars on polysiloxanes has also been achieved by using acetal as the linking function, by Ogawa [26]. A diethylacetal polysiloxane (diethoxypropylsiloxane) has been first prepared by hydrosilylation, by the reaction with acrolein diethylacetal. Then, acetalation of unprotected sugars was carried out in a DMF-dioxane solution, with activated clay or TsOH as catalyst, at 130°C for 2,5 h while ethanol, part of dioxane (and residual water) were distilled off (Fig. 15). After several tests for ensuring good solubilization of both the sugar and the acetal-functional polysiloxane, the DMF-dioxane mixture was selected. The reaction was tested with short siloxanes (trisiloxane), telechelic polysiloxanes (diethylacetal-terminated) and comb-like diethylacetal polysiloxanes.

When unprotected glucose was used, high molecular weight (with telechelic polysiloxanes) or insoluble cross-linked materials were obtained, due to the acetalation reaction occurring at both 4,6 and other OH positions, especially the 1-OH position. Acetalation is thus not selective enough in 4,6-OH position to afford a clean grafting of one glucose for each acetal side-group. On the contrary, reaction

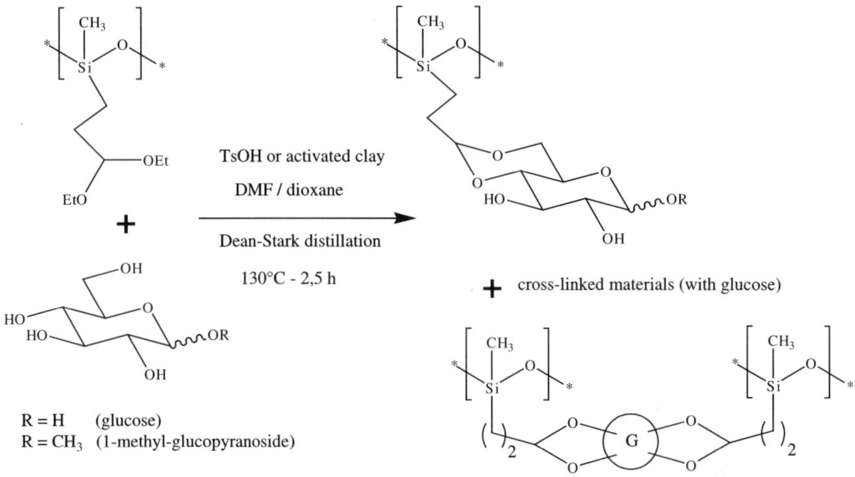

Fig. 15 Grafting of an unprotected sugar on an acetal-functionalized polysiloxane by acetalation

with 1-O-methylglucose afforded the expected monodimensional polysiloxane grafted through the cyclic acetal of glucose in the 4,6-OH position.

Concern about redistribution reactions of the polysiloxane backbone and the formation of small amount of aldehyde-functional polysiloxane as side-product led the authors to explore several acid catalysts with the lowest moisture content possible, with activated clay being finally preferred. Size-exclusion chromatography was performed on some of the resulting polysiloxanes.

7.2 A second approach to bypass the lack of selectivity of the acetalation was described by the same author, by using a derivative of gluconolactone as the polyol to be grafted on the diethylacetal-polysiloxane [27]. This strategy gets rid of the competitive anomeric position of glucose that led to cross-linking in the former work (Section 7.1). No insoluble cross-linked material was observed, while two positions for the acetal formation, implying 6,5-OH's and 6,4-OH's were suggested. Conditions for acetalation were the same as described above. The polysiloxanes used were the same too (short, telechelic, comb-like).

7.3 The grafting of a sugar group to a telechelic (oligo)polysiloxane by an acetal function has also been described by Thiem et al. (see ref. [20] and also Section 5.4) (Fig. 16).

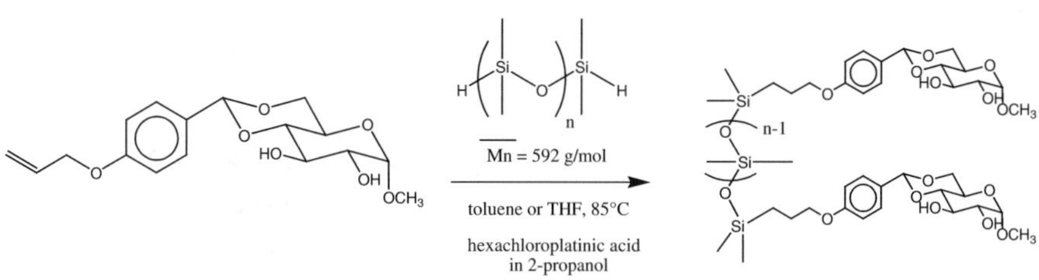

Fig. 16 Grafting by hydrosilylation of a sugar derivatized with double bond through acetalisation

In this case, an acetal-functional olefin was first linked to an unprotected methylglucoside in 4,6 position before performing hydrosilylation. The two free hydroxyles remaining on the glucoside unit did not seem to impair the hydrosilylation with hexachloroplatinic acid. Characterization of the resulting sugar-ended oligosiloxane was made by NMR and MALDI-TOF.

8 Enzymatic Grafting of Carbohydrates on Polysiloxanes

Enzymatic reactions have been used for functionalizing polysiloxanes with sugar units. Two strategies have been described, one by the group of R.Stadler (1995), the other one by the group of R.A.Gross (2005). The advantages of enzymatic reactions are their ability to provide regioselective reactions on sugars without the need of protective groups, and mild conditions that should preserve the polysiloxane backbone integrity.

8.1 The first approach used by the group of Stadler was to lengthen the size of an oligosaccharide already grafted to the polysiloxane. It was achieved by adding glucose units by a glycosylation reaction with glucose-1-phosphate catalysed by a phosphorylase [28, 29]. The starting polysiloxane was a maltoheptaoside-grafted or a maltoheptaonamide-grafted polysiloxane. The first one (Fig. 17) was prepared by hydrosilylation with the allylated and acetyl protected heptasaccharide followed by deprotection in dry MeOH/THF/K_2CO_3 at room temperature or reflux for 1–3 days. The second one was prepared by the opening of the terminal gluconolactone of an acetyl-protected heptasaccharide by an aminopropyl-polysiloxane, followed by the same deprotection step. Once obtained these starting sugar-grafted polysiloxanes, they were submitted to the phosphorylase-catalysed reaction, in the presence of glucose-1-phosphate, in aqueous citrate buffer at 37°C, pH 6.20. Up to 300 glucose units could be added to each heptaose unit of the starting glycopolysiloxane, and was determined by the titration of the released phosphate. The reaction outcome is related to the solubility of the starting glycopolysiloxane in the buffer solution. Thus, a lower grafting rate of 25–27 units has been obtained for the less-soluble, higher molecular weight malto-heptaonamidopropyl polysiloxanes. At the end of the reaction, after denaturation by heating, the coagulated enzyme was removed by filtration.

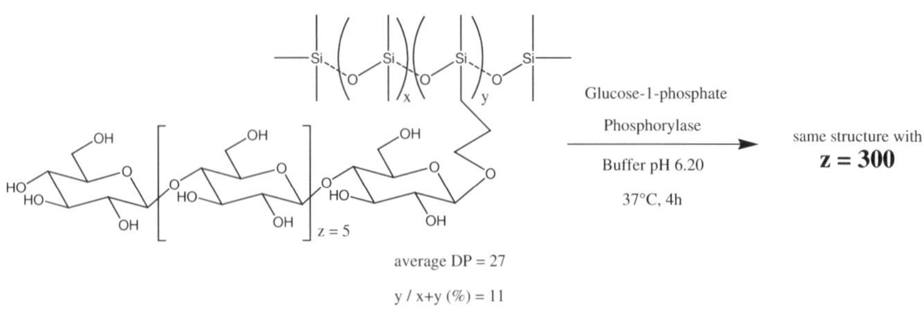

Fig. 17 Enzymatic lenghthening on an oligosaccharide-grafted polysiloxane

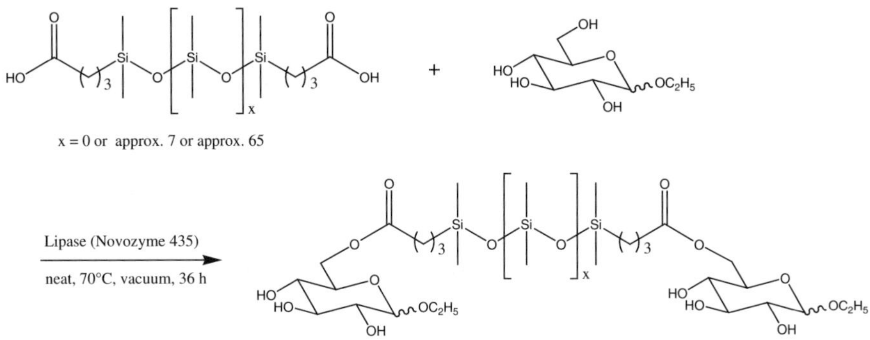

Fig. 18 Enzymatic esterification of glucoside with a telechelic carboxylic acid-terminated polysiloxane

This work showed that enzymatic glycosylation carried out in aqueous solution is compatible with the presence of polysiloxane and can lead to long-oligosaccharide grafts, if a small oligosaccharide precursor is already grafted as the starting point for the enzymatic reaction. However, after the enzymatic reaction, glycopolymers are hardly soluble in solvents, impairing full characterization.

8.2 Recently, the formation of ester bonds by enzymatic catalysis between a telechelic polysiloxane terminated by two carboxylic acid functions and ethyl-glucoside has been described by Gross et al. [30, 31]. The esterification reaction was catalysed by a lipase (Novozyme 435) in the absence of solvent, at 70°C for 36 h, with the continuous withdraw of the water formed during the reaction (Fig. 18). Three polysiloxanes were selected for the study, with $x=0$, $x\approx 7$ and $x\approx 65$. By NMR, they confirmed the selective grafting of ethylglucoside by its 6-OH position, at the terminal carboxylic groups of the telechelic polysiloxanes. Because of the use of commercial grade polysiloxanes, in which a non-negligible part was terminated by a cyano instead of carboxylic group, they obtained also unexpected monoesterified (and cyano) polysiloxanes. The good selectivity towards the 6-OH position is also confirmed by MS-ESI, by the absence of a mass peak corresponding to ethylglucoside units grafted with two siloxane chains. Applied to the longer polysiloxanes ($x\approx 7$ and $x\approx 65$), the mass spectrum displayed the expected distribution of mass, while SEC chromatograms showed the presence of peaks at the expected Mw, slightly increased by the addition of sugars units. Thus, polysiloxanes can react by enzymatic catalysis, while keeping the integrity of the siloxane bonds.

9 Main-Chain Carbohydrate-Polysiloxane Block Copolymers

An original approach for combining polysiloxanes and sugars has been reported by Thiem et al. [32] and Domschke et al. [33], with the preparation of main-chain carbohydrate-polysiloxane copolymers. In these structures, the sugar units are

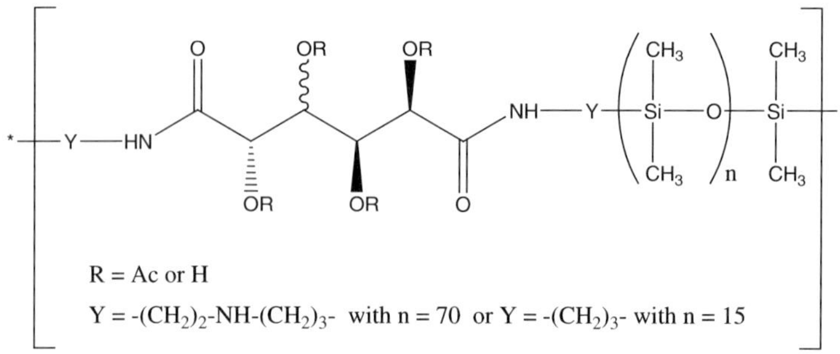

Fig. 19 Main chain carbohydrate – polysiloxane polyamides

included in the main chain of the polymer, through amide linkages, leading to "carbohydrate-segmented silicone polyamides" (Fig. 19).

In this case, the selectivity issue is resolved by using only difunctional units, both for polysiloxanes (telechelic polysiloxanes) and for sugar units. From the sugar chemistry point of view, the originality comes from the particular reactivity of sugar diacids esters (glucaric, galactaric, tartaric acid), able to react with amines in mild conditions, thanks to the cooperative activation by the neighbouring hydroxyles or thanks to the formation of intermediate reactive lactones.

The different segmented copolymers were obtained from the reaction of amino-terminated polysiloxane (DP ≈ 70 or 15) with either: acetyl-protected galactaroyl dichoride, unprotected dimethylgalactarate or unprotected dimethyl glucarate/lactones mixture. The reaction is carried out at 0°C, room temperature or 80°C for several days, in the presence of triethylamine. The solvent was choloroform (with protected aldaric acid derivatives) or isopropanol, methanol or DMSO (with unprotected aldaric esters). The reaction can thus be carried out in a homogeneous or heterogeneous way. Peak molecular weight (Mp) of the resulting polymers were around 20,000 g/mol for polymers prepared with unprotected sugars at room temperature, and much higher (Mw ≈ 100,000 g/mol) for polymers prepared with protected sugars at 0°C (with polysiloxane DP ≈ 70).

10 Miscellaneous Strategies

10.1 Nagase et al. described a second strategy for grafting a sugar-end group on their polysiloxanes, to make transdermal penetration enhancers (see ref. [16] and also Section 5.2). The grafting is achieved by the reaction of an acetyl-protected glucosyl-thio-pseudourea on an iodopropyl-ended polysiloxane. The deprotection with sodium methylate provides the desired product (Fig. 20).

10.2 A variation for grafting of sugar units on aminopolysiloxanes has been described, by using an acetyl-protected glucosyl isothiocyanate as the electrophilic carbohydrate reagent, followed by a deprotection step [39]. The aminopropylpolysiloxane is prepared by redistribution of aminopropyl-D_4 with decamethyltetrasiloxane,

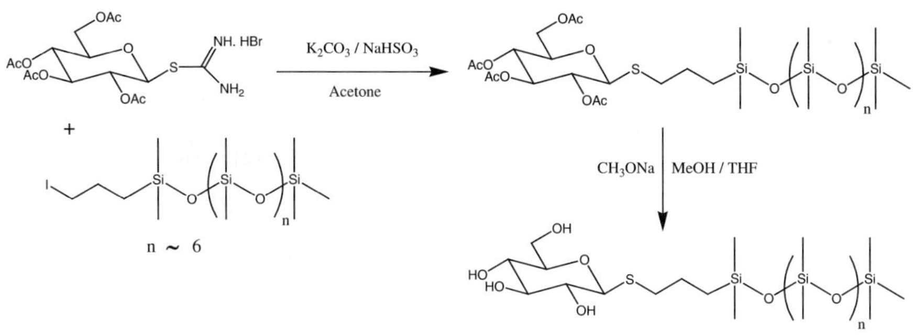

Fig. 20 Grafting of a glucosyl-thio-pseudourea on a iodopropyl-ended oligosiloxane

and then reacted with the glucosyl isothiocyanate in chloroform at room temperature and reflux. Deprotection of acetyl groups was made in methanol with anhydrous ammonia overnight at room temperature. The molecular weight obtained from size-exclusion chromatography are 5530, 5420, 5200 g/mol respectively after each step (redistribution, glucose isothiocyanate grafting, and deprotection).

The same group described the preparation of sugar-grafted polysiloxanes, by hydrosilylation of a trimethylsilyl-protected allylgluconamide, followed by deprotection with BF_3-Et_2O in dry dichloromethane at −10°C [40]. This strategy is similar to what was described by Stadler et al. At this stage, the status of the polysiloxane after the deprotection reaction was not specified.

11 What About the Properties of Glycosilicones?

The aim of this paper was mostly to give an overview of the synthetic pathways for making sugar-grafted polysiloxanes. Given that the syntheses are not always as easy as expected, quite few material would be finally available for making extensive physicochemical studies. However, in some papers, the authors have described some of their properties. Stadler et al. have provided quite a lot of results concerning solubility, aggregation in solution (light scattering) and glass transitions (DSC) of their sugar-grafted polysiloxanes [12–15, 28, 29]. Wagner et al. described mostly the surface-active properties, including the wetting properties of their sugar-polysiloxanes derivatives, since their aim was to make new sugar-silicone surfactants [6–8, 34–38]. Thiem et al. characterized the solubility, the thermal degradation and the biodegradability of their segmented sugar-polysiloxane polyamides block copolymers [32]. Ogawa et al. tested the solubility and emulsifying properties of their derivatives [26, 27]. Nagase et al., used them for enhancing transdermal drug permeation [16–18]. Hamaide et al. studied the solubility properties in water, the surface active properties and the application to the stabilisation of nanoparticles of their glycopolysiloxanes [21, 22]. Gross et al. studied their carbohydrate-ended polysiloxanes by TGA and DSC [30, 31].

In fact, a lot remains to be done for getting a clear knowledge of the relationships between structure and properties of sugar-grafted polysiloxanes. Both the set up of easier syntheses and a better knowledge of their properties would help their development.

12 Conclusion

This review on polysiloxanes grafted with sugars highlighted the most critical points of their synthesis. First of all, if the steps involved require even slightly acidic or basic media, the polysiloxane backbone may be affected. This point should be systematically checked in these syntheses. Because of this fragility, quite often, deceiving results were obtained when protection-deprotection strategies have been used. It seems however that, for deprotection reactions, results were not always consistent and reproducible. For example, deprotection seemed to proceed without chain scission in some cases, while the same conditions applied to another structure led to degradation. Probably, precise reaction conditions (especially, dryness or reaction time), the kind of structure (e.g. telechelic) and grafting amounts (especially lower ones), play an important role in the reaction outcome. To conclude, classical strategies for synthetizing "glycopolymers", performed easily and successfully with other kinds of polymers, failed in the case of polysiloxanes.

Besides the direct grafting of unprotected sugars is possible by taking advantage of the different reactivity of functions not in competition with the hydroxyl functions. In this case, cross-linking can be avoided. However, in these strategies, miscibility issues of both the polysiloxane and the sugar derivative have been quite systematically encountered. These strategies are the most promising from the point of view of a possible future development. The interest in these glycosilicones should increase, since they should display good toxicological profiles, advantageous interactions with biological material, tunable hydrophilic/lipophilic properties and tunable thermomechanical properties.

Acknowledgements The French Ministry of Education, Research and Technology (MENRT) is acknowledged for WM grant.

References

1. T.K. Lindhorst, "Essentials of Carbohydrate Chemistry and Biochemistry", 2007, Wiley VCH Ed., Chap 6, 213–237.
2. Patent A.J. O'Lenick, US 5,428,142 (27/06/1995)
3. J. Fitremann, Unpublished results.
4. e.g. patents WO 02/088456 (07/11/2002), WO 2006/127924 (30/11/2006).

5. K. Beppu, Y. Kaneko, J-I. Kadokawa, H. Mori, T. Nishikawa, Polymer J., 2007, 39(10), 1065–1070. "Synthesis of sugar-polysiloxane hybrids having rigid main-chain and formation of their nano aggregates"
6. R. Wagner, L. Richter, R. Wersig, G. Schmaucks, B. Weiland, J. Weissmueller, J. Reiners, Appl. Organometal. Chem., 1996, 10(6), 421–435. "Silicon-modified carbohydrate surfactants. I. Synthesis of siloxanyl moieties containing straight-chained glycosides and amides."
7. R. Wagner, L. Richter, B. Weiland, J. Reiners, J. Weissmüller, Appl. Organometal. Chem., 1996, 10(6), 437–450. "Silicon-modified carbohydrate surfactants. II. Siloxanyl moieties containing branched structures."
8. R. Wagner, L. Richter, B. Weiland, J. Weissmüller, J. Reiners, W. Kraemer, Appl. Organometal. Chem., 1997, 11(6), 523–538. "Silicon-modified carbohydrate surfactants.III. Cationic and anionic compounds."
9. T. Dietz, B. Gruning, P. Lersch, C. Weitemeyer, US 5,891,977, 06/04/1999. "Organopolysiloxanes comprising polyhydroxorganyl radicals and polyoxyalkylene radicals".
10. G. Torres, G. Wajs, FR 2646672, 11/09/1990. "Elastomère de silicone mouillable convenant à la fabrication de lentilles de contact".
11. R. Wersig, G. Sonnek, C. Niemann, Appl. Organomet. Chem. 1992, 6, 701–708. "Novel nonionic siloxane surfactants"
12. G. Jonas, R. Stadler, Makromol. Chem., Rapid Commun. 1991, 12, 625 "Polysiloxanes with statistically distributed glucose and galactose units, 1: Synthesis and thermal characterization"
13. G. Jonas, R. Stadler, Acta Polymer., 1994, 45, 14–20. "Carbohydrate modified polysiloxanes II. Synthesis via hydrosilation of mono-, di-and oligosaccharide allylglycosides".
14. K. Loos, G. Jonas, R. Stadler, Macromol. Chem. Phys. 2001, 202, 3210–3218, "Carbohydrate Modified Polysiloxanes, 3. Solution Properties of carbohydrate-Polysiloxane conjugates in Toluene".
15. V. von Braunmühl, R. Stadler, Polymer, 1998, 65, 1617. "Synthesis of aldonamide siloxanes by hydrosilylation".
16. T. Akimoto, K. Kawahara, Y. Nagase, T. Aoyagi, Macromol. Chem. Phys., 2000, 201(18), 2729–2734. "Preparation of oligodimethylsiloxanes with sugar moiety at a terminal group as a transdermal penetration enhancer".
17. T. Akimoto, K. Kawahara, Y. Nagase, T. Aoyagi, J. Control. Release, 2001, 77, 49–51. "Polymeric transdermal drug penetration enhancer: The enhancing effect of oligodimethylsiloxane containing a glucopyranosyl end group".
18. T. Akimoto, Y. Nagase, J. Control. Release, 2003, 88, 243–252. "Novel transdermal drug penetration enhancer: synthesis and enhancing effect of alkyldisiloxane compounds containing glucopyranosyl group".
19. M. Haupt, S. Knaus, T. Rohr, H. Gruber, J. Macromol. Sci. Pure Appl. Chem. 2000, A37(4), 323–341. "Carbohydrate modified polydimethylsiloxanes. Part 1. Synthesis and Characterization of carbohydrate silane and siloxane building blocks"
20. D. Henkensmeier, B. C. Abele, A. Candussio, J. Thiem, Macromol. Chem. Phys., 2004, 205, 1851–1857. "Synthesis and characterization of terminal carbohydrate modified polydimethylsiloxanes".
21. C. Racles, T. Hamaide, Macromol. Chem. Phys., 2005, 206, 1757–1768. "Synthesis and characterization of water soluble saccharide functionalized polysiloxanes and their use as polymer surfactants for the stabilization of polycaprolactone nanoparticles."
22. C. Racles, T. Hamaide, A. Ioanid, Appl. Organometal. Chem. 2006, 20(4), 235–245. "Siloxane surfactants in polymer nanoparticles formulation."
23. E. Fleury, Patent WO 2005 087843, 22/09/2005. "Grafted polymers comprising a polyorganosiloxane backbone and glycoside units".
24. ISPO International Symposium, Montpellier (Mèze, France), juin 2007.
25. E. Fleury, S. Halila, H. Driguez, S. Cottaz, T. Hamaide, S. Fort, FR 2 900 931, 16/11/2007. "Hybrid compounds based on silicones and at least one another polymer or nonpolymer molecular entity bonded to the silicone chains by nitrogen-containing rings, their preparation process, and their applications."

26. T. Ogawa, J. Polymer Sci., Part A: Polymer Chem. 2003, 41(21), 3336–3345. "Simplified synthesis of carbohydrate-functional siloxanes via transacetalation. I. Glucose-functional siloxanes".
27. T. Ogawa, Macromol. 2003, 36(22), 8330–8335. "Simplified Synthesis of Amphiphilic Siloxanes with Methyl Gluconyl Glycinate Functionalities via Transacetalation".
28. V. von Braunmühl, G. Jonas, R. Stadler, Macromol., 1995, 28, 17–24. "Enzymatic grafting of amylose from polydimethylsiloxanes".
29. V. v. Braunmühl, R. Stadler, Macromol. Symp. 1996, 103, 141–148. "Polydimethylsiloxanes with amylose side chains by enzymatic polymerization"
30. B. Sahoo, K.F. Brandstadt, T.H. Lane, R.A. Gross, Organic Lett. 2005, 7(18), 3857–3860. "Sweet Silicones": Biocatalytic Reactions to Form Organosilicon Carbohydrate Macromers".
31. B. Sahoo, K.F. Brandstadt, T.H. Lane, R.A. Gross, in "Polymer Biocatalysis and Biomaterials", ACS Symposium Series, 2005, 900, 182–190. "Sweet silicones": Biocatalytic reactions to form organosilicon carbohydrate macromers".
32. D. Henkensmeier, B. C. Abele, A. Candussio, J. Thiem, Polymer, 2004, 7053–7059, "Synthesis, characterisation and degradability of polyamides derived from aldaric acids and chain end functionalised polydimethylsiloxanes".
33. A. Domschke, D. Lohmann, J. Höpken, EP 0826158, 15/09/1999.
34. R. Wagner,; Y. Wu, L. Richter, T. Pfohl, S. Siegel, J. Weissmueller, J. Reiners, Appl. Organometal. Chem., 1996, 10(6), 421–435. "Silicon containing structures at interfaces. The wetting behavior of carbohydrate-modified Si surfactants on perfluorinated surfaces and the modification of rough metal surfaces by hydrophilic polysiloxane networks."
35. R. Wagner, L. Richter, Y. Wu, J. Weissmuller, J. Reiners, K.-D. Klein, D. Schaefer, S. Stadtmuller, in Organosilicon Chemistry III: From Molecules to Materials, N. Auner, J. Weis, Eds, 1998, 510–514."Carbohydrate-modified siloxane surfactants: The effect of substructures on the wetting behavior on non polar solid surfaces.
36. R. Wagner, L. Richter, Y. Wu, J. Weissmuller, A. Kleewein, E. Hengge, Appl. Organometal. Chem., 1998, 12(4), 265–276. "Silicon-modified carbohydrate surfactants. VII: Impact of different silicon substructures on the wetting behavior of carbohydrate surfactants on low-energy surfaces – distance decay of donor-acceptor forces."
37. R. Wagner, Y. Wu, L. Richter, S. Siegel, J. Weissmuller, J. Reiners, Appl. Organometal. Chem., 1998, 12(12), 843–853. "Silicon-modified carbohydrate surfactants IX: dynamic wetting of a perfluorinated solid surface by solutions of a siloxane surfactant above and below the critical micelle concentration."
38. R. Wagner, Y. Wu, L. Richter, J. Reiners, J. Weissmuller, A. De Montigny, Appl. Organometal. Chem., 1999, 13(1), 21–28. "Silicon-modified carbohydrate surfactants. VIII. equilibrium wetting of perfluorinated solid surfaces by solutions of surfactants above and below the critical micelle concentration-surfactant distribution between liquid-vapor and solid-liquid interfaces."
39. Eur. Polymer J., 2004, 40, 165–170.
40. Carbohydr. Polymers, 2006, 65, 321–326.

Part 2
Functional Polysilsesquioxanes

Silsesquioxane-Based Polymers: Synthesis of Phenylsilsesquioxanes with Double-Decker Structure and Their Polymers

Kazuhiro Yoshida, Takayuki Hattori, Nobumasa Ootake, Ryouji Tanaka, and Hideyuki Matsumoto

Abstract Double-decker shaped silsesquioxane was synthesized by hydrolysis and condensation of trimetoxyphenylsilane in the presence of sodium hydroxide and it led to macromonomers having carbon functionalized groups. Then, these were polymerized by hydrosilylation polymerization, and their properties were measured.

1 Introduction

Polyhedral oligomeric silsesquioxanes (POSS) are known molecules having various structures [1] as seen in Fig. 1. However double-decker shaped silsesquioxane has not been obtained. POSS is an interesting class of monomers for the preparation of silicon-based polymers. In recent years, silsesquioxane-based polymers have received much attention because of their useful properties, which include heat resistance, mechanical stability, low dielectric constant, etc. [2, 3].

2 Synthesis of Double-Decker Shaped Silsesquioxane and its Derivatives

Recently, sodium phenylsiloxanolate synthesis has been reported. For example, cyclo tetra siloxanolate $[(Na^+)_4C_6H_5Si(O)O^-]_4$ was synthesized by reacting tri-n-butoxyphenylsilane with sodium hydroxide in the presence of water [4], partial cage structure $[(Na^+)C_6H_5Si(O_{0.5})_2O^-]_3[C_6H_5Si(O_{1.5})]_4$ was obtained that reacts with trimethoxyphenylsilane and sodium hydroxide in the presence of water [5]. In 2003, we have

K. Yoshida
Goi research center, Chisso Petrochemical Corporation, Ichihara, Chiba 290-8551, Japan
e-mail: k.yoshida@chisso.co.jp

F. Ganachaud et al. (eds.), *Silicon Based Polymers*,

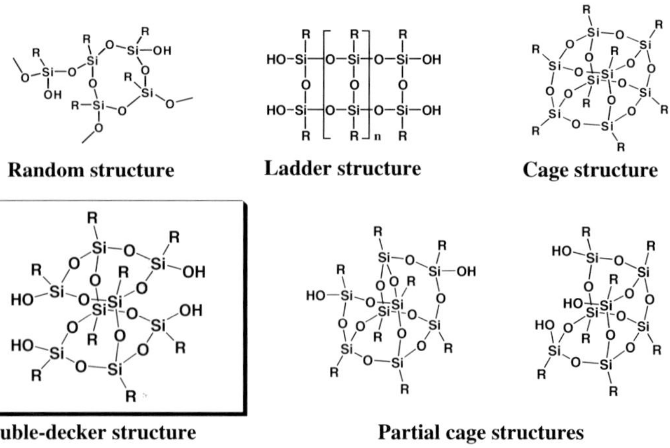

Fig. 1 Example of some POSS structures

reported the synthesis of the sodium phenylsiloxanolate $Ph_8Si_8O_{14}Na_4$ (DD-Na) and its derivatives [6–9], which are the first examples of the double-decker Si-O-Si frameworks (Scheme 1).

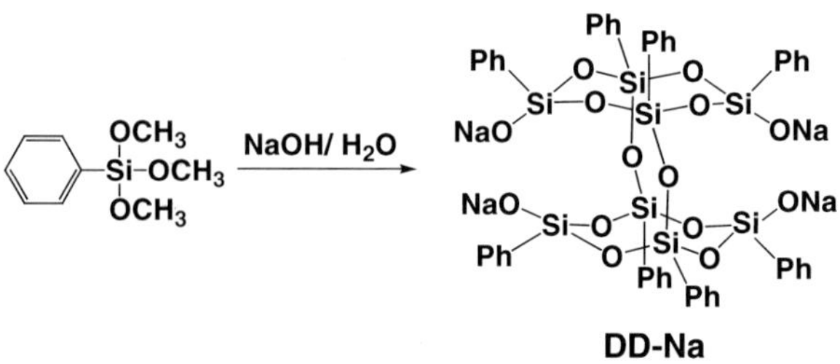

Scheme 1 Synthesis of double-decker shaped silsesquioxane

DD-Na is easily converted into the sila-functional silsesquioxane (DD-OH). The structure of DD-OH was confirmed by ^{29}Si-NMR and other spectroscopic analyses (Scheme 2) [10,11]. In quite recent time, Lee reported that the double-decker shaped silsesquioxane seems to be produced with the consumption of the partial cage and the cyclic siloxanes in the same condition [12].

The DD-Na is synthesized in a high yield (>70%) by the treatment of trimethoxyphenylsilane in the presence of sodium hydroxide in aqueous 2-propanol. We optimized the reaction conditions for the formation of DD-Na; it can be prepared by preparative scale (33 kg/batch). DD-Na give POSS derivatives (Scheme 3). For example it reacts with dichlorodimethylsilane to give a cage structure compound

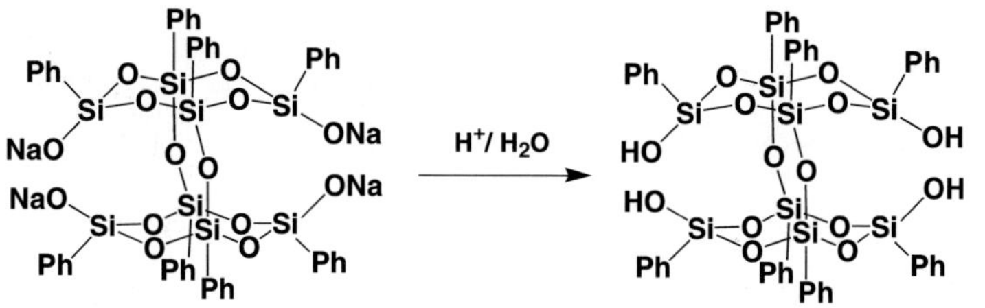

Scheme 2 Synthesis of sila-functional silsesquioxane

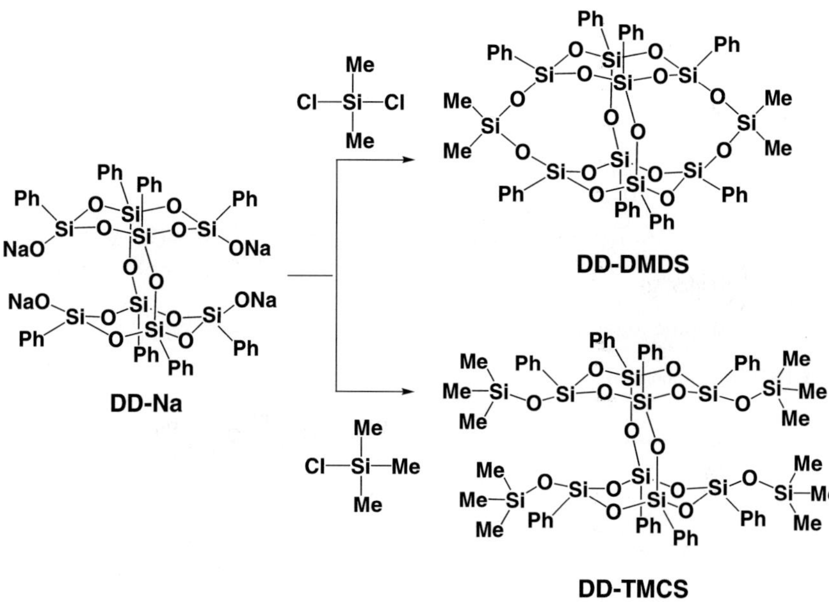

Scheme 3 Synthesis of POSS derivatives

(DD-DMDS): on the other hand, it reacts with chlorotrimethylsilane to give a partial cage structure compound (DD-TMCS).

The structures of these products were determined by X-ray crystallography as well as spectroscopic methods (Fig. 2).

Similarly, DD-Na lead to macromonomer by reaction with a variety of chlorosilanes bearing carbon-functional groups (Fig. 3). The substituent R on silicon is exemplified with hydroxyl, amino, epoxy, carboxyl, or vinyl functional group.

Therefore, bi-functional macromonomers reacts with various kinds of organic monomers and it makes organic-inorganic hybrid materials consisting of linear polymer. In addition, tetra-functional macromonomers makes network polymer by cross-linking (Fig. 4.)

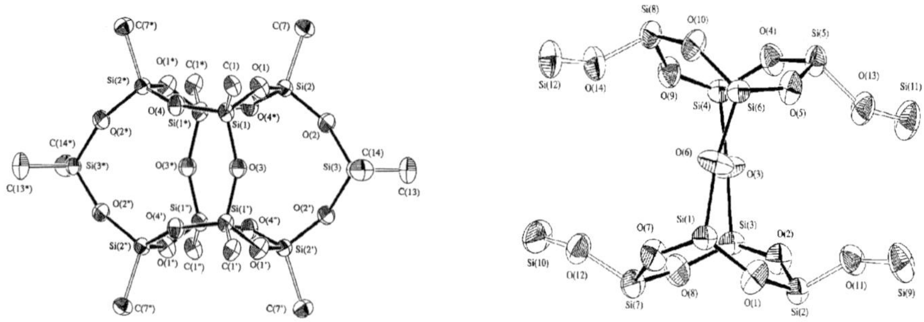

Fig. 2 ORTEP drawing, left is DD-DMDS and right is DD-TMCS

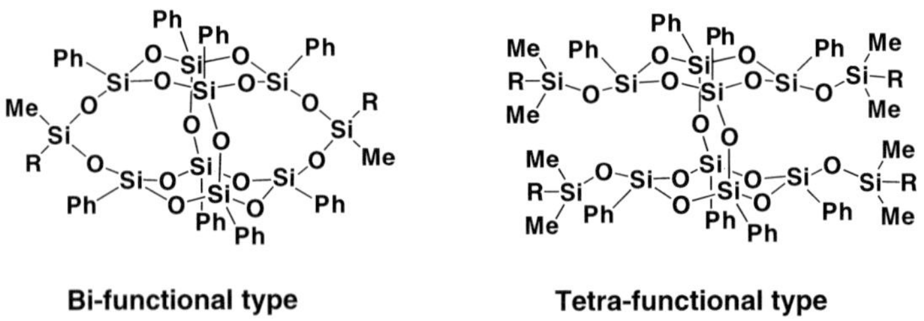

Fig. 3 PSQ-macromonomers

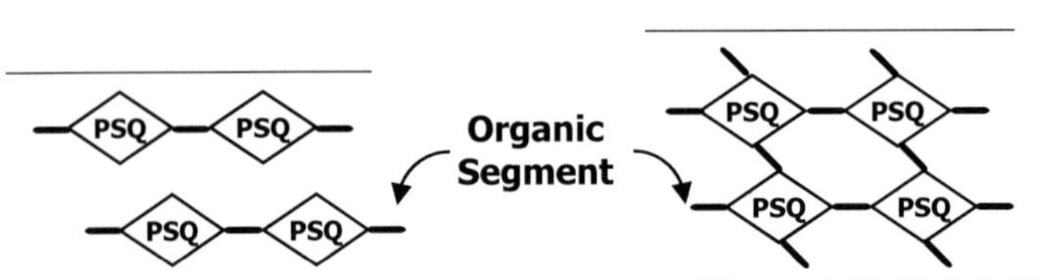

Fig. 4 Structure images of organic-inorganic hybrid materials

3 Synthesis of Silsesquioxane Based Polymers and Property of Their Polymers

We have obtained silsesquioxane-based polymer (HS-Polymer) by using the hydrosilylation reaction of vinyl-functionalized macromonomer (DD-2V) with some SiH-functionalized siloxanes (Scheme 4).

The resulting HS-polymers are colorless powder soluble in tetrahydrofuran and esters as well as aromatic hydrocarbons such as toluene, xylene and mesitylene. The HS-polymers can be cast into a transparent and flexible film as seen in Fig. 5.

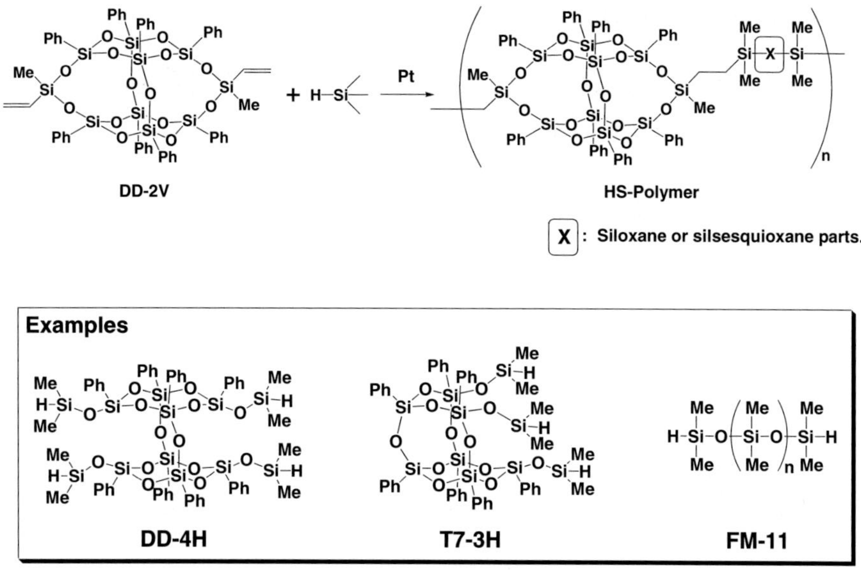

Scheme 4 Preparation of hydrosilylation polymers

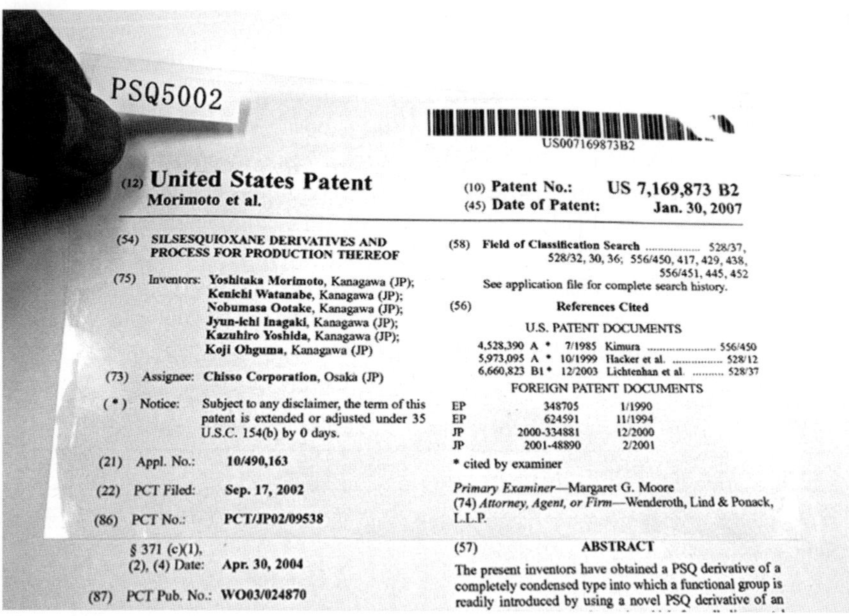

Fig. 5 Photograph of the HS-polymer film

4 Properties of HS-Polymers Film

An UV/Vis spectrum of HS-polymer film with a thickness of 100 micrometers was studied by UV/Vis spectrophotometer, and result is shown in Fig. 6. The film does not show absorption in the range of 400–800 nm, and cut off wavelength is 277 nm. As a result, HS-polymer film has high transparency.

Reflective index was studied by abbe refractometer, and result is 1.53 at 589 nm, and its birefringence is 0.0001, which indicate that this film is not anisotropic. And the film is insulant, since its volume resisitivity is 1×10^{17} Ωcm, and its dielectric constant is 2.8.

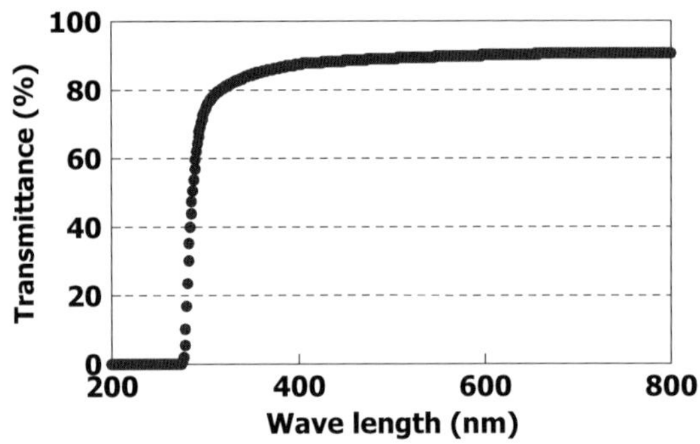

Fig. 6 UV-vis spectrum of the HS-polymer film

5 Weatherability

Accelerated weathering test was carried out by the exposure of the films to a Sunshine weather meter with Xenon arc type accelerated weathering machine. HS-polymer's film does not chalk nor yellow. However under the conditions employed, polyimide (PI) is led to chalking and polyethylenenaphtalate (PEN), polycarbonate (PC) and polyethylenetelephtalate (PET) are led to yellowing (Table 1).

Synergistic effect of transparency to ultra violet rays and visible light rays, the strong Si-O-Si framework, and chemical resistance gives the HS-polymer excellent weatherability as one of the promising features.

Table 1 Comparison for results of accelerated weathering test

Times	HS-Polymer	PI	PEN	PC	PET
250 hrs	OK	Chalking	Yellowing	Yellowing	Shrinkage
500 hrs	OK	↓	↓	↓	Yellowing
1000 hrs	OK	↓	↓	↓	Yellowing Cracks

* The symbol 'OK' indicates both non-chalking and non-yellowing.

6 Summary

We optimized the reaction conditions for the formation of Double-decker shaped silsesquioxane (DD-Na). As a result, it can be prepared in a high yield by industrial process (>70% yield and >33 kg/batch). Also, we have developed route to macromonomer of bi-functional and tetra-functional from DD-Na. The silsesquioxane macromonomers are versatile precursors for the preparation of organic-inorganic hybrid materials. For example, silsesquioxane-based polymer (HS-Polymer) obtained from vinyl terminated silsesquioxane macromonomer reacts with SiH terminated silsesquioxane macromonomer and the HS-Polymer's film have unique optical and electrical properties with high transparency, highly resistance for the ultraviolet rays and low dielectric constant. To our opinion, the material obtained in this work is one of the most promising materials as inorganic/organic hybrid precursor, which will open new area of applications.

References

1. Baney R H et al. (1995) Silsesquioxanes. Chem. Rev 95:1409–1430
2. Lichtenhan J D et al. (1993) Silsesquioxane-siloxane copolymers from polyhedral silsesquioxanes. Macromolecules 26:2141–2142
3. Seino M et al. (2006) Hydrosilylation polymerization of double-decker-shaped silsesquioxane having hydrosilane with diynes. Macromolecules 39:3473–3475
4. Shchegolikhina O et al. (2000) Synthesis and structure of sodium phenylsiloxanolate. Organometallics 19:1077–1082
5. Oikawa H et al. (2002) Synthesis and basic characteristics of novel T7-Phenylsilsesquioxane derivatives using hydrolytic condensation reaction. Polym Prep Jpn (Soc Polym Sci Jpn) 51:783
6. Yoshida K et al. (2003) Synthesis of Double-decker-shaped phenylsilsesquioxane derivative using hydrolytic condensation reaction. Polym Prep Jpn (Soc Polym Sci Jpn) 52:316
7. Ohguma K et al. (2003) Synthesis of novel siloxanes derivatives containing reactive organic functional groups. Polym Prep Jpn (Soc Polym Sci Jpn) 52:1317
8. Morimoto Y et al. (2007) Silsesquioxane derivatives and process for production thereof. United States Patent; US 7169873B2
9. Yoshida K et al. (2006) Silsesquioxane derivative and process for producing the same. United States Patent Application; 20060052623A1
10. Yoshida K et al. (2003) Synthesis of novel Double-decker-shaped phenylsilsesquioxane with reactive functional group of tetrasilanol. Polym Prep Jpn (Soc Polym Sci Jpn) 52:131
11. Yoshida K and Ootake N (2006) Organic silicon compound and production process for the same, and polysiloxane and production process for the same. United States Patent Application; 20060155091A1.
12. Lee D W and Kawakami Y (2007) Incompletely condensed silsesquioxanes: formation and reactivity. Polym J 39:230–238

Organosilica Mesoporous Materials with Double Functionality: Amino Groups and β-Cyclodextrin Synthesis and Properties

Stéphanie Willai, Maryse Bacquet, and Michel Morcellet

Abstract Our work concerns the preparation of new organofunctional mesoporous silica gels. The goal of this study is to combine in a unique material –on one hand β-cyclodextrin (β-CD), able to form stable inclusion complexes with many organic molecules, – and on the other hand aminogroups (from aminopropylsilane precursor) which are likely to chelate metallic cations. At the same time the formation of a mesoporous silica network would improve and regulate the access to functional sites. In order to tailor the functionalities content of our materials, a preparation in two steps has been chosen. First, a new silica precursor (β-CDAPS) has been synthesized from a β-CD derivative and (3-aminopropyl) trimethoxysilane (APS). A detailed characterization of the obtained product has helped to determine the structure of β-CDAPS repeating unit and quantify its functionality. Then this hybrid precursor has been co-condensed with tetraethylorthosilicate (TEOS) via a sol-gel process involving the use of surfactants by a S–I+ mechanism. Porous and chemical structures of a series of these functional mesoporous silicas were characterized. Finally, preliminary adsorption tests of aqueous model pollutants, carried out on these hybrid materials, have been discussed.

1 Introduction

Templated mesoporous silica materials have been discovered by Kresge et al. in 1992 [1, 2]. Since then, efforts have been made to produce by direct co-condensation synthesis hybrid materials functionalized by different types of reactive groups,

S. Willai
Laboratoire de Chimie Organique et Macromoléculaire, UMR CNRS 8009, Université des Sciences et Technologies de Lille, 59655 Villeneuve d' Ascq France

F. Ganachaud et al. (eds.), *Silicon Based Polymers*,

including amino groups [3–8] and cyclodextrin moieties [9–15]. However, mesoporous materials then obtained are functionalized by only one kind of functional group. Our purpose is to prepare by the direct co-condensation method new hybrid materials functionalized by both amino groups and β-cyclodextrin moieties. Such materials do not exist in the literature and would be useful for decontamination of mixed polluted effluents, where metallic and organic pollutants occur simultaneously. On the one hand, cyclodextrins (CDs) are cyclic oligosaccharides with a torus shape and are able to form stable inclusion complexes with many organic molecules. On the other hand, amino groups are likely to chelate metallic cations.

A preparation in two steps has been chosen: first a new hybrid silica precursor β-CDAPS, containing β-CD groups and amine functions, has been prepared and characterized. Then, this precursor has been co-condensed with tetraethyl orthosilicate (TEOS) via a sol-gel process involving the use of surfactant. We chose the anionic surfactant sodium dodecylsulfate (SDS). The chemical and structural characterization combined with adsorption tests of *p*-nitrophenol and lead nitrate led to the evaluation of the accessibility and the effectiveness of the binding functions in these hybrid materials.

2 Experimental Section

2.1 Materials and Methods

β-cyclodextrin (β-CD) was supplied by Roquette Frères (France). (3-aminopropyl) trimethoxysilane (97%) (APS), tetraethyl orthosilicate (99.999%) (TEOS) and N,N-dimethylacetamide (99%) (DMAc) were purchased from Aldrich and used without further purification. Sodium dodecylsulfate (>97%) (SDS) was obtained from Fluka. Para-nitrophenol (99%) (p-NP) was supplied by Acros Organics and lead(II) nitrate (99.99%) by Sigma.

ThermoGravimetric Analysis (TGA) was performed on TGAQ50 (TA Instruments) under air up to 800°C. FTIR spectra were obtained on a Spectrum One from Perkin Elmer Instruments. Elemental Analysis was performed by Service Central d'Analyse from CNRS located at Vernaison (France). Conductimetry back-titration of amine functions (after contact with HCL 0.1 N in excess) was performed using a Tacussel conductimetry probe by a 0.1 M NaOH solution. Porosimetry measurements were performed on a SORPTOMATIC 1990 from CE Instruments: specific surface areas were calculated using BET model between $P/P_0=0$ and $P/P_0=0.4$ and pore size repartitions were determined using BJH model. SEM photos were obtained on an Ultra 55 (Zeiss).

Adsorption capacities were studied -by Atomic Emission Spectroscopy (Vista-Pro ICP-OES from VARIAN) for lead analysis and -by UV Spectrometry (Nicolet Evolution 300 from ThermoElectron Corporation) for *p*-nitrophenol determination.

2.2 *Synthesis of Hybrid Precursor β-CDAPS Containing Amino Groups and β-CD Moieties*

First, Tosyl- -cyclodextrin (β-CDOTs) was synthesized according to reference [16]. RMN ^{1}H spectroscopy and Elemental analysis revealed that 91% molar of monotosyl-βCD and 9% molar of ditosyl-βCD were obtained. β-CDOTs contained about 6% w/w of water, determined by the Karl Fischer method.

Then, in a tricol, heated at 303 K under nitrogen, 5 g (0.004 mole) of β-CDOTs were dissolved in 125 mL of DMAc (kept previously on molecular sieves) until the solution cleared. 7 mL (0.04 mole) of (3-aminopropyl)trimethoxysilane (APS) were added dropwise. The solution was vigorously stirred for 1 hour and then precipitated in 1.4 L of acetone (distilled twice, on molecular sieves). The white precipitate was kept under stirring for 2 hours and filtered. The product obtained (β-CDAPS) was dried 2 days at 353 K under vacuum and crushed to form homogeneous yellow powder. The reaction scheme expected is presented on Fig. 1. Si, N, C, N, H contents were determined by TGA, amine titration and elemental analysis.

2.3 *Synthesis of Hybrid Silica Materials Using an Anionic Surfactant*

The synthesis was adapted from reference **[6]**. Sodium dodecylsulfate (SDS) was dissolved in water-ethanol (molar ratio 9:1) with a homogenizer (Ultra-Turrax T25) at 293 K. Then the precursor β-CDAPS (dispersed in a water-ethanol mixture) and tetraethyl orthosilicate (TEOS) were added in varying proportions. The resulting mixture was stirred 1 h at 293 K and kept statically for 2 days at 373 K. The precipitate obtained was filtered. Surfactant elimination was performed using Soxhlet extraction over ethanol and water (2×400 ml for each solvent) and for 2 days each. The obtained product was dried under vacuum at 333 K for 1 day. The initial molar

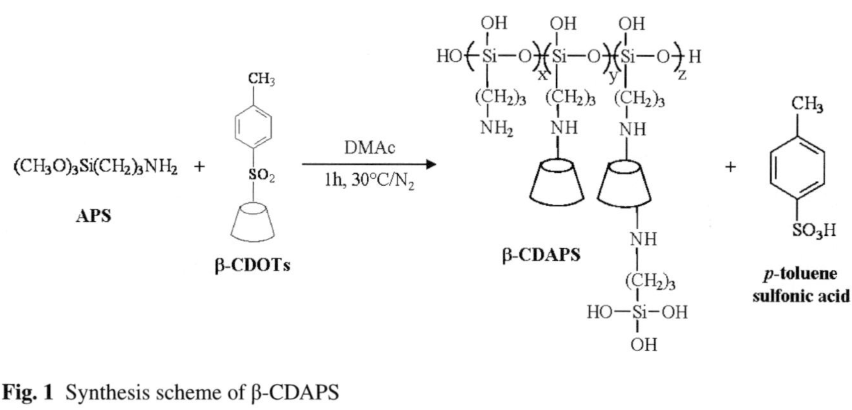

Fig. 1 Synthesis scheme of β-CDAPS

composition was 0.5 TEOS: x β-CDAPS: 0.1 SDS: 180 H_2O: 20 EtOH where x is equal to 0.5, 0.25 or 0.1. The obtained hybrid silica materials were designated TbSn, where n = 1, 2 and 3 corresponds to x = 0.5, 0.25 and 0.1 respectively.

2.4 Synthesis of Hybrid Silica Materials Without Surfactant

The synthesis was the same as the one presented above but without the use of surfactant. TEOS was added to a β-CDAPS solution (in a water-ethanol mixture with molar ratio 9:1) in varying proportions. The resulting mixture was stirred 1h at ambient temperature and then kept statically for 2 days at 373 K. The precipitate obtained was filtered and the powder dried under vacuum at 333 K. The initial molar composition was 0.5 TEOS: x β-CDAPS: 180 H_2O: 20 EtOH where x is equal to 0.5, 0.25 or 0.1. The obtained hybrid silica materials were designated Tbn, where n = 1, 2 and 3 corresponds to x = 0.5, 0.25 and 0.1 respectively.

2.5 Adsorption Tests

Batch tests in aqueous solutions at pH 5 were performed on hybrid silica particles. For Pb^{2+}, 30 mg of support were added to 10mL of an aqueous solution of $Pb(NO_3)_2$ at 100 ppm and stirred for 20 h. For p-NP, 20 mg of support were added to 10 mL of an aqueous solution of p-NP at 20 ppm and stirred for 4 h. Then the solutions were filtered and the supernatants were titrated.

3 Results and Discussion

3.1 Characterization of β-CDAPS Precursor

Native -cyclodextrin is not reactive enough to be condensed directly with APS, so a modified cyclodextrin is needed. Monotosyl-β-cyclodextrin is the most used to prepare mono-functionalized cyclodextrins.

The soft experimental conditions in the synthesis of β-CDAPS were optimized using experimental design. The hybrid precursor β-CDAPS has been characterized in order to determine its structure. FTIR spectra show the presence of amino groups (1564 cm^{-1}) on one hand and siloxane structure (1100–1000 cm^{-1}) on the other hand, confirmed by TGA measurements. This TGA gave a Si ratio of 5.80 ± 0.04%., confirmed by elemental analysis. N content was determined by elemental analysis at 2.98 ± 0.02%, which corresponds to 2.13 ± 0.04 mmol/g of amine functions, also deduced from amine titration.

A quantitative estimation of the β-CDAPS structure was undertaken, taking into account the chemical elements determination (Si, N, C, H) and the presence of primary and secondary amine, cyclodextrin moieties, and Si-O-Si bonds revealed by FTIR spectra.

So a model, consisted of three types of units with x, y, z as molar ratios for each one, was assumed (Fig. 1): y is fixed at 1, z at 0.1 (considering the amount of ditosyl β-CD) and x can vary. From this model, we calculated theoretical values of Si, C, N, H contents for different values of x and compared it with experimental contents. Finally, the best accordance led to a structure of the hybrid precursor β-CDAPS, close to 2.3 aminopropyl groups (x=2,3 in Fig. 1) for 1 β-CD group.

3.2 Characterization of Hybrid Silica Materials

Hybrid silica materials were prepared *via* a sol-gel pathway at pH 9. The influence of anionic surfactant (SDS) was studied by comparing templated materials (TbSn series) with hybrid materials obtained without surfactant (Tbn series). Two mechanisms of mesostructure formation can be considered as represented on Fig. 2. The pka of aminopropyl chain is about 10.6: in the reaction mixture propylamines are partially protonated. Electrostatic interactions between dodecylsulfate anion and NH_3^+ and sodium cation neutralization may then occur, resulting in the condensation of the silica structure around surfactant micelles and aminopropyl groups at the surface of the pores. The other mechanism is SDS chains complexation by β-CD cavity, which would result in β-CD groups located at the surface of the pores and aminopropyl less accessible, due to steric hindrance caused by β-CD bulky groups. A complete characterization of the products and adsorption capacities will help understanding the formation mechanism of mesoporous hybrid silica.

TGA curves of TbSn and SDS are presented on Fig. 3. The lack of deflection in TGA curves of TbSn at 200°C proved the absence of SDS in thses materials (confirmed by elemental analysis which gives a S content lower than 0.3%).The appearances of TbSn thermograms show the formation of the crosslinked silica structure: indeed, the higher the initial TEOS content, the larger the shift and SiO_2

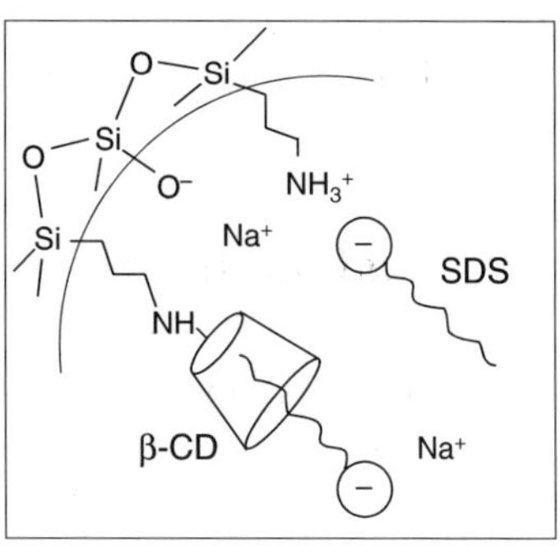

Fig. 2 Schematic representation of the two mechanisms of mesostructure formation expected: electrostatic interaction between SDS dodecylsulfate anion and protonated propylamine and SDS chain complexation by β-CD cavity

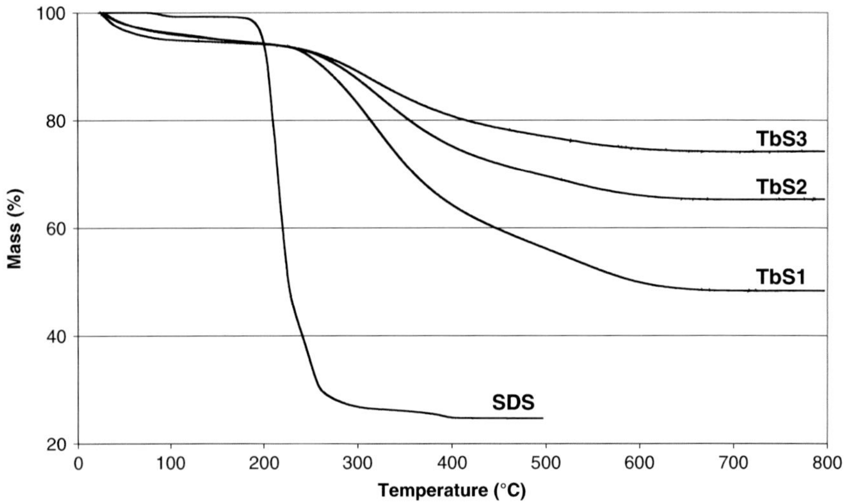

Fig. 3 TGA curves under air of SDS and TbSn materials

residue at 800°C in the TGA curve. Moreover, the anionic surfactant seems to allow a good incorporation of the hybrid precursor during the formation of the mesostructure and may play a role of compatibilizer between TEOS and β-CDAPS. The evolution of the quantity of β-CDAPS incorporated in the final product was as expected.

The presence of organic groups (attributed to β-CD moieties and amino groups) was shown from %C and %N determined by elemental analysis. These values for the different materials are gathered in Table 1.

The morphology of the hybrid silica materials was studied by SEM. Figure 4 shows photos obtained for samples TbS2 and Tb2. The structure of Tb2 is denser and is constituted of particles of 80 nm diameter whereas TbS2 presents a less compact texture with particles of 300 nm diameter.

Pore size distributions of TbSn samples are shown on Fig. 5. This series of hybrid materials present mesopores with a major family of pores with radius at about 2 nm. By comparison with radii of materials synthesized without surfactant (*Tbn, rp between 6 nm and 16 nm*) in Table 1, we proved the benefit of using SDS to obtain smaller pores with a narrower distribution.

3.3 Study of Adsorption Capacities

Results of adsorption capacities tests are gathered in Table 1. Samples prepared in the presence of SDS present good efficiency towards Pb^{2+} cations and *p*-nitrophenol. By comparing two samples with equivalent numbers of accessible amines (e.g TbS1 and Tb3), we can see that these samples adsorb the same quantities of Pb^{2+} cations.

Table 1 Proportions of total amino groups and β-CD (determined by elemental analysis), accessible amines (titration by conductimetry), specific surface areas, pore radii and adsorption capacities of lead cation and *p*-nitrophenol for the different samples

Initial molar composition				Physicochemical characterization					Adsorption results	
Sample	TEOS	β-CDAPS	SDS	S_{spe} (m^2/g)	r_p (nm)	Total amines (mmol/g)	Accessible amines (mmol/g)	β-CD (mmol/g)	Pb^{2+} (μmol/g)	p-NP (μmol/g)
Tb1	0.5	0.5	–	334	15 to 25	–	4.4	–	203	24
Tb2	0.5	0.25	–	179	6.4	3.1	3.3	0.35	102	10
Tb3	0.5	0.1	–	131	6.3	2.5	1.8	0.34	88	10
TbS1	0.5	0.5	0.1	257	2.0	3.3	1.8	0.63	87	61
TbS2	0.5	0.25	0.075	291	1.9	2.8	0.8	0.39	35	19
TbS3	0.5	0.1	0.06	217	1.9	2.0	0.7	0.21	26	10

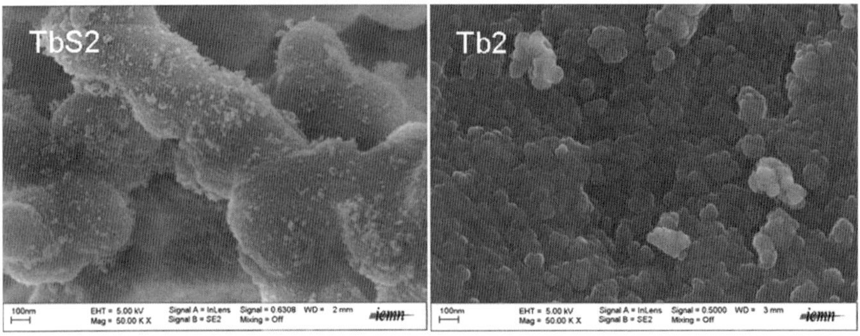

Fig. 4 SEM photos of hybrid materials prepared with SDS (TbS2) and without SDS (Tb2)

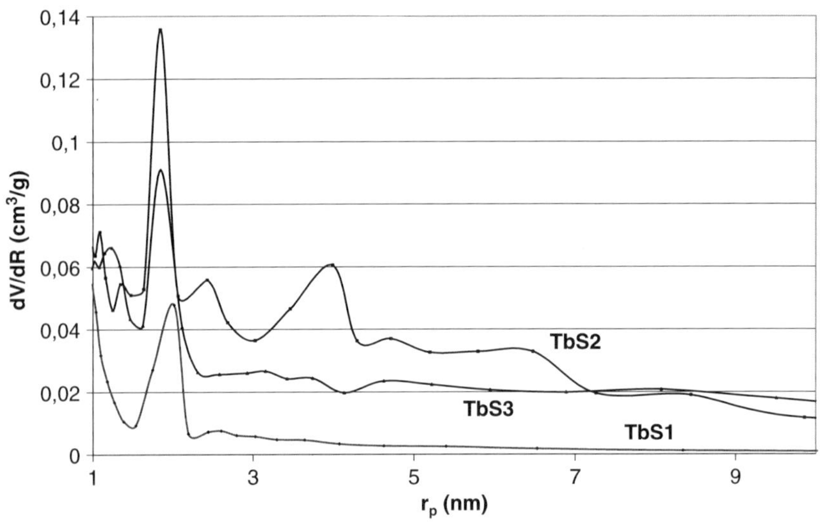

Fig. 5 Pore size distributions of hybrid materials prepared in presence of SDS (TbS1, TbS2 and TbS3)

On the other hand, the adsorption of *p*-nitrophenol is better for samples prepared in the presence of SDS. Indeed, for the same quantity of βCD, TbS2 is more efficient for p-NP adsorption than Tb2.

3.4 Determination of Anionic-Templated Formation Mechanism

In our synthesis conditions, the concentration of SDS during the preparation of sample TbS1 was lower than its Critical Micellar Concentration: this material will be excluded of the following discussion.

When the hybrid materials are synthesized in the presence of SDS, they possess lower pore sizes, due to the formation of SDS micelles. In these materials, a greater amount of β-CD groups has been incorporated (not necessarily all accessible) than in materials without template. However, the amount of accessible amino groups is lower and very similar for TbS2 and TbS3, although total amines are different. A plateau seems to be reached at around 0.8 mmol/g of accessible amino groups.

This may be explained as follows: the SDS interacts with the maximum of partially protonated aminopropyl groups, by electrostatic interactions between dodecylsulfate anion and NH_3^+ and sodium cation neutralization. The remaining amino groups are entrapped into the silica network. The amount of β-CD moieties incorporated depend on the initial molar ratio of β-CDAPS, and not of the presence or not of SDS. β-CD groups may not interact with SDS. We can say, that for these hybrid materials, the templated mechanism is a S^-I^+ one (S^- for anionic surfactant, I^+ for cationic aminosilica species).

4 Conclusion

Hybrid mesoporous materials have been prepared using the anionic-templating route by electrostatic interactions between SDS and aminopropyl groups. They possess accessible amino groups (up to 0.8 mmol/g) and β-CD moieties (up to 0.39 mmol/g). Preliminary adsorption tests showed that these materials are promising for environmental purposes.

Acknowledgements This work was cofinanced by CNRS and the Region Nord Pas de Calais (France)

References

1. Kresge C T, Leonowicz M E, Roth W J, Vartuli J C, Beck J S. (1992). Nature, 359:710.
2. Beck J S, Vartuli J C, Leonowicz M E, Kresge C T et al. (1992). J Am Chem Soc, 114:10834.
3. Fowler C E, Burkett S L, Mann S. (1997). Chem Commun:1769.
4. Macquarrie D J, Jackson D B, Mdoe J E. (1999). New J Chem, 23:539.
5. Huh S, Wiench J W, Yoo J C, Pruski M, Lin V S Y. (2003). Chem Mater, 15:4247.
6. Yokoi T, Yoshitake H, Tatsumi T. (2003). Chem Mater, 15:4536.
7. Yokoi T, Yoshitake H, Tatsumi T. (2004). J Mater Chem, 14:951.
8. Yokoi T, Yoshitake H, Yamada T, Kubota Y, Tatsumi T. (2006). J Mater Chem, 16:1125.
9. Huq R, Mercier L, Kooyman P J. (2001). Chem Mater, 13:4512.
10. Bibby A, Mercier L. (2003). Green Chem, 5:15.
11. Sawicki R, Mercier L. (2006). Environ Sci Technol, 40:1978.
12. Liu C, Lambert J B, Fu L. (2003). J Am Chem Soc, 125:6452.
13. Liu C, Wang J, Economy J. (2004). Macromol Rapid Commun, 25:863.
14. Liu C, Lambert J B, Fu L. (2004). J Org Chem, 69:2213.
15. Liu C, Naismith N, Economy J. (2004). J Chromatogr A, 1036:113.
16. Brady B, Lynam N, O'Sullivan T, Ahern C, Darcy R. (2004). Org Syn Coll, 10:686.

Direct synthesis of Mesoporous Hybrid Organic-Inorganic Silica Powders and Thin Films for Potential Non Linear Optic Applications

Eric Besson, Ahmad Mehdi, Catherine Réyé, Alain Gibaud, and Robert J. P. Corriu

Abstract Powders of mesoporous organosilica functionalized with a Non Linear Optical chromophore in the channel pore (azobenzene diethylphosphonate) were achieved in one step and were synthesized by templated-directed co-condensation of tetraethylorthosilicate (TEOS) and the functional organotriethoxysilanes. The materials were characterized by ^{13}C, ^{31}P and ^{29}Si NMR experiments, nitrogen gas adsorption, powder X-ray diffractrion (XRD). Optically transparent and highly ordered multifunctional mesostructured films obtained by evaporation induced self-assembly (EISA approach) approach were deposited on glass or silicon substrates by dip-coating. Thin films were monofunctionalized in the channel pores or bifunctionalized (channel pore/framework) and allowed us to evidence the salt effect induced by an organometallic complex on the structure of the mesostructured film. They were characterized by Grazing Incidence Small Angle X-ray Scattering (GISAXS) and X-ray reflectivity.

Keywords Hybrid materials · NLO · bifunctionality · thin films · mesoporous and GISAXS

The study of functionalized hybrid organic–inorganic materials is an expanding field of investigation, which should give rise to advanced materials [1].

Ordered mesoporous silicas are of particular interest for several applications, because of highly uniform porosity, mechanical stiffness, and thermal stability.

Such materials have often been prepared by post-synthetic grafting of an organotrialkoxysilane, $RSi(OR')_3$, onto the pore surface of ordered mesoporous silica [2–5].

A. Mehdi
Institut Charles Gerhardt, UMR 5253 CNRS, Chimie Moléculaire et Organisation du Solide, Université Montpellier II, Place E. Bataillon, 34095, Montpellier Cedex 5, France.
e-mail: ahmad.mehdi@univ-montp2.fr

F. Ganachaud et al. (eds.), *Silicon Based Polymers*,

This post-synthetic process has allowed the anchoring of a large number of organic groups, including bulky ones. However, this method does not allow to control the loading or the distribution of the functional groups on the inner pore surfaces [6, 7]. One alternative approach consisted of the introduction of organic groups during the synthesis of the material by the co-condensation of tetraethylorthosilicate (TEOS) and an organotriethoxysilane $RSi(OEt)_3$ in the presence of a structure-directing agent. In this case, the functional groups of the resulting materials are regularly distributed on the pore surfaces [8].

A variety of shapes, including thin films and monoliths, can be prepared, exhibiting good optical properties (transmission in visible range) and mechanical strength (easy machining) required for optical properties [9]. Then, numerous silica and siloxane-based hybrid organic–inorganic materials, mainly as coatings, have been developed in the past years, showing emission (solid-state dye lasers, electroluminescent devices), photochromism (optical switching, information storage), optical non linearity (second-order NLO, hole-burning) and sensing properties [10, 11].

As silica is transparent in the UV-visible range, these materials are ideal for optical applications. The mild synthesis conditions offered by the sol–gel process allow the incorporation of optically active organic molecules into the glassy matrix to form doped gels with specific optical properties.

Second-order non-linear optics continue to be an area of research because of its tremendous potential in the design of photon-based new materials for optical switching, data manipulation, and information processing [12].

Optimized NLO chromophores were incorporated into organic-inorganic hybrid materials through sol-gel process [7, 13]. This class of hybrid materials is very competitive for photonic application due to its peculiar properties, such as high transparency, low optical propagation losses and easy film processing. Such system containing covalent bonds between NLO chromophores and the host matrix network represents an interesting alternative to organic polymers, which exhibit poor reproducibility of optical quality and low Tg.

We now describe the synthesis of a mesoporous hybrid material containing silylated azobenzenediethylphosphonate ligand located in the channel pores, by hydrolytic polycondensation of the compound **1** (Scheme 1) in the presence

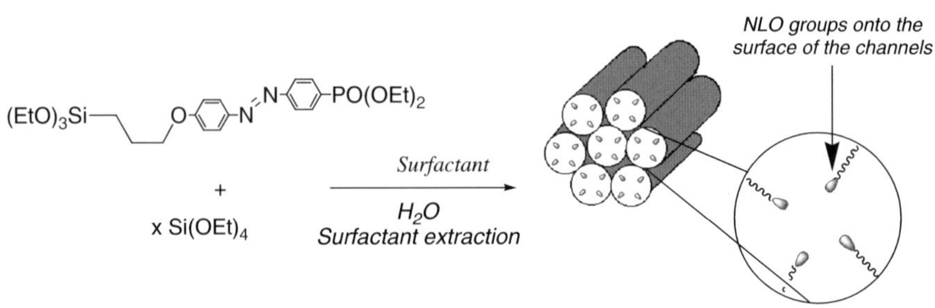

Scheme 1 General pathway for the preparation of meoporous bifunctionalized organo-silica thin films (See also Plate 16 in the Colour Plate Section)

of a structure-directing agent. The presence of chelating function on the NLO chromophore is expected to raise the NLO response by complexing a salt or coupling different properties as NLO and photoluminescence with a lanthanide salt.

As powders are unfavorable with respect to the potential applications in nonlinear optics and in the optical characterization of the composites, ordered mono and bifunctionalized organo-silica thin films were prepared through evaporation-induced self-assembly (EISA) and characterized.

The synthesis of the mono-silylated azobenzene diethylphosphonate **1** is outlined in Scheme 2: the 4-nitrophenyldiethylphosphonate **2** was synthesized thanks to a palladium catalyzed P–C cross coupling between diethylphosphonic acid and 4-bromonitrobenzene. This procedure [14] afforded **2** in 80% yield after purification. In order to perform the diazocoupling, nitro function of **2** was reduced by a Clemmensen type procedure to obtain **3** in a quantitative yield without further purification. The 4-(4-hydroxyphenylazo)phenyldiethylphosphonate **4** was synthesized by the diazocoupling reaction between **3** and the phenol. The silylation of **4** was accomplished by a Williamson type reaction with iodopropyl triisopropoxy-silane. It is worth noting that triisopropoxy groups were preferred to triethoxysilyl groups in silylation reactions, because the isopropoxy groups are poorly reactive especially towards hydrolysis. That allowed the purification by silica column chromatography in 71% yield. Finally, isopropoxide/ethoxide exchange was achieved in ethanol in presence of p-toluene sulfonic acid as catalyst to give compound **1** in high yield (Scheme 2). The organosilane **1** was fully characterized by ^{1}H, ^{31}P, ^{29}Si, ^{13}C NMR spectroscopies, and elemental analysis.

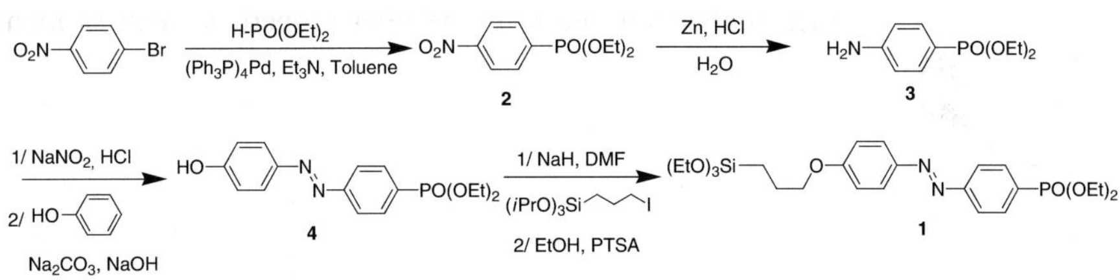

Scheme 2 Sequence of reaction for the preparation of the precursor **1**

The functional materials of **SBA-15** type were prepared by co-hydrolysis and polycondensation of **1** and x equivalents of tetraethylorthosilicate (TEOS) in presence of P123 as structure directing agent. The materials obtained were called **SBA1a** and **1b** for x=9 and 19 respectively. Our procedure was the following:

1.46 g of P123 was dissolved in 58 mL of a mixture ethanol/aqueous HCl solution pH=1.5 (75/25 v). This solution was added to a mixture of 3.1 g of TEOS (x=9) and 0.9 g of **1** at ambient temperature. The mixture was stirred for 3 h giving rise to a micro-emulsion. After heating this perfectly transparent solution at 60°C, a small amount of NaF (30.0 mg) was added with stirring to induce the polycondensation. The mixture was stirred at 60°C for 48 h. The resulting solid was filtered off and washed. The surfactant was then removed by hot ethanol extraction in a Soxhlet apparatus affording 1.5 g of the **SBA1a** as an orange powder in 88% yield. The molar composition of the reaction mixture was: 0.04 F^-: 1 TEOS: 0.11 **1**: 0.02 P123: 0.03 HCl: 54 H_2O: 50 EtOH.

Small-angle X-ray scattering patterns exhibit a weak single diffraction peak at $q = 0.68$ and $0.66\,nm^{-1}$ for **SBA1a** and **SBA1b** respectively, characteristic of the wormlike structure (Fig. 1) with a regular pore-pore distance [15]. The corresponding *d* value was calculated to be 9.2 and 9.5 nm respectively.

The N_2 adsorption-desorption measurements (Fig. 2) showed a type IV isotherm, characteristic of mesoporous materials with BET surface areas of 303 and 608 $m^2.g^{-1}$ for **SBA1a** and **SBA1b** respectively, and narrow pore size distributions centered at 3.7 and 5.9 nm.

The incorporation of **1** in the mesoporous materials and the removal of surfactant were shown by solid-state NMR spectroscopies. ^{13}C CP-MAS NMR spectrum of **SBA1a** (Fig. 3) exhibits two signals at 63.13 (**CH_2**-O) and 15.94 ppm (**CH_3**-CH_2), revealing the presence of diethylphosphonate groups. The spectrum exhibits three additional signals attributed to the propyl spacers at 69.86 (**CH_2**O), 22.89 (**CH_2**), and 7.99 ppm (**CH_2**Si). Seven signals in the aromatic area were characteristic of diazodiphenyl unit.

Solid-state ^{29}Si CP-MAS NMR spectrum of **SBA1a** displayed signals at −101.32 and −110.07 ppm attributed to the substructures Q^3 and Q^4, respectively, denoting high cross-linking of the silica. One additional peak at −66.58 ppm assigned to the substructure T^3 [C-Si(OSi)$_3$], indicates the high cross-linking of the organosilsesquioxane species.

The solid state ^{31}P NMR spectrum of **SBA1a** and **SBA1b** displays one signal at 19.20 ppm proving that the diethylphosphonate groups were not modified during the sol–gel process.

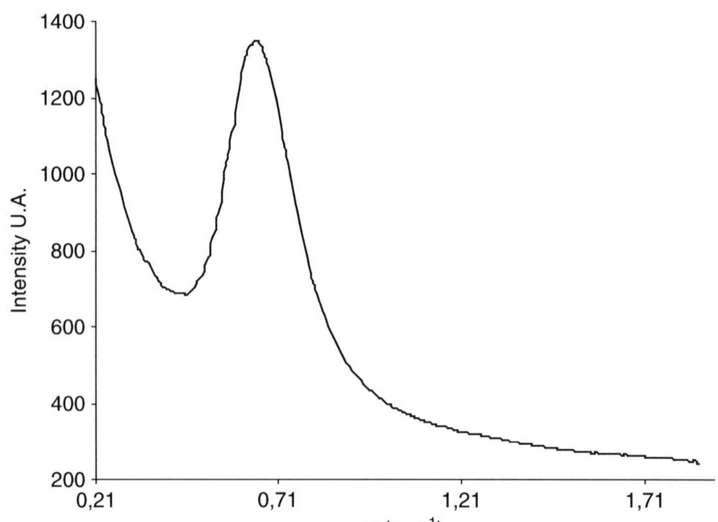

Fig. 1 SAXS pattern of **SBA1b**

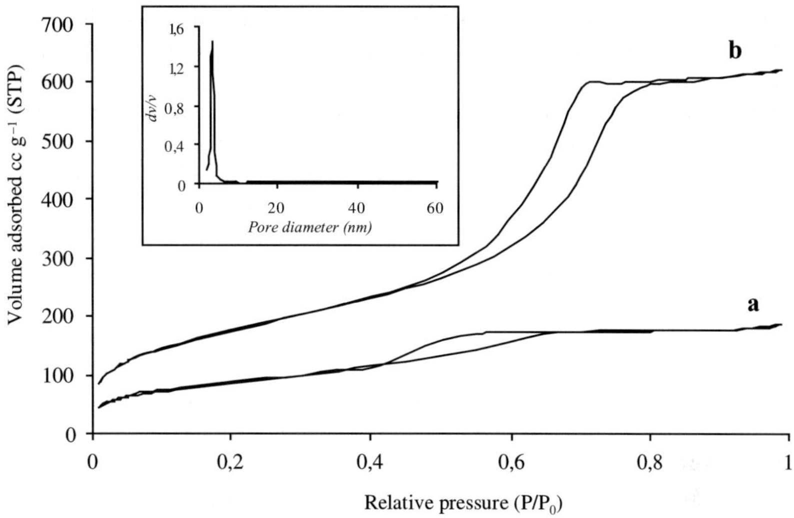

Fig. 2 N_2 adorption-desorption isotherm of **SBA1a** (**a**) and **SBA1b** (**b**). The inset shows the BJH pore size distribution plot for **SBA1a**.

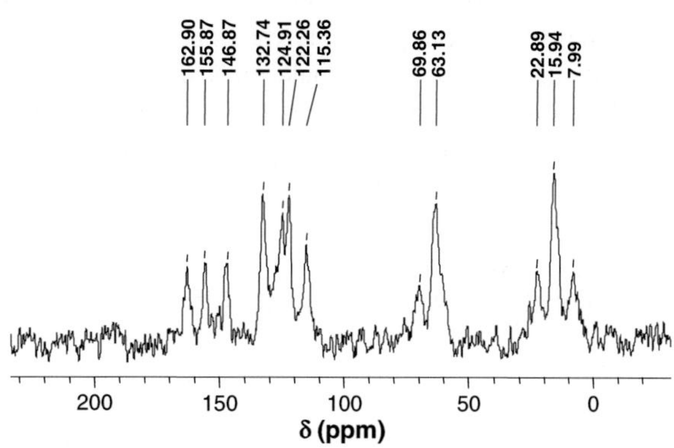

Fig. 3 ^{13}C CP/MAS NMR spectrum of **SBA1a**

The results of the elemental analysis of the final material led to the experimental formula $C_{19}H_{29}N_2PSi_{10}$, the calculated formula being $C_{19}H_{24}N_2PSi_{10}$, which indicates that the loading in azobenzene diethylphosphonate is close to the expected value.

A standard procedure [16] was adapted for the preparation of functional ordered organosilica thin films. The initial sol was prepared in two times. First, a pre-hydrolysed hybrid sol was prepared by mixing 1.66 g of TEOS (7.98 mmoles), 0.23 g (0.42 mmoles) of **1**, 1.0 g of ethanol and 1.25 g of H_2O (pH = 1.5). The resulting mixture was stirred at room temperature for 1h. A second solution containing 0.6 g of P123 and 27.2 g of EtOH was then added to the sol. After 2 h of stirring, 2.0 g of H_2O (pH = 1.5) was added. Thin films were then prepared by dip-coating float glass substrates in the solution at a constant withdrawal speed of 14 cm *per* minute in a sealed cabinet in which the relative humidity was maintained at 70% and the temperature at 25°C. The resulting thin film called $\mathbf{F}_{SBA1b}$ was obtained after 10 min of aging time under the same conditions. The surfactant template was selectively removed by solvent extraction using Soxhlet apparatus and EtOH under reflux for 6 h.

The structure of the thin films was characterized by both GISAXS measurements and X-ray reflectivity (XRR). GISAXS measurements were carried out at Brookhaven National Laboratory at the X22B beam line of the NSLS. The incident beam had a fixed energy, E = 8 keV. It was impinging on the surface of the sample at an angle of incidence of about 0.25° (slightly above the critical angle of incidence of the glass substrate) to fully penetrate inside the thin film. The GISAXS patterns were collected on a MARCCD detector located 73 cm after the sample and the specularly reflected beam was attenuated with a beam stop to avoid detector saturation. XRR measurements were performed on the Philips reflectometer of LPEC laboratory (Université du Maine, Le Mans). The X-ray beam coming out of a copper tube was monochromatized with a 0002 graphite monochromator.

Figure 4b shows the GISAXS pattern measured after surfactant extraction. Surprisingly, it can be seen that the pattern exhibits a lamellar symmetry.

The XRR (Fig. 4.a) shows the scattering features along the specular direction with a dynamical range of more than six orders of magnitude. All the films exhibit Bragg reflections located at almost the same wave vector transfer. A rough estimation from the Bragg peak positions yields d_{100} = 4.8 nm and d_{200} = 2.4 nm confirming the lamellar structure while the bulk material was not organized.

Then, we attempted to prepare a 2D hexagonal mesoporous organosilica thin film containing **1** in the channel pores and an erbium complex **5Er** (the synthesis of which was previously described [17]) in the framework (Scheme 3).

Indeed, inorganic salts have been used to improve the hydrothermal stability [18–20], control the morphology [21–23], extend the synthesis domain [24], and tailor the framework porosity [25, 26] during the formation of mesoporous

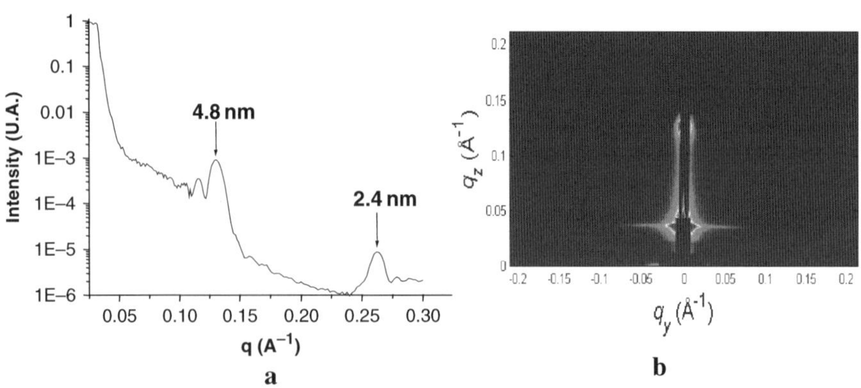

Fig. 4 X-ray reflectivity curve (**a**) and GISAXS pattern (**b**) showing the lamellar structure of the film $\mathbf{F}_{SBA1b}$

Scheme 3 Erbium complex **5Er** introduced in the framework

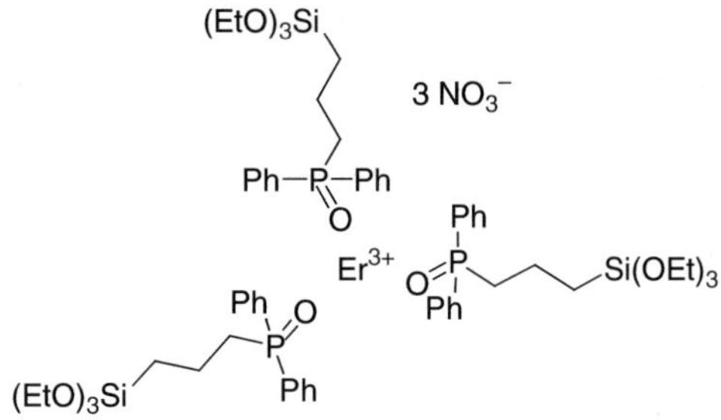

materials. Our goal was to combine the inorganic salt effect and the control of the structure of the organosilica in order to obtain a bifunctionnal (channel pore/framework) mesoporous organosilica thin films. The thin film named $\mathbf{F}_{\mathbf{SBA1,5Er}}$ was prepared from precursors **1** and **5** according to the same procedure as this used for $\mathbf{F}_{\mathbf{SBA1b}}$.

Figure 5b shows the GISAXS pattern of $\mathbf{F}_{\mathbf{SBA1,5Er}}$ measured after surfactant extraction. It can be seen that this pattern exhibits the 2 D hexagonal symmetry characteristic of the *p6m* space group. The XRR (Figure 5.a) shows the scattering features. A rough estimation from the Bragg peak positions yields $d_{10}=6.5$ nm and $d_{20}=3.4$ nm confirming the 2D hexagonal structure.

These results can be attributed to the specific effect of the erbium salt on the self-assembly interaction between surfactant head-groups and inorganic species [27].

In non-ionic surfactant templating systems, the effect of ionic strength has also been studied [28, 29]. Generally, ionic interactions are emphasized in these reports, which leads to better order either in mesostructure or morphology [30]. In our case, we obtained the same result, but this is the first time to our knowledge, that an organometallic complex acts as an inorganic salt in the structuration of an organosilica.

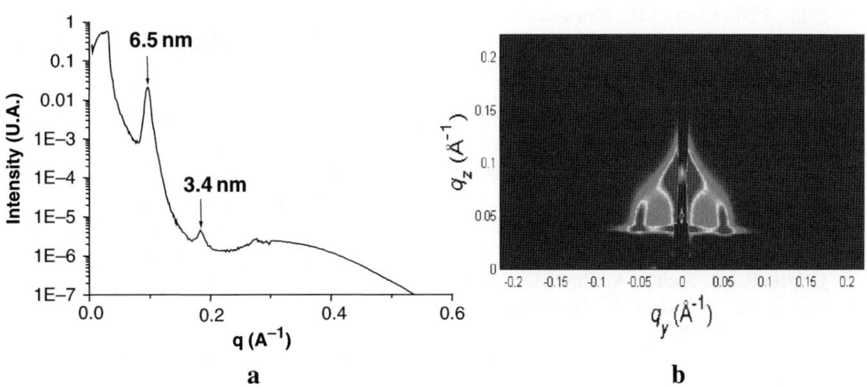

Fig. 5 X-ray reflectivity curve (**a**) and GISAXS pattern (**b**) showing the 2D-hexagonal structure of the film $\mathbf{F}_{\mathbf{SBA1,5Er}}$

In conclusion, we first described the one step synthesis of nanoporous organosilica powders functionalized with a Non Linear Optical chromophore in the channel pore (azobenzene diethylphosphonate). These powders have been totally characterized and present a meso-porosity with a wormlike structure. Thin films monofunctionalised with NLO chromophore in the channel pores or bifunctionalized (NLO in the channel pores/erbium salt in the framework) were also prepared in one step and characterized by Grazing Incidence Small angle X-ray Scattering (GISAXS) and X-ray reflectivity. Interestingly, while the monofunctional thin film $\mathbf{F}_{\mathbf{SBA1b}}$ presents a lamellar structure, the bifunctional one $\mathbf{F}_{\mathbf{SBA1,5}}$ (channel pores/framework) presents a 2D-hexagonal phase having p6m symmetry. Thus, the GISAXS experiments highlighted a salt effect on the final structure of the thin films. This is a clear evidence that a complex could act as an inorganic salt in the structuration of mesoporous films. Further investigations coupling other functional organic groups and complexes are now in progress in order to attempt a generalization of this effect. Our synthetic approach is a route to produce interactive materials in thin films, i.e. materials simultaneously coupling two properties at the nanometric scale.

Acknowledgements The authors thank Dr Philippe Dieudonné (LCVN, URM 5587 Montpellier, France) for powder SAXS measurements, the CNRS and the 'Université Montpellier II' for financial support. Work at Brookhaven National Laboratory was supported by U.S. DOE Contract No. DE-AC02–98CH10886.The authors wish to thank B. Ocko and the members of the X22 beam line for their support during the GISAXS measurements.

References

1. Sanchez, C., Julian, B., Belleville, P., Popall, M. *J. Mater. Chem.*, **2005**, 15, 3559.
2. Zhang, W., Froba, M., Wang, J., Tanev, P.T., J. Wong, J., Pinnavaia, T.J. *J. Am. Chem. Soc.*, **1996**, 118, 9164;
3. Mercier, L., Pinnavaia, T.J. *Adv. Mater.*, **1997**, 9, 500.
4. Cauvel, A., Renard, G., Brunel, D. *J. Org. Chem.*, **1997**, 62, 749.
5. Price, P.M., Clark, J.H., Macquarrie, D.J. *J. Chem. Soc., Dalton Trans.*, **2000**, 101.
6. Macquarrie, D.J., Jackson, D.B., Mdoe, J.E.G., Clark, J.H. *New J. Chem.*, **1999**, 23, 539.
7. Walcarius, A., Delacôte, C. *Chem. Mater.*, **2003**, 15, 4181.
8. Mouawia, R., Mehdi, A., Reyé, C., Corriu, R.J.P. *New J. Chem.*, **2006**, 1, 1077.
9. Brinker, C. J., Scherer, G.W. *Sol-Gel Science*, Academic Press: London, **1990**.
10. Sanchez, C., Lebeau, B., Chaput, F., Boilot, J-P. *Adv. Mater.*, **2003**, *15* (23), 1969.
11. Penard, A.L., Gacoin, T., Boilot, J.P. *Acc. Chem. Res.*, **2007**, *40*, 895.
12. Samyn, C., Verbiest, T., Persoons, A., *Macromol. Rapid Commun.*, **2000**, 21, 1.
13. Cui, Y., Qian, G., Chen, L., Wang, Z., Wang, M. *Macromol. Rapid Commun.*, **2007**, 28, 2019.
14. Hirao, T., Masunaga, T., Yoshiro, O., Agawa, T. *Synthesis*, **1981**, 56.
15. Corriu, R.J.P., Mehdi, A., Reye, C., Thieuleux, C. *Chem.Mater.*, **2004**, *16*, 159.
16. Mehdi, A., Dourdain, S., Bardeau, J.-F., Reyé, C., Corriu, R.J.P., Gibaud, A. *J. Nanosci. nanothecnol.*, **2006**, 377.
17. Besson, E., Mehdi, A., Reye, C., Corriu, R. J. P. *J. Mater. Chem.*, **2006**, 16(3), 246.
18. Ryoo, R., Jun, S. *J. Phys. Chem. B*, **1997**, *101*, 317.
19. Kim, J. M., Kim, S. K., Ryoo, R. *Chem. Commun.*, **1998**, 259.
20. Kim, J. M., Jun, S., Ryoo, R. *J. Phys. Chem., B* **1999**, *103*, 6200.
21. Zhao, D., Yang, P., Chmelka, B. F., Stucky, G. D. *Chem. Mater.* **1999**, *11*, 1174.

22. Zhao, D., Sun, J., Li, Q., Stucky, G. D. *Chem.Mater.*, **2000**, *12*, 275.
23. Yu, C.; Tian, B.; Fan, J.; Stucky, G. D.; Zhao, D. *J. Am. Chem. Soc.* **2002**, *124*, 4556.
24. Yu, C., Tian, B., Fan, J., Stucky, G. D., Zhao, D. *Chem. Commun.*, **2001**, 2726.
25. Newalkar, B. L., Komarneni, S. *Chem. Mater.*, **2001**, 13, 4573.
26. Newalkar, B. L.; Komarneni, S. *Chem. Commun.* **2002**, 1774.
27. Lukens, W. W., Schmidt-Winkel, P., Zhao, D., Feng, J., Stucky, G. D. *Langmuir*, **1999**, 15, 5403.
28. Zhang, W., Glomski, B., Pauly, T.R., Pinnavaia, T.J. *Chem. Commun.*, **1999**, 1803; Prouzet, E., Cot, F., Nabias, G., Larbot, A., Kooyman, P., Pinnavaia, T.J. *Chem. Mater.*, **1999**, 11, 1498.
29. Bagshaw, S.A. *J. Mater. Chem.*, **2001**, 11, 831.
30. Olkhovyk, O., Jaroniec, M. *J. Am. Chem. Soc.*, **2005**, 127, 60.

Self-Association in Hybrid Organic-Inorganic Silicon-Based Material Prepared by Surfactant-Free Sol-Gel of Organosilane

Bruno Boury

Abstract This brief review presents some of the important results obtained in the field of a control of the organisation of hybrid materials prepared by the Sol-Gel process of precursors of the general formula $RSi(O_{1.5})_n$. The different strategies involved are based on the used of weak supramolecular interactions during the hydrolysis/polycondensation process.

1 Introduction: The Sol-Gel Process for Silicon-Based Organic-Inorganic Materials

The Sol-Gel process is a well-known route of inorganic polycondensation now taught at the Master level [1–3]. Using the concept of "chimie douce", this general approach can be applied to silicon based-material. Indeed, the mild conditions required for the formation of the Si-O-Si framework allows the presence of organic species which can be introduced as a doping agent, then, the material can be assimilated to a nanocomposite (Scheme 1) [4–6]. Alternatively the organic part can be a specific moiety covalently bonded to the silicon with the general formula $R[SiX_3]n$ (X=Cl, OR, H,... and n>1). A homogeneous distribution of the organic and inorganic part is always achieved because the covalent bonds linking the Si-O-Si network and the organic parts are introduced via the structure of the monomer [6].

Development of the organosilicon chemistry allows the access to very different monomers for this approach. In all cases, the objective is to combine the qualities of the Si-O-Si network (vitreous structure, transparency, thermal and chemical stability) with those of the organic entities [7].

B. Boury
Institut Charles Gerhardt Montpellier, UMR 5253 CNRS-UM2-ENSCM-UM1, CMOS, Place E. Bataillon, 34095 Montpellier, France
e-mail: boury@univ-montp2.fr

F. Ganachaud et al. (eds.), *Silicon Based Polymers*,

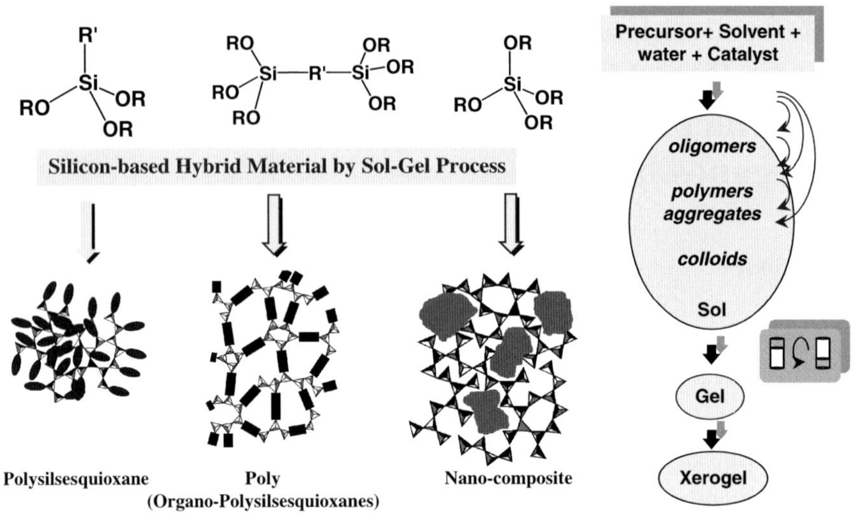

Scheme 1 General overview of the different possibilities that offers the Sol-Gel process for the preparation of silicon-based hybrid material (See also Plate 17 in the Colour Plate Section)

For a given precursor, the characteristics of the solid (composition, structure, porosity and morphology) rely on the choice of the experimental parameters that can control the kinetics of the reactions, therefore these solids are considered as kinetically controlled materials [8–11]. Concerning their organization and structure, it is obvious that the presence of an organic moiety in the precursors has brought into this domain of researches an important dose of supramolecular chemistry. Probably because silica prepared by Sol-Gel chemistry is an amorphous material, these silicon-based hybrid materials were considered initially amorphous too. Meanwhile, the presence of the R group attached to the silicon, not only may introduce important modifications of the reactivity compared to TEMOS or TEOS, it also comes with the possibility of weak interactions between these organic parts. Thus, compared to amorphous silica, the hybrid materials made of R-$[SiO_{1.5}]_n$ units have the possibility to form systems with different organization either anisotropic or isotropic, at a long or a short range order (Scheme 2). This short review focuses on the structure and organization process of such hybrid inorganic-organic materials, and is a non-exhaustive presentation of the progress of the concepts developed in this field. Note that we are talking here of materials prepared in the absence of any surfactant or structure directing agent, this type of materials have been recently reviewed by Inagaki et al [12].

The chemistry involved in the preparation of hybrid materials by the Sol-Gel process is *a priori* similar to the one used to prepare silica gel by the same process (Scheme 3) and it involves necessarily the formation of a Si-O-Si network and silanol groups.

The Si-O-Si linkage has very specific properties. For example, in polysiloxanes [13–16], the structural parameters that determine their general behavior are (a) the relatively high length of the Si-O bond (1.64 Å), (b) an important free volume by segment (c) the conformational flexibility of the polysiloxane bridge (Si-O-Si can vary

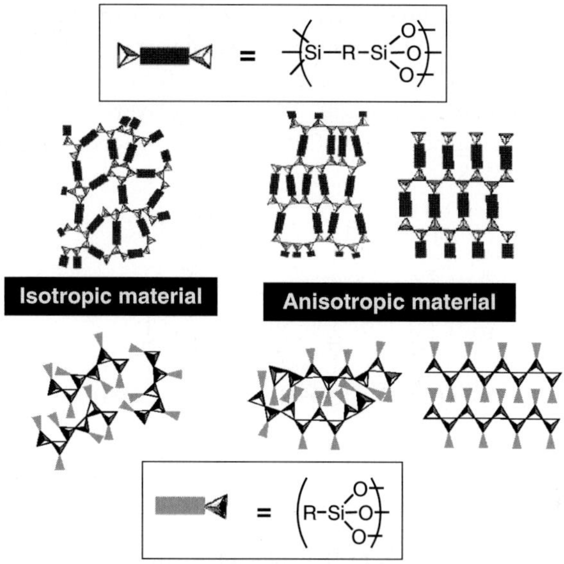

Scheme 2 Various possibilities of organisation and structure for silicon-based hybrid materials (See also Plate 18 in the Colour Plate Section)

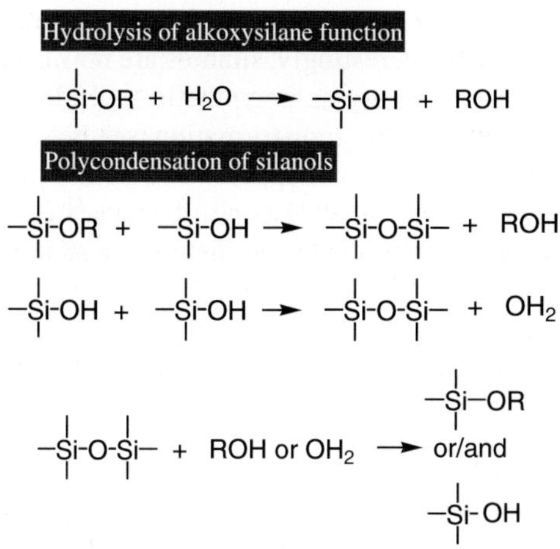

Scheme 3 General chemistry at silicon involved during the Sol-Gel process

from 104° to 180°) and (d) the weak interactions between segments (the activation energy for viscous flow is below 40 KJ.mol^{-1}). In this situation, almost all the di-substitued polysiloxanes exhibit thermotropic liquid crystal behavior, revealing their ability to self-organize as mesomorphous phases [17]. This liquid crystal behavior is not only observed when mesogenic groups are used as substituents at the silicon, forming the so-called side-chain and main-chain polysiloxanes. Indeed, it is also observed for these polysiloxanes with “simple” alkyl groups and it is proposed to adopt a columnar structure. On this subject see also the article by Hurduc et al. in this book.

Molecules with silanol functions are known to form supramolecular arrays in the solid state (linear polymeric chains, 2D-sheets or layers, 3D-networks) based on intermolecular H-bonding between Si-OH groups like the one described in Scheme 4 for dialkyl-silanediol [18].

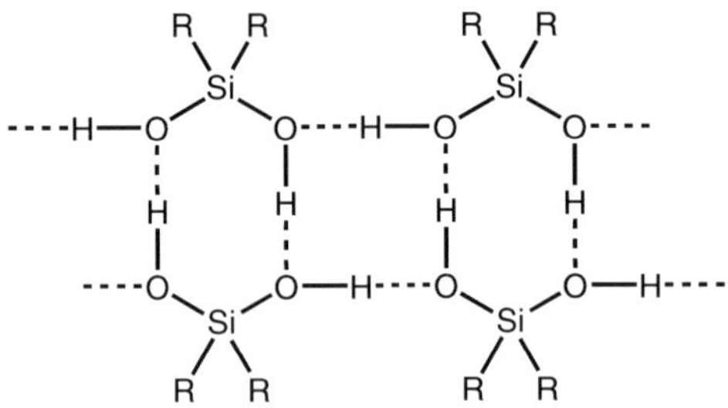

Scheme 4 Example of H-bonging network in disilanol structure

Recently, the possibility of Si-O-H... phenyl hydrogen-bonds was discovered in arylsilanols [19]. However, the nature and the size of the organic group attached to silicon can limit the development of such networks and lead to an association of a limited number of units forming dimers, trimers,...

Therefore, silanol species can be a driving force of the resulting organization in the hybrid material. More interestingly, silanols are now used as *reactive supramolecular functions*, that is to say as a synthetic tool allowing the association of molecular units and their further transformation, see below. On the other hand, these self-assembling properties do not appear to be clearly predictable, especially because in the Sol-Gel process other H-bonding species (mainly alcohol, water and sometimes solvent) are present and can disturb the silanol self-association.

2 The case of Polysilsesquioxanes R-$[SiO_{1.5}]_n$ n=1

Polysilsesquioxanes of general formula R-$[SiO_{1.5}]_n$ n=1 are generally prepared by hydrolysis/polycondensation of the corresponding precursors $RSiX_3$ (X=Cl, OR, H). Four basic structures can be envisaged: molecular compound with a cage structure, polymer with a regular ladder structure, polymer with a 3-D disordered structure and lamellar solid.

2.1 Cage-like structures (POSS)

Cage-like structures (POSS) are attractive building blocks (Scheme 5), their organization in the solid is controlled by the shape of these building blocks and mostly by the supramolecular interactions between the organic groups attached to the silicon atoms, some of them exhibiting thermotropic liquid crystal behavior [20,21]. Now

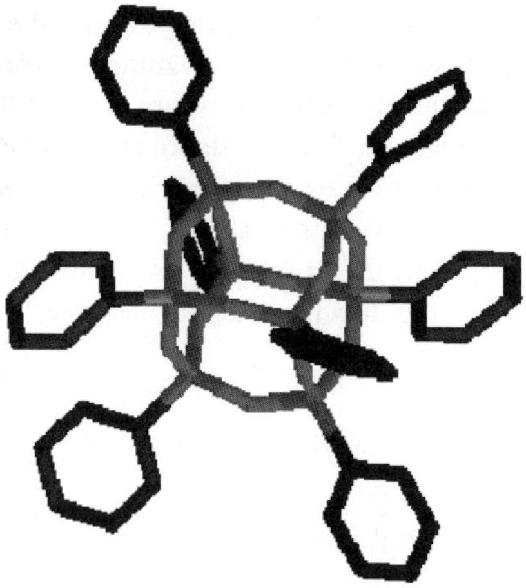

Scheme 5 Cage-like structure of T_8 Cube polysilsesquioxane (See also Plate 19 in the Colour Plate Section)

research in this field is directed towards the preparation of cages with different structures [22] (see also the article of Yoshida et al. in this book).

2.2 Ladder silsesquioxane polymers

Ladder silsesquioxane polymers are very unique in the polymer family since they can be considered as an intermediate structure between a 1D macromolecular chain and a 2D lamellar structure (Scheme 6). Indeed, theoretically, polysilsesquioxanes can adopt such structures. However, their preparation in high yield and leading unambiguously to the precise ladder structure is very recent although they have been postulated frequently in the past. Matsumoto et al. succeeded in their

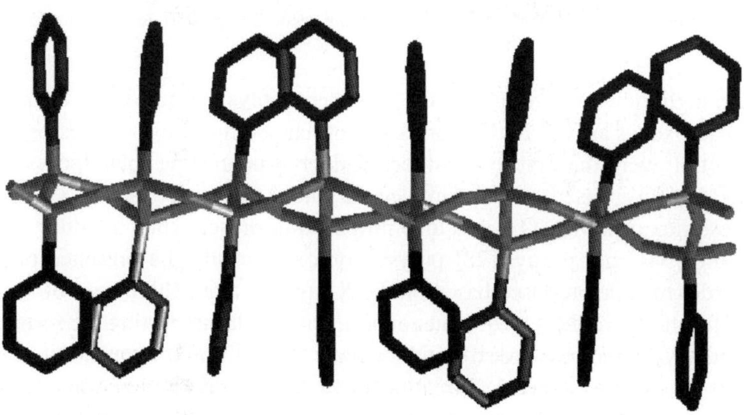

Scheme 6 Ladder-like structure of polysilsesquioxane (See also Plate 20 in the Colour Plate Section)

preparation by oxidation of polysilanes [23] or by a step by step process [24]. They have also reported the possibility to take advantage of the reactivity of silanol-containing precursors whose silanol functions are condensed [25].

Zang et al used such a reactivity of the silanols but while also taking advantage of the self-organization promoted by this function in the solid-state structure of the precursor. Combined with the presence of organic groups with strong self-organization ability at silicon, they obtained the solid-state condensation of the pre-organized precursor leading directly to the formation of the ladder structure [26].

2.3 *Polymers with a 3-D disordered structure*

Polymers with a 3-D disordered structure are systems described as resins with a disordered cross-linked polymeric structure (Scheme 7).

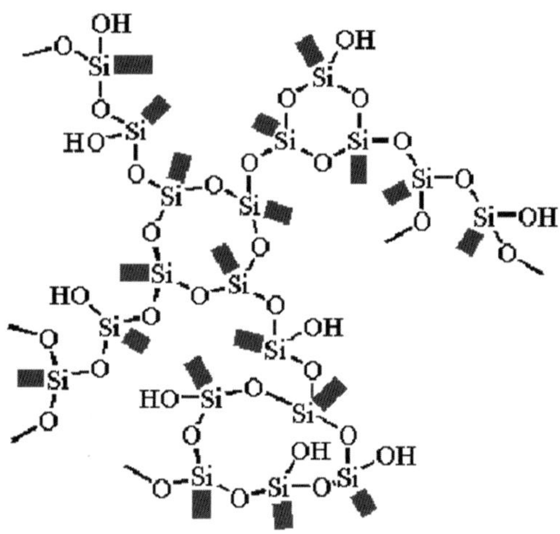

Scheme 7 Idealized structure for polysilsesquioxane resin (See also Plate 21 in the Colour Plate Section)

Such high molecular weight solids like phenylpolysilsesquioxanes or methylpolysilsesquioxanes are often used for matrix of nanocomposites or as resins [27–30]. Brown et al. were the first to consider a ladder structure for phenylpolysilsesquio xanes $C_6H_4SiO_{1.5}$ based on macromolecular and X-ray analysis [31–33]. Methylp olysilsesquioxanes $MeSiO_{1.5}$ was also proposed to present such a ladder structure. For hydro [34], and propyl [28] polysilsequioxanes only the presence of a short range order may be proposed based on the X-ray data that exhibit a set of few broad signals for these products [30]. Other experimental data supporting the formation of these laddered structures are still under debate. In addition the preparation of these products require some chemical treatment and "equilibration" reactions. It suggests that a reorganization process is required and the mixture of macromolecular species associating cage and ladder structures is apparently always obtained [28]. Loy et al. have extensively studied the ability of this system to form gels, resins or crystals [36].

2.4 Lamellar structures

Lamellar structures (Scheme 8) obtained by hydrolysis/polycondensation of organo-silicon precursors are generally the result of a segregation between polar and non polar part of a precursor already set up in it: hybrophobic alkyl chain on one hand and alkoxy or halogeno as "masked-silanol" on the other hand. That was the case with the first polysilsesquioxane organized on a long-range order reported by Allara et al. using liquid octadecyltrichlorosilane $C_{18}H_{37}SiCl_3$ [37].

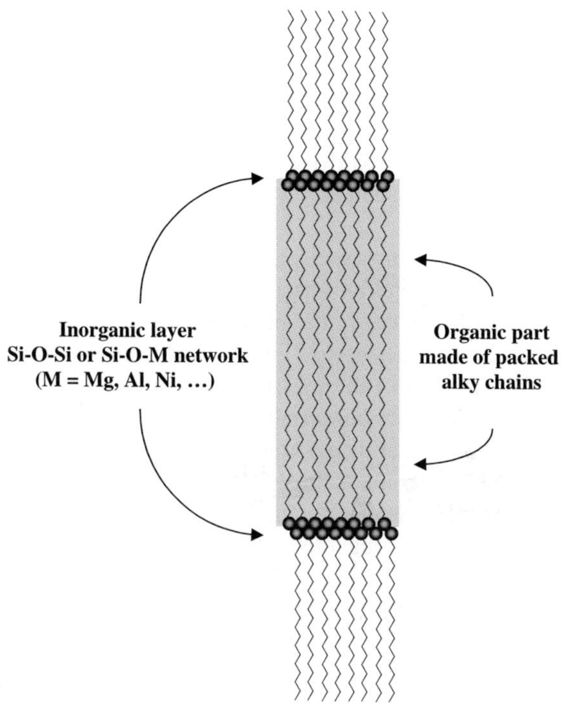

Scheme 8 Idealised lamellar structure of polysilsesquioxanes (See also Plate 22 in the Colour Plate Section)

Its hydrolysis in the absence of solvent by direct contact with water leads to the formation of flakes of octadecylpolysilsesquioxane with a highly organized lamellar structure. It consists of periodically alternated layers, an"organic" one made of alkyl chains and an "inorganic"one made of siloxy backbones plus residual silanols and physi-sorbed water. The high level of organization in these solids is proven by the positioning of the alkyl chains between each other in the "organic layers", this packing is close to the one reported for alkyl chains of crystallized octadecane. Similarly, Da Fonseca et al demonstrated that layered structures based on a polysilsesquioxane of the formula R-$SiO_{1.5}$ but combined with Si-O-M network, (M=Mg [38–41], Al [38,39], Cu [41] or Ni [42]), can be obtained by mixing the organosilane precursor with metal salts.

Numerous contributions by Kuroda et al. have established that a layered 2D-network can be obtained by hydrolysis/polycondensation of a trialkoxysilane with long alkyl chains [43–50]. Like for other layered structures, the architecture of the solid consists of a packing of alternated layers, one made of a siloxy backbone to which is covalently attached the other one, the "organic layer", made of octadecyl groups. Here again, we found the silanols as one of the driving force of the organization of the system. The formation during the hydrolysis/polycondensation process leads to $RSiX_n(OH)_{(3-n)}$ (X=OMe or Cl) entities with a "surfactant" character that is at work for promoting the resulting structure. The formation of a layered structure is assumed to occur either before or simultaneously with the polycondensation of the silanol group. Kuroda's group is currently working on this approach.

Layered materials can also be prepared from more functional precursors of the general formula A-CH_2-CH_2-CH_2-Si$(X)_3$ (X=Cl or OR) where A can be for example amino groups. White et al reported the formation of fibrous materials using H_2N-CH_2-CH_2-CH_2-Si$(OEt)_3$ (APS) [51]. The formation of carbamates by reaction with CO_2 from air and their association with Na^+ ions is assumed to result in the observed lamellar structure.

Based on the formation of ammonium carbamate salts, Corriu et al. have obtained the formation of layered material by self-assembling of monosilylated monomers. Starting from H_2N-$(CH_2)_n$-Si$(OMe)_3$, the method consisted of using the reversible reaction between CO_2 and amines groups giving rise to ammonium carbamate salts. The hydrolytic sol–gel process of these carbamates provided hybrid materials with lamellar structure containing ammonium carbamate salts (Scheme 9). Subsequently, the elimination of CO_2 upon heating generated materials with free amino groups in which the long-range order was maintained. The structuration appears dependent on both the ionic interaction between ammonium carbamate pairs and also on van der Waals interactions between the long alkylene chains [52,53].

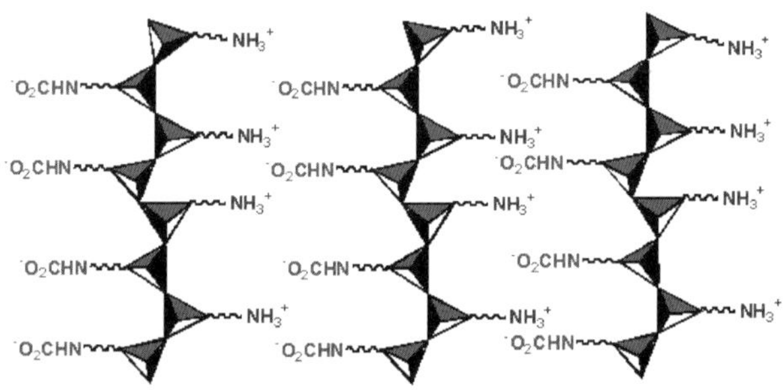

Scheme 9 Idealised structure of the assembly of polysilsesquioxanes layers through carbamate/ammonium interactions (See also Plate 23 in the Colour Plate Section)

Corriu's group has reported another possibility to form layered materials by direct hydrolyis/polycondensation of functional monosilylated organosilanes. In the case of CN-$(CH_2)_n$-Si$(OEt)_3$ hydrolysis/polycondensation of the ethoxy groups simultaneously with the formation of the carboxylic functions from the cyano groups lead to a 2D lamellar structure (Scheme 10) [54]. Strong hydrogen-bonding between carboxylic groups along with the Sol-Gel process is assumed to promote the structuration. Similar systems have been developed that combined the presence of diazo-goups **[55]**.

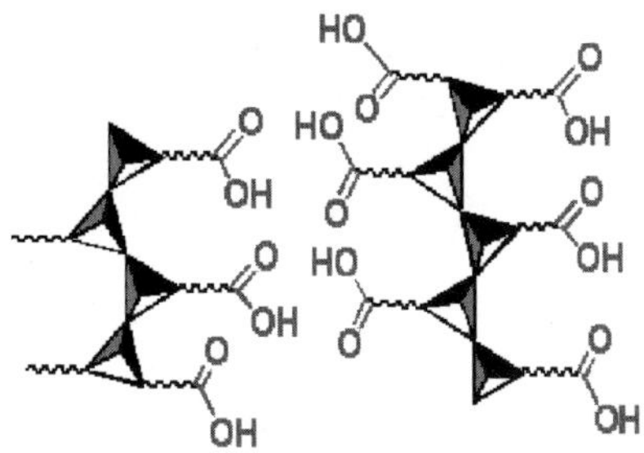

Scheme 10 Idealised structure of the assembly of polysilsesquioxanes layers through H-bonds between carboxylic acid groups (See also Plate 24 in the Colour Plate Section)

Moreau et al. took advantages of hydrophobic interactions and hydrogen-bonding between adenine-containing and thymine-containing monosilylated precursors (Scheme 11). Lamellar structures are obtained with an interlayers distance in agreement with the length of the sum of the organic moieties [56].

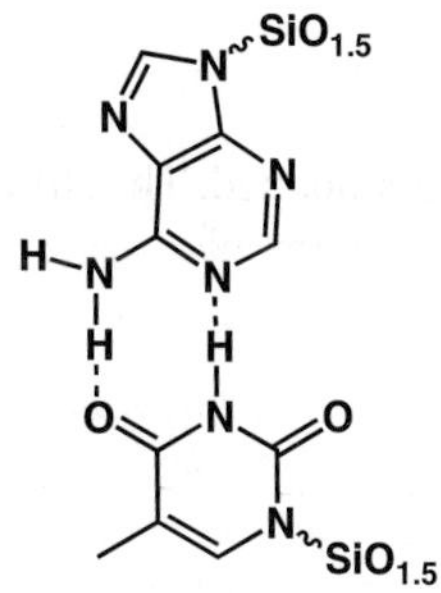

Scheme 11 Self association through H-bonds between adenine-containing and thymine-containing monosilylated precursors

By a similar approach based on H-bonding supramolecular interaction between an adenine- and an uracil-containing precursors, Barboiu et al obtained materials with a micrometric plate-like morphology with a 1-dimensional periodicity of 3.5 Å [57].

3 Organo-Polysilsesquioxanes R-$[SiO_{1.5}]_n$ n>1

The organo-polysilsesquioxane solids have a polysilsesquioxane-based structure. These systems with precursor containing at least 2 –SiX_3 groups were first developed by Shea et al.[58], and by Corriu et al [59]. and many reviews have covered the topic of their synthesis and reactivity [7, 60–62].

Initially, the anisotropic organization in these materials was evidenced by three main sets of data. Organization in some of these solids was suggested by their reactivity in the solid state (polymerization [63–65] or complexation [66–72]).

Secondly, the broad signals observed by X-ray powder diffraction of the xerogel support the idea of an anisotropic organization of $R[SiO_{1.5}]_n$ ($n>1$), whose position and intensity depend on the nature of the precursors (Scheme 12) [73–75].

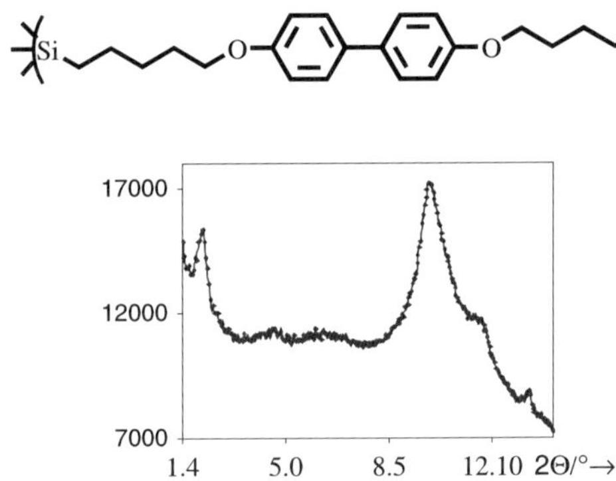

Scheme 12 X-ray diffraction patern of a powder of bridged-polysilsesquioxane with mesogenic group as a bridge

Finally, at the mesoscopic level, we reported the birefringence observed by microscopy in polarized light of these xerogels that demonstrates unambiguously the anisotropic organization of the medium as exemplified by Fig. 1 [73, 76–78].

A rather high level of organization can be achieved this way in certain cases, for example when carborane-containing precursors are used [79], or with biphenyl-containing precursors [78,80]. However, an organization at the long-range order in these solids requires a better control of the supramolecular forces than can self-associate the precursor during the Sol-Gel process to be achieved. For this, two points are important, first to avoid the use of an organic solvent as much as possible in order to favor any type of hybrophobic or aromatic interactions and secondly to introduce as much as possible a strong interaction between precursors.

Fig. 1 Images by microscopy in polarized light of three bridged-polysilsesquioxanes (See also Plate 25 in the Colour Plate Section)

In this idea, hydrophobic interactions alone require long alkyl chains to be efficient enough. Corriu et al have shown the effect of C12 to C20 chains on the formation of lamellar structures and the ability to achieve 2D hexagonal structures with C30 chains [81,82].

Other weak interactions like ππ-stacking interactions between aromatic groups can be used to promote self-association between precursor units, as such or as subunits of a mesogenic organic group [76]. In the case of an aromatic moiety exhibiting photophysical properties the organization of the solid after Sol-Gel process can result in preservation or very important modification of such properties compared to the monomer [83–85].

Obviously, H-bonds are also very efficient weak interactions to promote the self-association during the Sol-Gel process. Ureido groups appeared very efficient in doing this job as demonstrated by Moreau et al. [86–91] or Brinker et al. [92] The self-organization of the ureido-containing precursors (Scheme 13) during the Sol-Gel process leads to lamellar structured materials whose proposed structures are in agreement with MEB and X-ray powder diffraction analyses.

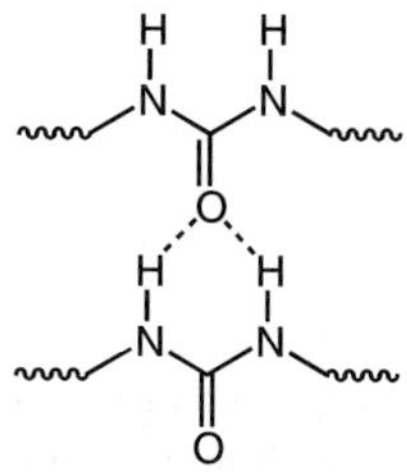

Scheme 13 H-bonds between subunits in an ureido-containing bis-silylated precursor or material

Alternatively, formation of highly organized materials is achievable by a new approach using a solvent-free reaction and taking advantage of the crystalline structure of the precursor. We investigated that approach by considering the hydrolysis/polycondensation of bis(trichlorosilyl)bibenzyle in the solid state by direct contact with water vapour (or liquid) [93]. The high reactivity of the Si-Cl function and the formation of HCl as a by-product prevent dissolution during the process. The crystalline structure of the precursor is not preserved during the process but it apparently serves as a scaffold for the building of the crystalline structure of the hybrid material. This is also achievable with crystallized alcoxysilanes [94].

Such approach of the synthesis by solid-state reaction is especially attractive with precursors that present silanol functions. Here again, the supramolecular association promoted by these functions combined with their reactivity allows to directly transform by a simple thermal treatment a molecular precursor to a polymeric hybrid material. Liu, et al. have reported on the possibility to polycondense silanols in the solid state [95], and Corriu et al. have obtained the formation of layered materials by condensation of bis-trisilanol precursors [96,97].

4 Conclusion

At the crossover between supramolecular and synthetic chemistry the hybrid materials represent a major field of science exploration and applications. In contrast with pure silica, the presence of the organic groups in R-$[SiO_{1.5}]_n$ (n>1) systems questions the possibility to have organization based on the presence of hydrophobic and hydrophilic parts.

The key to control the organization of the solid during the Sol-Gel process, i.e.: the design of the monomeric building blocks and the conditions in which they are assembled, are in the hands of the chemist. In these materials prepared in the absence of any template or structure directing agent, weak forces (H-bond, ππ-stacking, hydrophobic ...) are the driving forces of the organization of the materials. Similarly, these forces are also at work in mesoporous hybrid materials prepared by Sol-Gel process of organosilanes [98,99].

References

1. Brinker, C. J., Scherer, G. W. "*Sol-Gel Science*, Academic Press, Inc: Boston, 1990.
2. Jolivet, J.-P. "*De la solution à l'oxyde*, InterEditions: Paris, 1994.
3. Livage, J., Sanchez, C. *J. Non-Cryst. Solids* **1992**, *145*, 11.
4. Sanchez, C., Ribot, F. *New J. Chem.* **1994**, *Special issue*, 1007.
5. Sanchez, C., Lebeau, B. *Pure Appl. Opt.* **1996**, *5*, 689.
6. Judenstein, P., Sanchez, C. *J. Mater. Chem.* **1996**, *6*, 511.
7. Shea, K. J., Moreau, J. J. E., Loy, D. A., Corriu, R. J. P., Boury, B. in "*Functional Hybrid Material*", Gomez-Romero, P.Sanchez, C. Eds., Wiley-VCH, Weinheim, 2004.
8. Corriu, R. *Actualite Chimique* **2005**, *290–291*, 17.
9. Cerveau, G., Corriu, R. J. P., Framery, E. *Chem. Mater.* **2001**, *13*, 3373.
10. Cerveau, G., Corriu, R. J. P., Framery, E. *Chemical Communication* **1999**, *20*.
11. Cerveau, G., Corriu, R. J. P., Fischmeister-Lepeytre, C. *J. Mater. Chem.* **1999**, *9*, 1149.
12. Inagaki, S., Fujita, S. *Chem. Mater.* **2008**, *20*, 891.
13. Gemmell, P. A., Gray, G. W., Lacey, D. *Mol. Cryst. Liq. Cryst.* **1985**, *122*, 205.
14. Percec, V., Pugh, C. in "*Side Chain Liquid Crystals*", McArdle, C. B. Ed., Blackie, Glasgow, 1989.
15. Tschierske, C. *J. Mater. Chem.* **1998**, *8*, 1485.
16. Teyssié, D., Boileau, S. in "*Silicon-Containing Polymers*", Jones, R. G.Ando, W.Chojnowski, J. Eds., Kluwer Academic Publishers, Dordrecht, 2000, p 593.
17. Molenberg, A., Siffrin, S., Möller, M. *Macromol. Symp.* **1996**, *102*, 199.
18. Haiduc, I., Edelmann, F. T. "*Supramolecular Organometallic Chemistry*, Wiley-VCH: Weiheim, 1999.
19. Al-Juaid, S. A., Al-Nasr, A. K. A., Eaborn, C., Hitchcock, P. B. *J. Organomet. Chem.* **1992**, *C9*, 429.
20. Hanssen, R. W. J. M., van Santen, R. A., Abbenhuis, H. C. L. *Eur. J. Inorg. Chem.* **2004**, *4*, 675.
21. Laine, R. M. *J. Mater. Chem.* **2005**, *15*, 3725.
22. Hagiwara, Y., Shimojima, A., Kuroda, K. *Chem. Mater.* **2008**, *20*, 1147.
23. Unno, M., Tanaka, R., Tanaka, S., Takeuchi, T., Kyushin, S., Matsumoto, H. *Organometallics* **2005**, *24*, 765.
24. Unno, M., Suto, A., Matsumoto, H. *J. Am. Chem. Soc.* **2002**, *124*, 1574.
25. Unno, M., Matsumoto, T., Matsumoto, H. *J. Organomet. Chem.* **2007**, *692*, 307.

26. Zhang, Z.-X., Xie, P., Shen, Z., Jiang, J., Zhu, C., Li, H., Zhang, T., Han, C. C., Wan, L., Yan, S., Zhang, R. *Ang. Chem. Int. Ed. Engl.* **2006**, *45*, 3112.
27. Avnir, D., Klein, L. C., Levy, D., Schubert, U., Wojcik, A. B. in "*The chemistry of organic silicon compound*", Pappoport, Z.Apeloig, Y. Eds., John Wiley & Sons Ltd, 1998, Vol. 2, p. 2317.
28. Baney, R. H., Cao, X. "*in Silicon-Containig Polymers*, Kluwer Academic Publishers: Dordrecht, **2000**.
29. Baney, R. H., Cao, x. in "*Silicon-Containing Polymers"*, Jones, R. G.Ando, W.Chojnowski, J. Eds., Kluwer Academic Publishers, Dordrecht, **2000**, p. 157.
30. Baney, R. H., Itoh, M., Sakakibara, A., Suzuki, T., inclues, e. r. b. *Chem. Rev.* **1995**, *95*, 1410.
31. Brown, J. F., Prescott, P. I. *J. Am. Chem. Soc.* **1964**, *86*, 1402.
32. Brown, J. F. *J. Poly. Sci. C* **1963**, *1*, 86.
33. Brown, J. F., Vogt, L. H., Katchman, A., Eustance, J. W., Kaiser, K. M., Krantz, K. W. *J. Am. Chem. Soc.* **1960**, *82*, 6194.
34. Xie, Z., He, Z., Dai, D. *Chinese. j. Polym. Sci.* **1989**, *7*, 183.
35. Li, Z., Cao, X., Xu, H. *rect. Funct. Polym.* **2001**, *to be published.*
36. Loy, D. A., Baugher, B. M., Baugher, C. R., Schneider, D. A., Rahimian, K. *Chem. Mater.* **2000**, *12*, 3624.
37. Parikh, A., Schivley, M. A., Koo, E., Seshadri, K., Aurentz, D., Mueller, K., Allara, D. L. *J. Am. Chem. Soc.* **1997**, *119*, 3135.
38. Ukrainczyk, L., Bellman, R. A., Anderson, A. B. *J. Phy. Chem. B* **1997**, *101*, 531.
39. Silva, C. R., da Fonseca, M. G., Barone, J. S., Airoldi, C. *Chem. Mater.* **2002**, *14*, 175.
40. Jaber, M., Miehe-Brendle, J., Le Dred, R. *Chem. Lett.* **2002**, 954.
41. da Fonseca, M. G., Airoldi, C. *J. Mater. Chem.* **2000**, *10*, 1457.
42. da Fonseca, M. G., Silva, C. R., Barone, J. S., Airoldi, C. *J. Mater. Chem.* **2000**, *10*, 789.
43. Shimojima, A., Sugahara, Y., Kuroda, K. *J. Am. Chem. Soc.* **1998**, *120*, 4528.
44. Shimojima, A., Kuroda, K. *Chem. Lett.* **2000**, 1310.
45. Shimojima, A., Mochizuki, D., Kuroda, K. *Chem. Mater.* **2001**, *13*, 3603.
46. Shimojima, A., Kuroda, K. *Langmuir* **2002**, *18*, 1144.
47. Shimojima, A., Liu, Z., Ohsuna, T., Terasaki, O., Kuroda, K. *J. Am. Chem. Soc.* **2005**, *127*, 14108.
48. Shimojima, A., Wu, C., Kuroda, K. *J. Mater. Chem.* **2007**, *17*, 658.
49. Fujimoto, Y., Shimojima, A., Kuroda, K. *J. Mater. Chem.* **2006**, *16*, 986.
50. Fujimoto, Y., Heishi, M., Shimojima, A., Kuroda, K. *J. Mater. Chem.* **2005**, *15*, 5151.
51. Cabibil, H. L., Pham, V., Lozano, J., Celio, H., Winter, R., White, J. M. *Langmuir* **2000**, *16*, 10471.
52. Alauzun, J., Besson, E., Mehdi, A., Reye, C., Corriu, R. J. P. *Chem. Mater.* **2008**, *20*, 503.
53. Alauzun, J., Mehdi, A., Reye, C., Corriu, R. J. P. *J. Am. Chem. Soc.* **2005**, *127*, 11204.
54. Mouawia, R., Mehdi, A., Reye, C., Corriu, R. *J. Mater. Chem.* **2007**, *17*, 616.
55. Besson, E., Mehdi, A., Chollet, H., Reye, C., Guilard, R., Corriu, R. J. P. *J. Mater. Chem.* **2008**, *18*, 1193.
56. Moreau, J. J. E., Pichon, B. P., Arrachart, G., Wong Chi Man, M., Bied, C. *New J. Chem.* **2005**, *29*, 653.
57. Arnal-Herault, C., Barboiu, M., Pasc, A., Michau, M., Perriat, P., Van der Lee, A. *Chem. A Eur. J.* **2007**, *13*, 6792.
58. Shea, K. J., Loy, D. A., Webster, O. *J. Am. Chem. Soc.* **1992**, *114*, 3307.
59. Corriu, R. J. P., Moreau, J. J. E., Thépot, P., Wong Chi Man, M. *Chem. Mater.* **1992**, *4*, 1217.
60. Shea, K. J., Loy, D. A. *Chem. Mater.* **2001**, *13*, 3307.
61. Shea, K. J., Loy, D. A. *Acc. Chem. Res.* **2001**, *34*, 707.
62. Boury, B., Corriu, R. J. P. *Chem. Comm.* **2002**, 795.
63. Corriu, R. J. P., Moreau, J. J. E., Thépot, P., Chorro, C., Lèreporte, J. P., Sauvajol, J. L., Wong Chi Man, M. *Synth. Met.* **1994**, *62*, 233.
64. Corriu, R. J. P., Moreau, J. J. E., Thépot, P., Wong Chi Man, M. *Chem. Mater.* **1996**, *8*, 100.
65. Corriu, R. J. P., Moreau, J. J. E., Thépot, P., Wong Chi Man, M., Chorro, C., Lèreporte, J. P., Sauvajol, J. L. *Chem. Mater.* **1994**, *6*, 640.

66. Dubois, G., Reyé, C., Corriu, R. J. P., Chuit, C. *J. Mater. Chem.* **2000**, *10*, 1091.
67. Franville, A. C., Zambon, D., Mahiou, R., Chou, S., Troin, Y., Cousseins, J. C. *J. All. Comp.* **1998**, *275–277*, 831.
68. Franville, A.-C., Zambon, D., Mahiou, R. *Chem. Mater.* **2000**, *12*, 428.
69. Broudic, J.-C., Conocar, O., Moreau, J. J. E., Meyer, D., Wong Chi Man, M. *J. Mater. Chem.* **1999**, *9*, 2283.
70. Brethon, A., Hesemann, P., Rejaud, L., Moreau, J. J. E., Wong Chi Man, M. *J. Organomet. Chem.* **2001**, *627*, 239.
71. Raehm, L., Mehdi, A., Wickleder, C., Reye, C., Corriu, R. J. P. *J. Am. Chem. Soc.* **2007**, *129*, 12636.
72. Brandes, S., David, G., Suspene, C., Corriu, R. J. P., Guilard, R. *Chem. A Eur. J.* **2007**, *13*, 3480.
73. Boury, B., Corriu, R. J. P., Delord, P., Nobili, M., Le Strat, V. *Ang. Chem. Int. Ed. Engl.* **1999**, *38*, 3172.
74. Boury, B., Corriu, R. J. P., Delord, P., Le Strat, V. *J. Non-Cryst. Solids* **2000**, *265*, 41.
75. Schaefer, D. W., Beaucage, G., Loy, D. A., Shea, K. J., Lin, J. S. *Chem. Mater.* **2004**, *16*, 1402.
76. Ben, F., Boury, B., Corriu, R. J. P. *Adv. Mater.* **2002**, *14*, 1081.
77. Ben, F., Boury, B., Corriu, R. J. P., Delord, P., Nobili, M. *Chem. Mater.* **2002**, *14*, 730.
78. Ben, F., Boury, B., Corriu, R. J. P., Le Strat, V. *Chem. Mater.* **2000**, *12*, 3249.
79. Gonzalez-Campo, A., Boury, B., Teixidor, F., Nunez, R. *Chem. Mater.* **2006**, *18*, 4344.
80. Le Strat, V. *PhD dissertation* **2000**, Montpellier (France).
81. Alauzun, J., Mehdi, A., Reye, C., Corriu, R. J. P. *J. Mater. Chem.* **2005**, *15*, 841.
82. Alauzun, J., Mehdi, A., Reye, C., Corriu, R. J. P. *Chem. Comm.* **2006**, *3*, 347.
83. Nam, H., Granier, M., Boury, B., Park, S. Y. *Langmuir* **2006**, *22*, 7132.
84. Nam, H., Boury, B., Park, S. Y. *Chem. Mater.* **2006**, *18*, 5716.
85. Dautel, O. J., Wantz, G., Almairac, R., Flot, D., Hirsch, L., Lere-Porte, J.-P., Parneix, J.-P., Serein-Spirau, F., Vignau, L., Moreau, J. J. E. *J. Am. Chem. Soc.* **2006**, *128*, 4892.
86. Bantignies, J.-L., Vellutini, L., Sauvajol, J.-L., Maurin, D., Man, M. W. C., Dieudonne, P., Moreau, J. J. E. *J. Non-Cryst. Solids* **2004**, *345&346*, 605.
87. Bantignies, J.-L., Vellutini, L., Maurin, D., Hermet, P., Dieudonne, P., Wong, C. M. M., Bartlett, J. R., Bied, C., Sauvajol, J.-L., Moreau, J. J. E. *J. Phys. Chem. B* **2006**, *110*, 15797.
88. Moreau, J. J. E., Luc, V., Wong, C. M. M., Bied, C., Dieudonne, P., Bantignies, J.-L., Sauvajol, J.-L. *Chem. A Eur. J.* **2003**, *11*, 1527.
89. Moreau, J. J. E., Pichon, B. P., Wong, C. M. M., Bied, C., Pritzkow, H., Bantignies, J.-L., Dieudonne, P., Sauvajol, J.-L. *Ang. Chem. Int. Ed. Engl.* **2004**, *43*, 203.
90. Moreau, J. J. E., Vellutini, L., Wong Chi Man, M., Bied, C. *J. Am. Chem. Soc.* **2001**, *123*, 7957.
91. Moreau, J. J. E., Vellutini, L., Wong Chi Man, M., Bied, C. *J. Am. Chem. Soc.* **2001**, *123*, 1509.
92. Liu, N., Yu, K., Smarsly, B., Dunphy, D. R., Jiang, Y.-B., Brinker, C. J. *J. Am. Chem. Soc.* **2002**, *124*, 14540.
93. Boury, B., Ben, F., Corriu, R. J. P. *Ang. Chem. Int. Ed. Engl.* **2001**, *40*, 2853.
94. Boury, B., Muramatsu, H., Corriu, R. J. P. *J. Am. Chem. Soc* **2003**, *125*, 854.
95. Liu, L. Y., Xu, L., Hou, Z. J., Xu, Z. L., Chen, J., Wang, W. C., Li, F. M. *Phys. Let. A* **1999**, *262*, 206.
96. Cerveau, G., Corriu, R. J. P., Dabiens, B., Le Bideau, J. *Ang. Chem. Int. Ed. Engl.* **2000**, *39*, 4533.
97. Cerveau, G., Chappellet, S., Corriu, R. J. P., Dabiens, B. *Can. J. Chem.* **2003**, *81*, 1213.
98. Corriu, R. J. P., Mehdi, A., Reye, C. *J. Mater. Chem.* **2005**, *15*, 4285.
99. Fujita, S., Inagaki, S. *Chem. Mater.* **2008**, *20*, 891.

Part 3
Polysilanes

The Synthesis, Self-Assembly and Self-Organisation of Polysilane Block Copolymers

Simon J. Holder and Richard G. Jones

Abstract Block copolymers containing polysilane blocks are unique in that the polysilane components possess electro-active properties and are readily photodegradable. This review will discuss and assess the two major approaches to the synthesis of polysilane block copolymers via pre-formed polymer chain coupling and living polymerisation techniques. The self-organisation of polysilane block copolymers and the morphologies adopted in thin films are reviewed. Amphiphilic polysilane-containing block copolymers self-assemble in solvents selective for one block and a number of examples are highlighted. The versatility of these materials is highlighted by recent significant applications including the preparation of hollow cross-linked micellar aggregates in aqueous solutions and in patterned thin film generation subsequently employed as templates for the growth of cell cultures and $CaCO_3$.

Glossary

AFM = atomic force microscopy
ATRP = atom transfer radical polymerisation
PEO = poly(ethylene oxide)
PHEMA = poly(hydroxyethyl methacrylate)
PI = polyisoprene
PMAA = poly(methacrylic acid)
PMHS = poly(1,1-dimethyl-2,2-dihexyldisilene)
PMMA = poly(methyl methacrylate)
PMPS = polymethylphenylsilane

S. J. Holder
Functional Materials Group, School of Physical Sciences, University of Kent,
Canterbury, Kent. CT2 7NH. UK.

F. Ganachaud et al. (eds.), *Silicon Based Polymers*,

POEGMA = poly[oligo(ethylene glycol methyl ether) methacrylate]
PS = polystyrene
PTrMA = poly(triphenylmethyl methacrylate)
SCM = shell cross-linked micelle
SEM = scanning electron microscopy
TEM = transmission electron microsopy
TEMPO = 2,2,6,6-tetramethyl-1-piperidinyloxy

1 Introduction

Polysilanes are linear polymers with a backbone of catenated silicon atoms that are usually substituted with aryl and/or alkyl groups [1]. Polysilanes display electronic delocalization within the σ-bonded framework of the Si-Si backbone as a consequence of sp3 orbital interactions between Si atoms (the smaller dimensions of the hybrid orbitals of the carbon atoms preclude this delocalization in analogous unsaturated C-C polymer backbones). This electron delocalization plays a large part in determining the properties of polysilanes, in particular the spectroscopic and electroactive properties. As a consequence they have potential applications as semiconducting [2], photoconducting [3, 4], electroluminescent [5, 6], and non-linear optical materials [7]. They have found applications as precursors to β-SiC fibres [8, 9], as resists in microlithography [10, 11], and as photoinitiators of radical polymerization [12]. However, their mechanical properties are relatively poor, adversely affecting their processability and limiting their exploitation. In order to exploit their properties, attempts have been made to combine polysilanes with organic polymers with complementary characteristics in order to optimise mechanical properties. Whilst many structural properties can be optimised through the incorporation of two or more polymer components in a copolymer structure, block copolymers are also being increasingly studied as self-organising and self-assembling materials. The microphase separation of block copolymers gives a wide range of 3-dimensional morphologies in the bulk state [13] and the self-assembly of block copolymers in different liquid media give a variety of aggregate structures such as vesicles [14], micelles [15], and micellar fibres [16]. A number of applications have been demonstrated for block copolymer aggregate structures including their use as encapsulants for drug delivery [17–19] and templates for colloid synthesis [20]. It has been demonstrated, that the self-assembly and/or self-organisation of block copolymers might be used for the imposition of increased order to fine-tune the performance of conjugated polymers and facilitate the preparation of nano-scale devices [21–28]. Polysilanes as σ-conjugated polymers are therefore ideal targets for investigation as self-organizing and assembling block copolymer systems. Polysilanes are rod-like, they have semi-rigid, segmented backbones that ensue from their σ -conjugated chains and helical conformations [29, 30]. Thus, they offer a significant potential for self-alignment that tends to be similar to that for π-conjugated carbon based block copolymers. They also offer the possibility of bringing to block copolymer structures the characteristic properties that result from their σ-conjugation and, hence, access to new structure-property combinations through self-organization in bulk or self-assembly through aggregation. This paper will review the methodologies that have been developed for the synthesis of polysilane block copolymers, the resultant morphologies in thin films and their self-assembly in solution, and highlight some of their more interesting properties and applications.

2 Synthesis

2.1 Synthesis of Polysilane Blocks

Most but not all strategies for the synthesis of block copolymers based on polysilanes require pre-syntheses of the polysilanes that will comprise the blocks. Thereafter, the other copolymer component blocks are attached to the polysilane (a polymer coupling approach) or else grown from its appropriately functionalized chain ends (a living polymerization approach). There are four known procedures for the synthesis of polysilanes (Fig. 1), the Wurtz-type reductive-coupling of dichloroorganosilanes [31] (Reaction 1), the ring-opening of cyclosilanes [32] (Reaction 2), the anionic polymerization of 'masked' disilenes [33] (Reaction 3) and the catalytic dehydrogenation of primary silanes [34] (Reaction 4). For detailed descriptions of these procedures the reader is referred to recent reviews [35]. Of these approaches the catalytic dehydrogenation has yet found no application in the synthesis of block copolymers. This is principally because high molecular weight polysilanes are

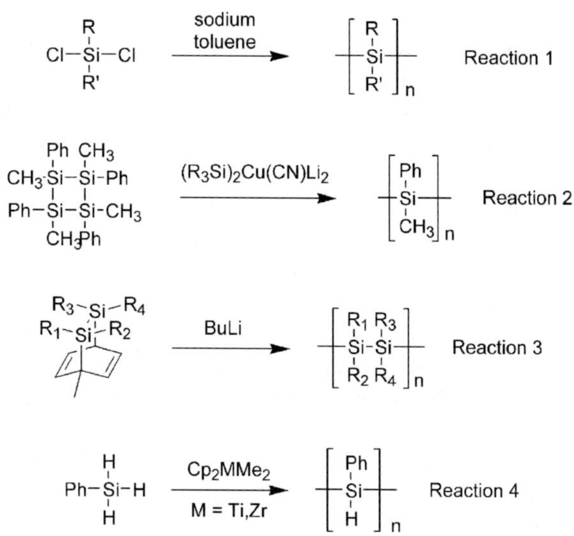

Fig. 1 Polysilane syntheses: the four main techniques

only obtained from the dehydrocoupling of tertiary organosilanes and the resultant polymers possess no discrete end-group functionalities for subsequent polymer coupling or polymerization initiation.

Both the anionic polymerization of masked disilenes and the ring-opening polymerization are living polymerizations under the appropriate conditions and hence after full monomer conversion they retain silyl anion end-functions suitable for block copolymer synthesis. Both of these polymerizations can give quantitative yields of polysilane with narrow molecular weight distributions and molecular weight parameters controlled by the ratio of monomer to initiator. However in both cases the monomers require careful and time-consuming multi-step syntheses. In contrast the Wurtz-reductive coupling polymerization under the appropriate conditions, can give polysilanes in high yields utilizing commercially available monomers and reagents [36, 37]. Furthermore the end-groups are predominantly Si-Cl bonds which are ideal for subsequent use in block copolymer syntheses. The principal disadvantage of the Wurtz-reductive coupling polymerization is that whilst it can be controlled to give monomodal samples, it gives polysilanes with high polydispersities (typically 1.5–2 under optimum conditions) with yields no higher than 50–80% under optimum conditions.

2.2 Synthesis of Polysilane Block Copolymers

2.2.1 Polymer Coupling Reactions

The use of polysilanes as photoinitiators of radical polymerization was one of the first means whereby they were incorporated within block copolymer structures [38–40], albeit in an uncontrolled fashion. However the resulting block copolymer structures were poorly defined and interest in them principally lay in their application as compatibilisers for polystyrene (PS) and polymethylphenylsilane blends PMPS. The earliest synthetic strategies for relatively well-defined copolymers based on polysilanes exploited the condensation of the chain ends of polysilanes prepared by Wurtz-type syntheses with those of a second prepolymer that was to constitute the other component block. Typically, a mixture of AB and ABA block copolymers in which the A block was polystyrene (PS) and the B block was polymethylphenylsilane (PMPS) was prepared by reaction of anionically active chains ends of polystyrene (e.g. polystyryl lithium) with Si-X (X=Br, Cl) chain ends of α,ω-dihalo-polymethylphenylsilane an example of which is shown in Fig. 2 [43, 44, 45]. Similar strategies were subsequently used to prepare an AB/ABA copolymer mixture in which the A block was poly(methyl methacrylate) (PMMA) [46] and also a multi- block copolymer of PMPS and polyisoprene (PI) [47].

A particularly interesting block copolymer made by the coupling approach was a multi-block copolymer of PMPS and poly(ethylene oxide) (PEO). This was prepared by reacting the Si—X chain ends of PMPS with the hydroxyl chain ends of well-defined commercial sample of poly(ethylene glycol) [47] (Fig. 3). Although

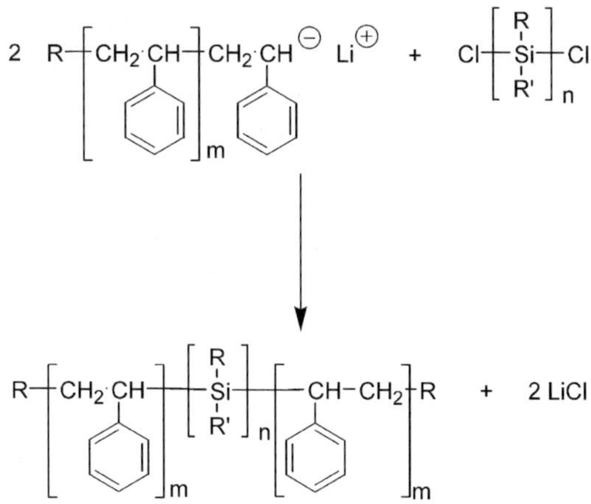

Fig. 2 Typical synthesis of a polystyrene-*block*-polysilane-*block*-polystyrene by a polymer coupling reaction

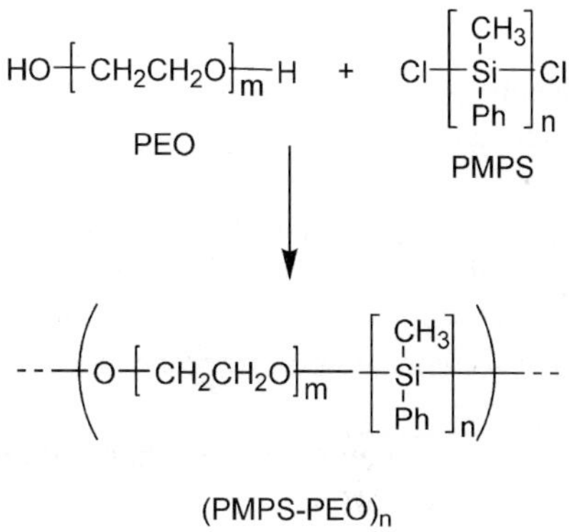

Fig. 3 Synthesis of a polymethylphenylsilane-poly(ethylene oxide) multi-block copolymer

the former had a normal molecular weight distribution, the latter was of a uniform distribution. As determined by size exclusion chromatography, the resultant copolymer consisted of all the $(AB)_n$ and $(AB)_nA$ structures from PMPS-PEO through to $(PMPS\text{-}PEO)_{16}$.

2.2.2 Living Polymerizations

The polymer coupling approach to block copolymer synthesis is seriously disadvantaged by the need to ensure stoichiometric equivalence of the reactive functional groups. These are in low concentration relative to the main chain units of the polymer chains and they are usually sensitive to trace impurities, particularly water (e.g. Si-Cl rapidly converts to Si-OH, Si^- rapidly converts to Si-H). Hence obtaining stoichiometric equivalents of the chain ends is extremely difficult and leads to poor reproducibility without scrupulous care using high vacuum line procedures. In contrast, the living polymerization approach either (i) uses the reactive chain end of a preformed polysilane to initiate polymerization of a vinyl monomer; (ii) uses the reactive chain end of a preformed polysilane to functionalize the chain with a suitable initiator for a subsequent living polymerization; (iii) uses a polymeric carbanion (e.g. polystyryl lithium) is to initiate the polymerization of the silane monomer. The resulting block copolymer structures are usually more defined than those prepared by polymer coupling and copolymers based on polysilanes with both AB and ABA structures have been synthesized these ways. The first such living polymerization syntheses were achieved using the living silyl-anionic chain ends on poly(1,1-dihexyl-2,2-dimethylsilane) and poly(1-butyl-1,2,2-trimethylsilane), prepared using the masked disilene procedure [33], to initiate the polymerization of methyl methacrylate, trimethylsilyl methacrylate and 2-(trimethylsilyloxy) ethyl methacrylate [48, 49, 50] (Fig. 4a). A further example of the synthetic utility of this approach came with the synthesis of poly(1,1-dimethyl-2,2-dihexyldisilene)-*b*-poly(triphenylmethyl methacrylate) (PMHS-*b*-PTrMA) [51]. The PTrMA block was synthesized in the presence of (-)-sparteine which induced the adoption of a helical conformation in the methacrylate block, i.e. a chiral one-handed helical chain. When the temperature was reduced to –20°C the polysilane chain was induced by this block to adopt itself a one-handed helix and become chiral and optically active.

PMPS-*b*-PS and PMPS-*b*-PI were synthesized by the anionic ring-opening polymerization of tetramethyltetraphenylcyclotetrasilane (prepared from commercially available octaphenyltetrasilane) initiated by the living anionic chain ends of polystyrene and polyisoprene [52, 53] (Fig. 4b).

Over the past 10 years the advent of controlled radical polymerization has resulted in an explosion of interest in the synthesis of block copolymer systems that were hitherto inaccessible [54]. The most commonly used methods of controlled radical

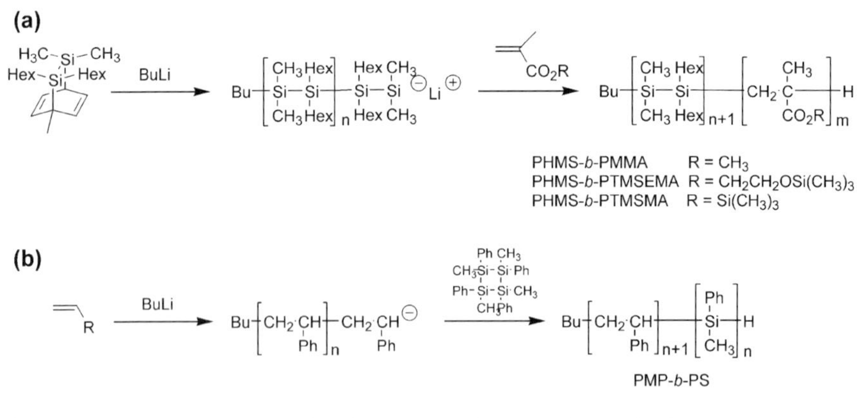

Fig. 4 Approaches to the synthesis of polysilane block copolymers by living polymerization techniques: (**a**) via anionic polymerization of masked disilenes; (**b**) via anionic ring-opening polymerization of cyclotetrasilanes

polymerization of vinyl monomers are nitroxyl mediated (e.g. TEMPO), reversible addition fragmentation (RAFT) and atom transfer radical polymerization (ATRP) [55, 56, 57]. Both TEMPO and ATRP based syntheses of polysilane block copolymers have been reported. Poly(styrene-*block*-methylphenylsilane-*block*-styrene) has been synthesized by a TEMPO-mediated polymerization from an end functionalized PMPS macromolecular initiator (Fig. 5) [58]. The first inorganic–organic hybrid copolymer system synthesized via ATRP was that of polystyrene grafts grown from a bromomethylated PMPS sample [59]. A more generally useful approach has been the end-functionalization of PMPS with an active ester alkyl halide 2-bromo-2-methyl propanoate followed by its application in the copper catalyzed ATRP of methyl methacrylate, hydroxyethyl methacrylate and oligo(ethylene glycol methyl ether methacrylate) (Fig. 6) [60, 61].

In these cases the precursor α,ω-dihaloPMPS of a broad polydispersity (typically 1.6–2.1) was reacted with hydroxyethyl 2-bromo-2-methylpropanoate

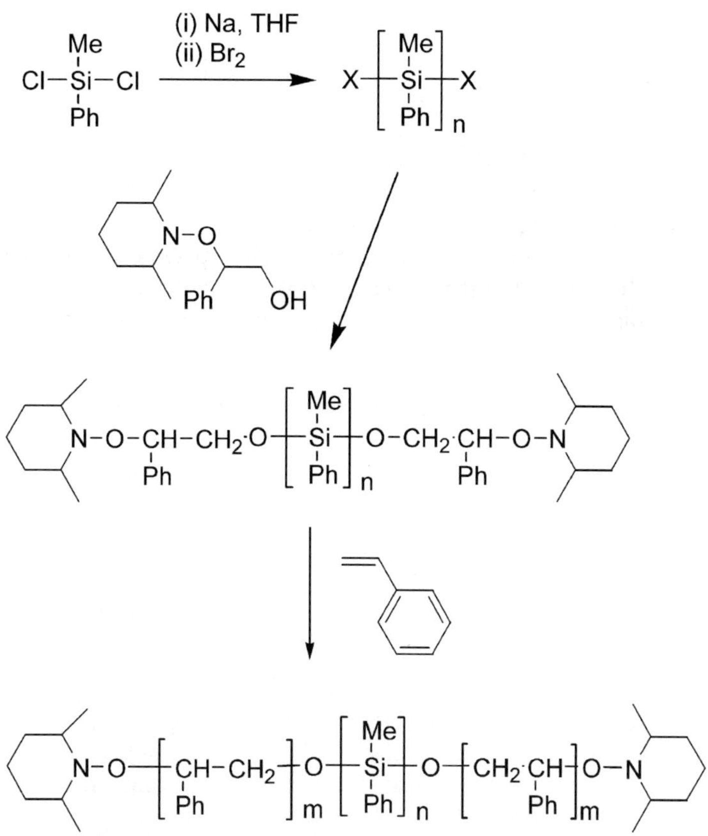

Fig. 5 Synthesis of polystyrene-*block*-polymethylphenylsilane-*block*-polystyrene by a TEMPO mediated macroinitiator approach

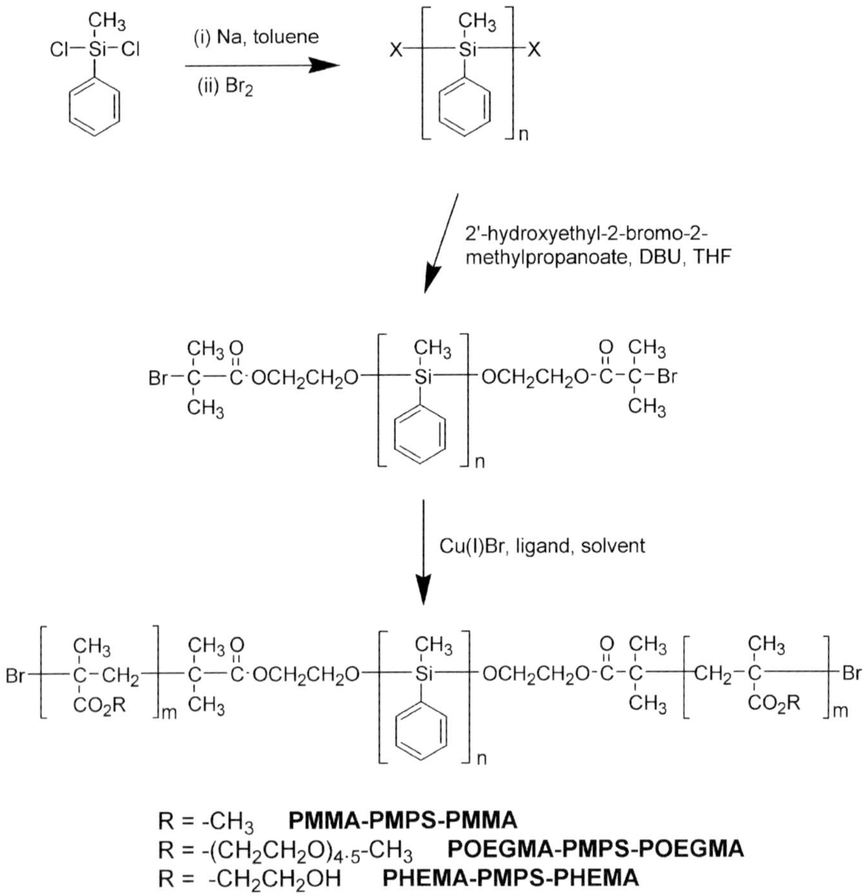

Fig. 6 Synthesis of various polymethylphenylsilane block copolymers by an ATRP based macroinitiator approach

to give end-functionalized PMPS (characterized by ^{1}H NMR). A variety of PMPS molecular weights were prepared in this manner and the appropriate vinyl monomers were polymerized using a Cu(I)Br catalyst and a bidentate N-containing ligand (e.g 2,2'-bipyridine) in a solvent (e.g. THF). The resultant ABA block copolymers were of lower polydispersities than the precursor PMPS blocks as expected from the incorporation of narrow distribution blocks at the chain ends of a sample with a formerly broad distribution. The molecular weight parameters measured by SEC were in broad agreement with the theoretical M_n values predicted from consideration of the monomer to initiator ratios and were observed to grow in a manner consistent with a controlled chain growth polymerization (Fig. 7a). Kinetic analyses ($\ln[M_0]/[M]$ versus time) of the polymerization of MMA from a PMPS macroinitiator clearly demonstrated 1st order behaviour as expected from ATRP (Fig. 7b).

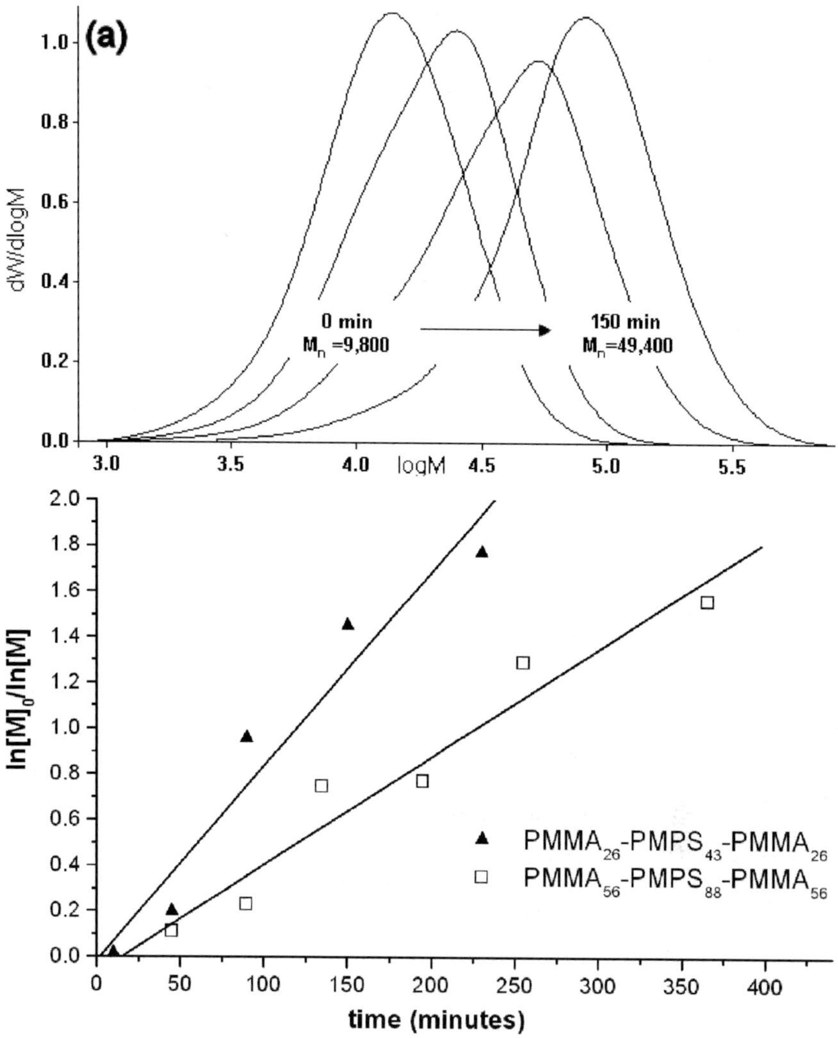

Fig. 7 ATRP of MMA from a Br-PMPS-Br macro-initiator: (**a**) SEC traces showing growth of a PMMA chain; (**b**) kinetic plots for the polymerisation of MMA demonstrating living nature eof polymerization

3 Self-Assembly and Self-Organization of Polysilane Block Copolymers

3.1 Self-Organization in Thin Films

The ability of copolymers consisting of chemically distinct polymeric segments to undergo microphase separation as a result of enthalpically driven segregation has led to a remarkable range of nanostructured morphologies being catalogued and studied [62]. Consequently, such materials have been the subject of intense study for over ten years [63, 64, 65]. Block copolymer thin films show many of the morphologies displayed by the bulk materials, but substrate and surface effects can play a much more pronounced role in the self-organization, particularly for very thin films. A large number of potential applications for these self-organizing thin films have been proposed and demonstrated. Examples include applications as lithographic masks [66, 67], photonic materials [68, 69, 70], and nanostructured membranes [71, 72].

Demoustier-Champagne et al. used atomic force microscopy (AFM) to observe microphase separation within cast films of PS-PMPS-PS/ PS-PMPS block copolymer mixture [43] that were used to compatibilize a blend of PMPS and PS. The fracture surface of blend films with the block copolymer incorporated show a far finer dispersion of particle sizes than those without. Matyjaszewski et al. studied PMPS-PS thin films by SFM (scanning force microscopy) and TEM (transmission electron microscopy) and Fig. 8 shows a TEM picture of a thin section of a film which was prepared by slow evaporation from THF, which is slightly selective for the polystyrene block [73].

The dark areas were assigned to the polysilane domains as a result of the stronger electron scattering in silicon rich regions compared to carbon rich regions. The domains were poorly defined but multiple images demonstrated that the morphology was real. The wormlike dark domains were consistent with a cylindrical morphology of the polysilane block in a matrix of polystyrene. The cylinders had approximately the same size throughout the entire sample with a diameter of 7±2 nm. This was roughly half of the extended chain length of 14.8 nm of the PMPS component (M_n = 9,000 g.mol^{-1}) of the block copolymer. Intriguingly by exposing a thin film of the PMPS-PS block copolymer to UV light of 360 nm the PMPS could be selectively degraded. Whilst initial SFM analysis of the film revealed no change in texture over the surface after UV exposure bundles of the PI became apparent (Fig. 9). A broad distribution (M_w/M_n = 2.4) multi-block copolymer of predominant structure (PMPS-*b*-PI)$_3$ has been demonstrated to form self-supporting films that are optically clear, strong, and flexible [47]. They were characterized using both AFM and neutron scattering [74] and despite relatively

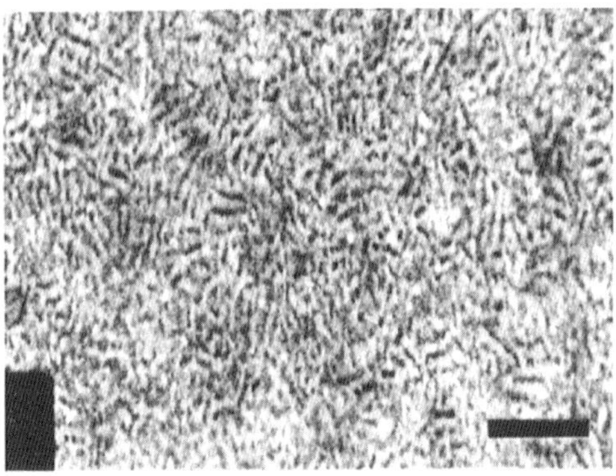

Fig. 8 TEM picture of a thin section of a film of a PMPS-b-PS copolymer prepared by slow evaporation from THF. Scale bar = 150 nm. Reproduced with permission from [73], Fossum et al., Macromolecules (1997) 30:1765–1767. ©American Chemical Society

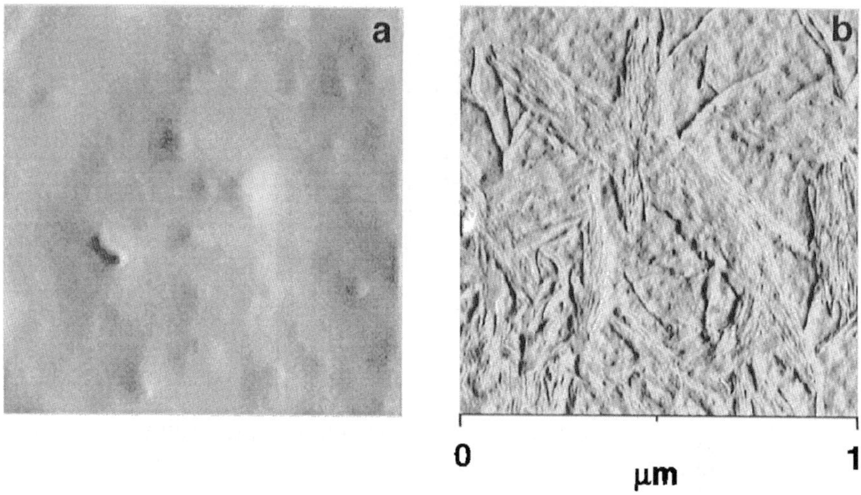

Fig. 9 SFM micrographs of a thick film of polystyrene-*block*-PMPS with Mn = 18,700 and 9,000 and overall polydispersity Mw/Mn = 1.22 (**a**) before and (**b**) after degradation with 360 nm light. Reproduced with permission from [73], Fossum et al., Macromolecules (1997) 30:1765–1767. © American Chemical Society

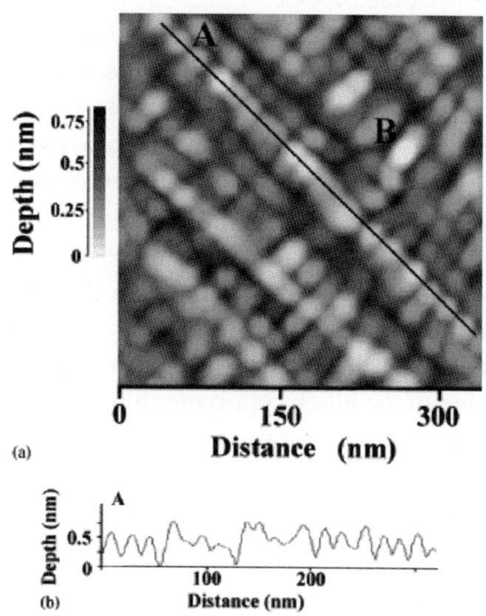

Fig. 10 (**a**) Representative AFM image of a PMPS-*b*-PI surface (root mean square roughness over imaged area, 0.071 nm). (**b**) Profile of PMPS-*b*-PI surface along line drawn in (**a**). Maximum distance between domain peaks: 22.0 nm; minimum distance: 16.8 nm. Reproduced from [74], Hiorns and Martinez (2003) Synth Met 139:463

high polydispersities in both component blocks (respectively 1.64 and 1.34 for PMPS and PI), a regular modulated morphology was observed with average domain repeat units of 18.8 and 18.6 nm at the surface and in the bulk respectively (Fig. 10). The regularity was further shown to be entirely consistent with the extended lamella-like structure shown in Fig. 11 rather than the thermodynamically less-favored structure within which the chains fold and reverse direction at the coiled PI segments.

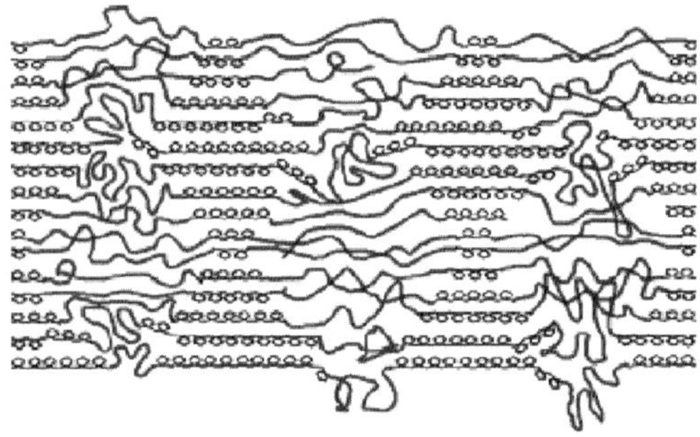

Fig. 11 Tentative, idealised, sectional representation of microphase separation and organization in a PMPS-*b*-PI film

The PMMA-*b*-PMPS-*b*-PMMA triblock copolymers prepared by the macroinitiator approach using ATRP [60] were only characterized using differential scanning calorimetry. The glass transition temperature (T_g) of PMPS is usually difficult to observe but within the copolymers it was clearly evident at 125–130°C. The T_gs of the PMMA blocks increased with block length in a manner consistent with the variation with chain length for homopolymers of PMMA and were also clearly visible by DSC. The presence of two T_gs provides strong evidence for microphase separation of the blocks.

Films of the POEGMA-*b*-PMPS-*b*-POEGMA series of copolymers synthesized by ATRP [61] were cast on glass, silicon, silver, and gold substrates and were investigated by a number of techniques [75]. The water contact angles at the surfaces of the block copolymers were observed to be directly related to the nature of the underlying substrate; e.g. hydrophilic glass substrate gave a low contact angle (~35°) and a hydrophobic gold substrate gave a relatively high contact angle (~90°) (see Fig. 12).

Selective delamination from the hydrophilic surfaces (glass and ozone-treated silicon) was observed for those copolymers with a high POEGMA content (weight ratios of POEGMA:PMPS > ~ 1.3, below this no delamination was observed). I.e. the nature of the substrate (hydrophilic vs. hydrophobic) directly controlled the adhesion of the block copolymer film to the substrate in an aqueous environment. Similar behavior was observed for a corresponding series of POEGMA-*b*-PS-*b*-POEGMA copolymers (in which the central block is polystyrene) of POEGMA:PS weight ratios > ~ 1). Thus the behavior was not intrinsic to polysilane block copolymers. Water contact angles of ~85° were observed for the PHEMA-*b*-PMPS-*b*-PHEMA and PMMA-*b*-PMPS-*b*-PMMA copolymers and a poly methylmethacrylic acid-*block*-polymethyl methacrylate-*block*-polymethyl methacrylic acid copolymer and no delamination was observed upon immersion in water. It was therefore proposed that the presence of POEGMA blocks is a critical factor in addition to appropriate hydrophobic:hydrophilic weight ratios.

The tapping mode AFMs of Fig. 13 are examples of the height and phase images of the POEGMA-*b*-PMPS-*b*-POEGMA films coated on glass. They reveal smooth surfaces with lateral microphase separation. Films of the copolymers coated on gold surfaces (hydrophobic) are similar, as typified by the phase images for both systems

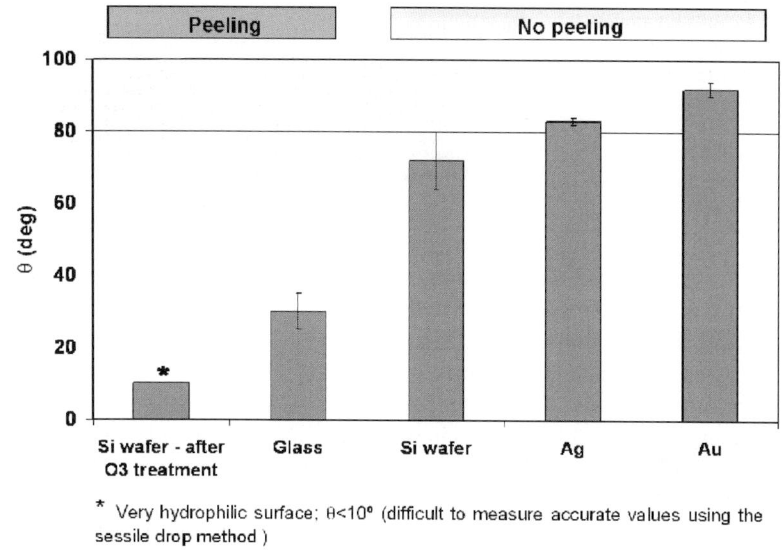

Fig. 12 Water contact angles (□) of different substrates. The block copolymer films of POEGMA-*b*-PMPS-*b*-POEGMA and POEGMA-*b*-PS-*b*-POEGMA selectively delaminate only from the substrates with a hydrophilic surface upon immersion in H2O. Reproduced with permission from [75] Popescu et al., (2004) Macromolecules 37:3431–3437. © American Chemical Society

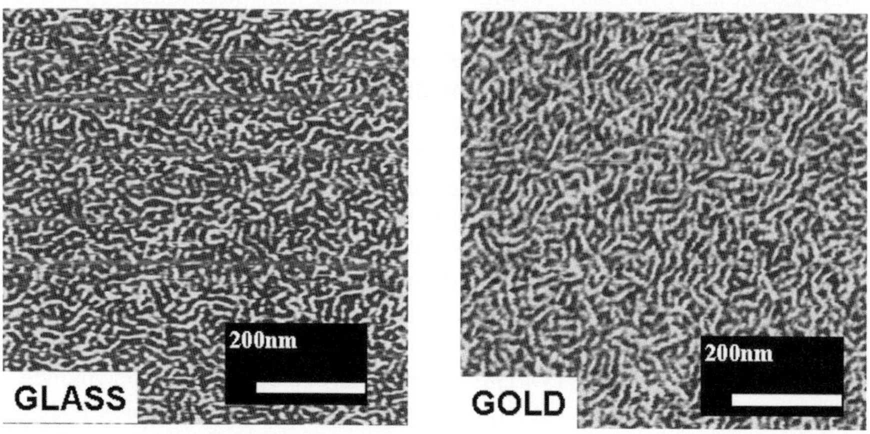

Fig. 13 Tapping mode AFM phase images of films of POEGMA-*b*-PMPS-*b*-POEGMA spin-coated on a gold and on a glass substrate. Reproduced with permission from [75] Popescu et al., (2004) Macromolecules 37:3431–3437. © American Chemical Society

shown on the same scale in Fig. 13. Despite this the contact angles of water at these surfaces differed considerably. X-ray photoelectron spectroscopy, XPS, was used to probe the effects of immersion in water of the hydrophilically unstable films of POEGMA-*b*-PMPS-*b*-POEGMA coated on gold and glass. Prior to immersion, the

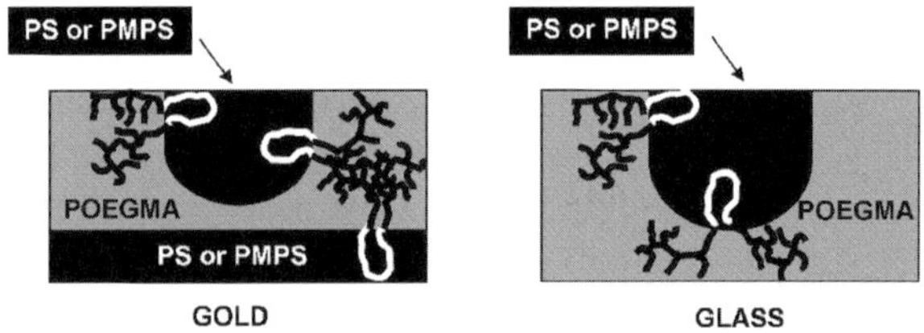

Fig. 14 Postulated models for the self-organization of the delaminating POEGMA-*b*-PMPS-*b*-POEGMA and POEGMA-*b*-PS-*b*-POEGMA block copolymers over gold and glass substrates (cross section) [75]

spectra indicated a slight accumulation of the PMPS on the outer surface of the films. The film coated on gold was unaffected by immersion but that coated on glass indicated a clear shift to a higher POEGMA content at the outer surface. Thus XPS indicated that a rearrangement of the surface morphology took place when the films over glass came into contact with water whereas no rearrangement occurred for films over gold. In contrast, the stable (non-delaminating) films coated on glass have relatively rough, granular surfaces with a micellar-like structure. The thin film cross sections of Fig. 14 represent models for the self-organization of the delaminating ABA block copolymers over glass and gold substrates that are based on the above observations.

The amphiphilic block copolymers PHEMA-*b*-PMPS-*b*-PHEMA and POEGMA-*b*-PMPS-*b*-POEGMA have found applications as templates for the patterning of cell growth [76] and the patterning of biomimetic crystallization processes [77]. In the first case the selective delamination of POEGMA-*b*-PMPS-*b*-POEGMA was utilized [76]. Samples were spun-cast onto patterned gold-electrodes (20 nm thick, 100–600 μm wide) on a glass substrate. The resulting films were immersed in a cell culture medium for 3 days over which time the copolymer detached from the glass surface but remained on the gold (the delamination was significantly faster using the cell culture medium rather than water) as illustrated schematically in Fig. 15. Similar effects were observed for the POEGMA-*b*-PS-*b*-POEGMA samples but in the case of the POEGMA-*b*-PMPS-*b*-POEGMA samples delamination could be monitored by UV-vis spectroscopy by noting the disappearance of the characteristic UV absorption peak due to the PMPS (Fig. 15). When the copolymer films were exposed to a cell culture medium in the presence of C2C12 mouse myoblasts (undifferentiated muscle cells) within 24 h a pattern of aligned myoblasts was visible on the glass lanes between the gold electrodes as a consequence of the delamination of the copolymer (Fig. 16). The copolymer film on the gold electrodes remained free of cells as a consequence of the POEGMA component of the copolymer (poly(ethylene oxide) compounds are resistant to protein adsorption and cell adhesion). It should be noted that this delamination process occurs within films forming a continuous thin films over the glass and gold surfaces. Thus it remains a remarkably simple means of protecting/covering the hydrophobic parts of patterned surfaces through a simple application of a polymer film followed by immersion and rinsing.

The PHEMA-*b*-PMPS-*b*-PHEMA amphiphilic ABA block copolymers were used to generate patterned calcium carbonate films with dimensions of several hundreds of microns using the photolithographic properties of the polysilane component [77]. PHEMA-*b*-PMPS-*b*-PHEMA was spin cast from THF solution onto glass substrates. On this polymer layer continuous films of calcium carbonate,

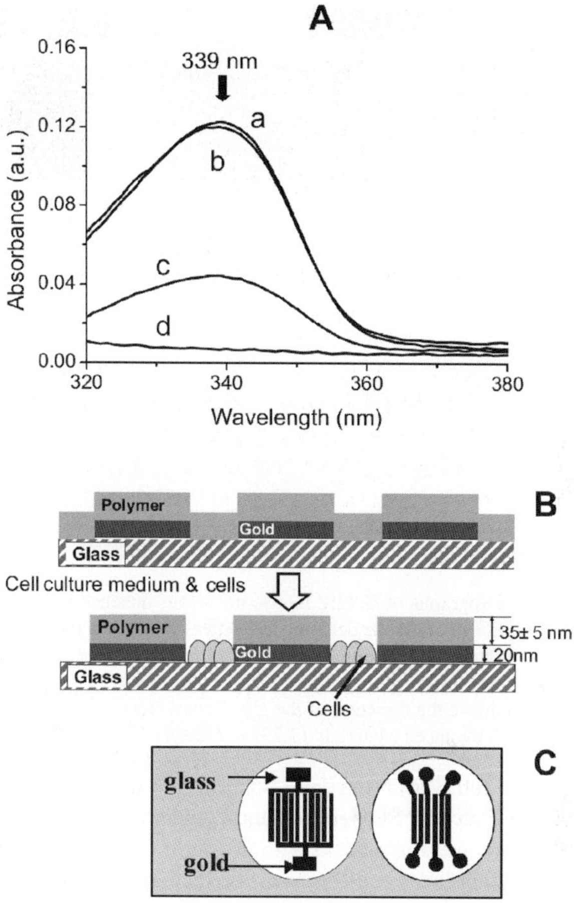

Fig. 15 Selective delamination process of POEGMA-PMPS-POEGMA films from glass. (**A**) UV spectra of POEGMA-PMPS-POEGMA films deposited on glass, as spincoated (a – *black*), after 3 days exposure to H_2O (b – *green*), after 5 days exposure to H_2O (c – *red*) and after 3 days exposure to C2C12 cell culture medium (d – *blue*). (**B**) Cross sectional schematics of the process of selective delamination upon exposure to cell culture medium and cell seeding on the substrates resulting in the formation of a pattern of non –adhesive copolymer-coated gold lanes next to clean glass lanes where the cells can attach. (**C**) Schematic gold electrode arrangements on glass surface. Reproduced with permission from [76] Popescu et al. (2005) Advanced Materials 17:2324–2329. © Wiley-VCH Verlag GmbH & Co

$CaCO_3$, ~ 1 µm thick were grown by immersing them in an aqueous $CaCl_2$ solution containing poly(acrylic acid) and allowing CO_2 vapor from $(NH_4)_2CO_3$ to diffuse into these solutions (Fig. 17).

Surface profilometry showed that the average thickness of the $CaCO_3$ films was ~1 µm. Optical microscopy, FT-IR and SEM all demonstrated that the films were amorphous with few embedded crystalline spherulites. Upon standing, the film

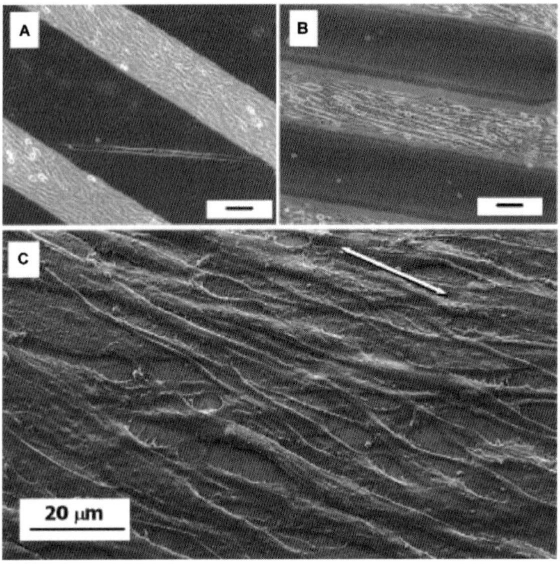

Fig. 16 (**A**, **B**) Optical micrographs of C2C12 mouse myoblasts attached to the glass lanes where the POEGMA-PMPS-POEGMA film has delaminated. After A) 24 h at a high cell seeding density, (**B**) 8 days, after differentiation into myotubes. Scale bars: 100 μm. (*dark gray* – gold lanes; *light gray* – glass lanes). (**C**) SEM image of the aligned C2C12 myotubes (13 days after seeding on the substrates). The arrows indicate the direction of the glass lanes. Reproduced with permission from [76], Popescu et al. (2005) Advanced Materials 17:2324–2329. © Wiley-VCH Verlag GmbH & Co

crystallized as demonstrated by SEM, FT-IR and Powder X-Ray Diffraction (PXRD). When the polymer films were irradiated with UV light (360 nm) through a mask for 2 h, the irradiated polymer lanes could be removed selectively by washing with ethanol, resulting in a pattern of polymer lanes 200 μm wide and ~30 nm high (as determined by surface profilometry). When these patterned films were subjected to the mineral film formation process, $CaCO_3$ was deposited both on the polymer lanes and on the areas from which the polymer had been removed. However, when the mineral was deposited onto a pre-exposed but undeveloped polymer film, the $CaCO_3$ layer grown on the irradiated lanes could be selectively removed upon immersion of the film into ethanol, resulting in the formation of a patterned $CaCO_3$ film (Fig. 18).

The non-patterned $CaCO_3$ films could be observed to crystallize within 1 h by optical microscopy. However the patterned films stayed amorphous for 2–3 h under ambient conditions and were only completely crystalline after 24 h, which is probably due to the use of ethanol in the patterning procedure, as this is known to stabilize ACC (amorphous calcium carbonate). Subsequently cell culture experiments were performed and the results indicated that the $CaCO_3$ substrates support rat bone marrow stromal cell attachment, proliferation and differentiation into osteoblast and osteoclast-like cells. Moreover, mineral formation by the osteoblast-like cells was favored on the $CaCO_3$ films compared to the developed polymer films. Also, the osteoclast-like cells can degrade the $CaCO_3$ films. Therefore, these patterns of $CaCO_3$ films can be regarded as suitable 2D model substrates for bone cells.

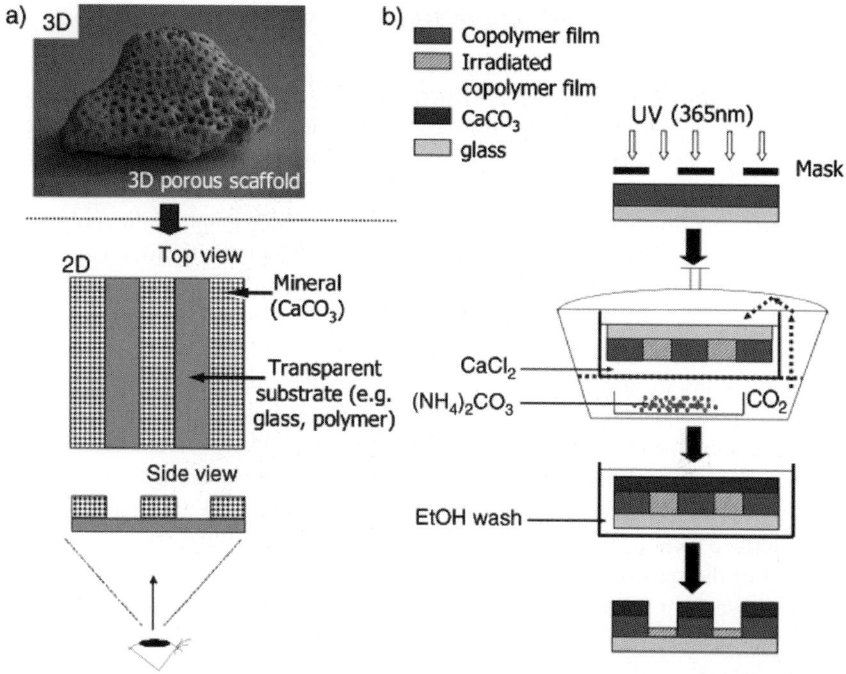

Fig. 17 Schematic representation of (**a**) a 2D model substrate consisting of alternating lanes of mineral ($CaCO_3$) and non-mineralized lanes and (**b**) of the experimental procedure for the generation of patterns of $CaCO_3$. Dimensions are not to scale. Reproduced with permission from [77] Popescu et al., (2006) Angew Chem Int Ed 45:1762–1767. © Wiley-VCH Verlag GmbH & Co

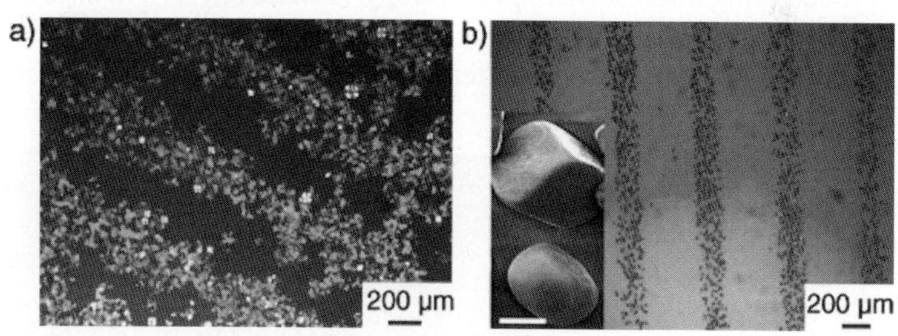

Fig. 18 Optical micrographs under cross-polarized light of (**a**) a pattern of $CaCO_3$ discrete crystals, (**b**) a pattern of $CaCO_3$ film, grown on patterned thin films of PHEMA-*b*-PMPS-*b*-PHEMA. Reproduced with permission from [77] Popescu et al., (2006) Angew Chem Int Ed 45:1762–1767. © Wiley-VCH Verlag GmbH & Co

In addition, the patterning method presented here is not restricted only to glass substrates unlike the use of patterned SAMs (self-assembled monolayers), where the choice of substrates is limited. In general this method would allow for the photogeneration of patterns of $CaCO_3$ on a variety of substrates, including e.g. conducting polymers, which would be beneficial for electrical stimulation of cells to enhance their proliferation and differentiation.

3.2 Self-Assembly of Polysilane Block Copolymers in Solution

The first self-assembling block copolymers were PS-*b*-PMPS-*b*-PS synthesised by Matyjaszewski and Möller. They observed micellar aggregates by AFM after casting dilute dioxane solutions (a solvent selective for the PS block) of the copolymer. The observed micelles were taken to have internal PMPS cores and were measured at 25–30 nm in diameter [73]. The first self-assembling amphiphilic polysilane block copolymers to be investigated was the PMPS-PEO multi-block copolymer with normal distribution PMPS blocks and uniform low polydispersity PEO blocks. After dialysis aqueous dispersions of this copolymer formed micellar as well as vesicular structures [78, 79] as shown in Fig. 19.

Encapsulation of the water soluble 5-carboxyfluoroscein dye confirmed the formation of vesicles. A pressure-area isotherm was recorded for a monolayer of the copolymer at the air-water interface to investigate the likely orientation of chains in the vesicle walls. A lift-off area of 30 nm^2 $molecule^{-1}$ corresponding to the approximate cross sectional area of three PMPS chains oriented perpendicular to the air-water interface were observed. More detailed studies [80] of the surface viscoelastic properties of the spread films at the air-water interface have revealed complex relaxation processes that follow none of the simple models that might be expected. This was attributed to the rigid nature of the polysilane blocks. The UV-vis spectra of the aqueous dispersions display a σ-σ^* transition associated with a λ_{max} = 342 nm compared to molecularly dissolved PMPS with λ_{max} = 337–340 nm, indicating that within the aggregates the polysilane segments were more extended than in solution. Based on these observations a structural model of the vesicles could be constructed (Fig. 19).

The aggregation of the PMPS-PEO copolymers in solution and dispersion was furthered probed using fluorescence spectroscopy. A small red shift in the fluorescent emission maximum was observed on increasing the water content, attributable to more effective energy transfer from shorter to longer PMPS segments. This indicated

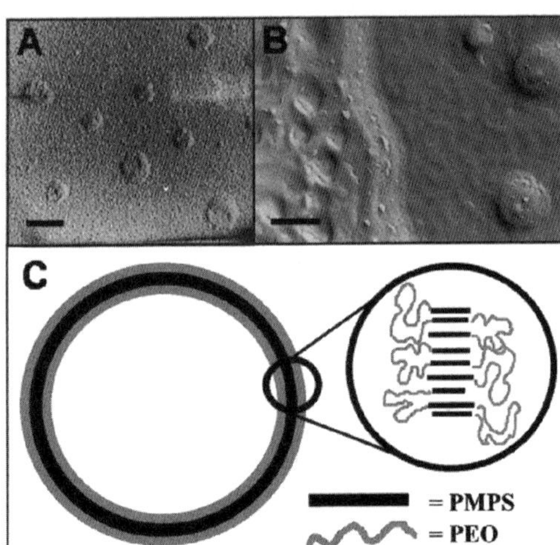

Fig. 19 (**a**, **b**) TEM images of vesicle dispersions of PMPSn-PEOm showing (**a**) a replica of intact (freeze fracturing) and (**b**) collapsed vesicles (platinum shadowing); bars represent 200 nm. (**c**) Schematic representation of the proposed structure of the vesicles showing the hydrophobic PMPS interior (*black*) shielded from the aqueous phase by the hydrophilic PEO layers (*gray*). Reproduced with permission from [79], Sommerdijk et al. (2000) Macromolecules 33:8289. © American Chemical Society

that higher water contents induced the alignment of PMPS segments. Absorption spectra showed similar trends. Below 40% water content the block copolymer was molecularly dissolved as evidenced by TEM and dynamic light scattering. However, from 40% to 80% water content, micellar fibers with diameters of 20 nm and up to several microns in length were observed. TEM images of samples both unstained and stained with uranyl acetate are shown in Figs 20a–c.

The unstained sample highlights the hydrophobic PMPS segments and the stained sample also shows the polar PEO segments as a protective hydrophilic sheath around the core of PMPS as depicted in the schematic of Fig. 20d. At water concentrations above 80%, both right- and left-handed helical aggregates were observed. These superstructures had lengths of 1–2 μm, widths up to 0.2 μm, and a pitch of approximately 0.15 μm (Fig. 21). The helicity of the aggregates was attributed to the known helical conformation of the polysilane segments [29, 30] arising form the close-packing of helical segments of the same screw-sense. This was in analogy to the close packing of the rigid helical poly(isocyanide) segments in polystyrene and peptide based poly(isocyanide) block copolymers.

The masked disilene procedure was used by Sakurai and co-workers to synthesize two samples of diblock copolymers of 1,1-dimethyl-2,2-dihexylsilane (MHS) and 2-(trimethylsilyloxy)ethyl methacrylate, which differed only in the relative lengths of their blocks. Hydrolysis of the trimethylsilyl protecting groups gave the corresponding amphiphilic diblock copolymers, poly(1,1-dimethyl-2,2-dihexyldisilene)-*b*-poly(2-hydroxyethyl methacrylate) (PMH S-*b*-PHEMA), depicted in Fig. 22 [48].

In the solid state at room temperature, PMHS has a λ_{max} at 334 nm in which it takes on an ordered conformation (originally thought to be *trans* but most probably helical) but in toluene has a λ_{max} of 307 nm (resulting from a disordered conformation). The PMHS-*b*-PHEMA copolymer (copolymer 1) in methanol exhibits absorption with a

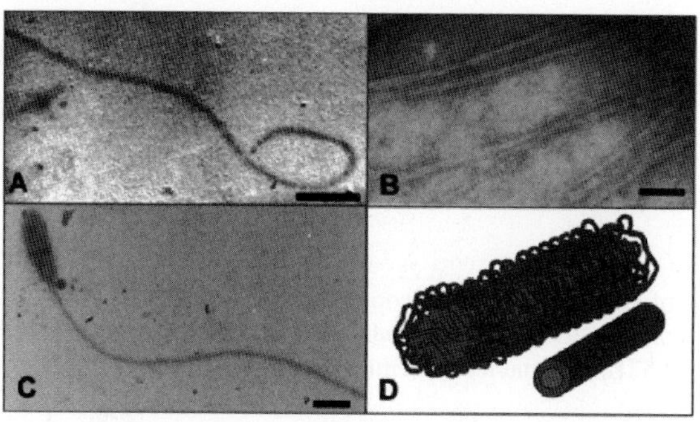

Fig. 20 Micellar fibers of PMPSnPEOm in mixtures of THF and water (25/75 by volume). TEM images (**a**) visualizing the polysilane core of micellar fibers (unstained, bar represents 250 nm); (**b**) revealing the PEO shell using uranyl acetate staining, (**c**) showing an example of the bulges found for many of these fibers. (**d**) Schematic representation of the structure of the micellar fibers showing the PMPS core and the PEO shell. Reproduced with permission from [79], Sommerdijk et al. (2000) Macromolecules 33:8289. © American Chemical Society

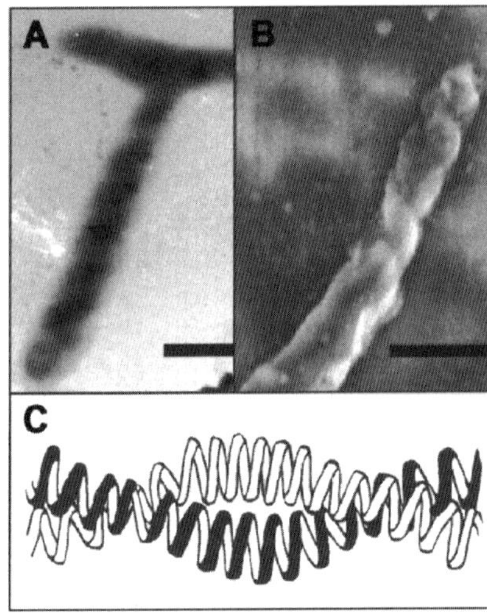

Fig. 21 Helical aggregates of PMPSnPEOm found in a water/THF mixture of 90/10 (v/v). (**a**) TEM image (unstained, bar represents 250 nm) of a right-handed helix and (**b**) SEM image (uncoated, bar represents 250 nm) of a left-handed helix. (**c**) Schematic representation of the formation of a superhelix from the coiling of two helical stands. Reproduced with permission from [79], Sommerdijk et al. (2000) Macromolecules 33:8289. © American Chemical Society

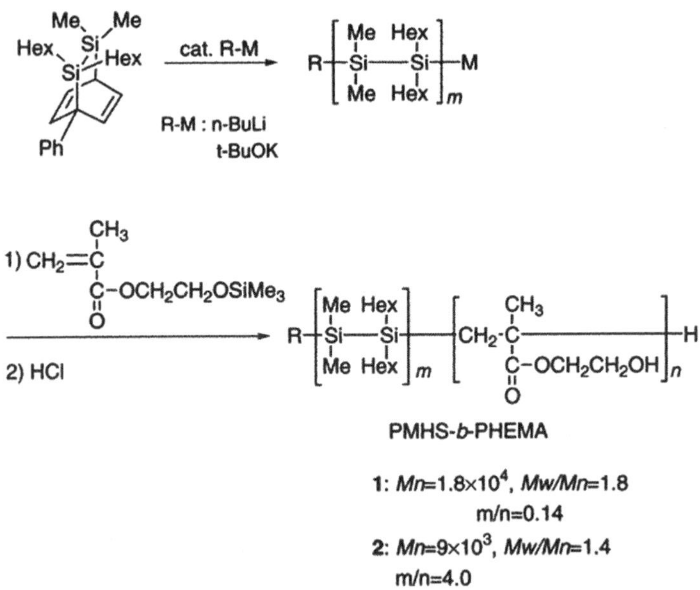

Fig. 22 Synthesis of amphiphilic diblock copolymers of poly(1,1-dimethyl-2,2-dihexyldisilene)-*b*-poly(2-hydroxyethyl methacrylate). Reproduced with permission from [48] Sanji et al. (1999) Macromolecules 32:5718. © American Chemical Society

λ_{max} at 334 nm so it was concluded that the PMHS blocks existed in a hydrophobic micellar core as a solid surrounded by the hydrophilic PHEMA blocks. In toluene solution, copolymer 2, with relatively shorter HEMA blocks, exhibited a λ_{max} of 307 nm. Thus, the PMHS blocks exist either in a corona or are molecularly dissolved. In addition, copolymer 1 showed solvatochromism related to the change between micelles and unimers. Copolymer 1 forms kinetically frozen micelles as the component polysilane block has a glass transition temperature (T_g) higher than room temperature so the morphology can be observed directly using AFM operating in the tapping mode. A cast solid film on mica-coated from a methanol solution indicated ellipsoidal micellar structures with a size of 50–60 nm in agreement with the observations of static light scattering experiments (Fig. 23).

Poly(1,1-dimethyl-2,2-dihexyldisilene)-*block*-poly (methacrylic acid) (PHMS-*b*-PMAA) with a PMHS:PMAA molar monomer unit ratio of 1:20 was also prepared by a sequential anionic polymerization of 'masked' disilenes and trimethylsilyl methacrylate, followed by hydrolysis of the trimethylsilyl protecting group [49]. PMHS-*b*-PMAA was soluble in water and self-assembled to form polymer micelles with an average diameter of 170 nm in water at a concentration of 0.2 g L-1 at 25°C. Subsequently methacrylic acid block was reacted with 2,2-(ethylenedioxy)diethylamine to form shell cross-linked micelles (SCM) (Fig. 24). ^{1}H-NMR signals from the PHMS core could not be detected in D_2O solution because of very long relaxation times arising from the solid core, however, upon the addition of an excess of THF-d8 to the solution, signals from the core were observed as the PMHS blocks became solvated. Solid state CP-MAS ^{29}Si NMR demonstrated the presence of two signals at −27.6 and −35.7 ppm, assignable to dihexylsilylene and dimethylsilylene units, respectively. DLS studies indicated the intensity-averaged diameter of the particles to be 160 nm with mono-dispersed spheres, consistent with the size of the micelles of the parent PMHS-*b*-PMAA, though slightly shrunken. AFM images revealed spherical particles of about 50 nm diameter in the dry state.

The polysilane core part within the shell cross-linked micelles was photodegraded by UV irradiation (≥280 nm) and dialysis against water produced nanometer-sized

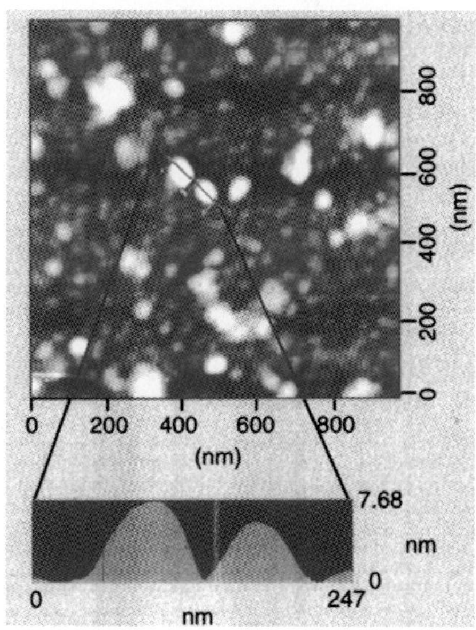

Fig. 23 Tapping mode AFM images and vertical profile of poly(1,1-dimethyl-2,2-dihexyldisilene)-*b*-poly(2-hydroxyethyl methacrylate) in the solid film on mica coated from a methanol solution (0.043 g/L). Reproduced with permission from [48] Sanji et al. (1999) Macromolecules 32:5718. © American Chemical Society

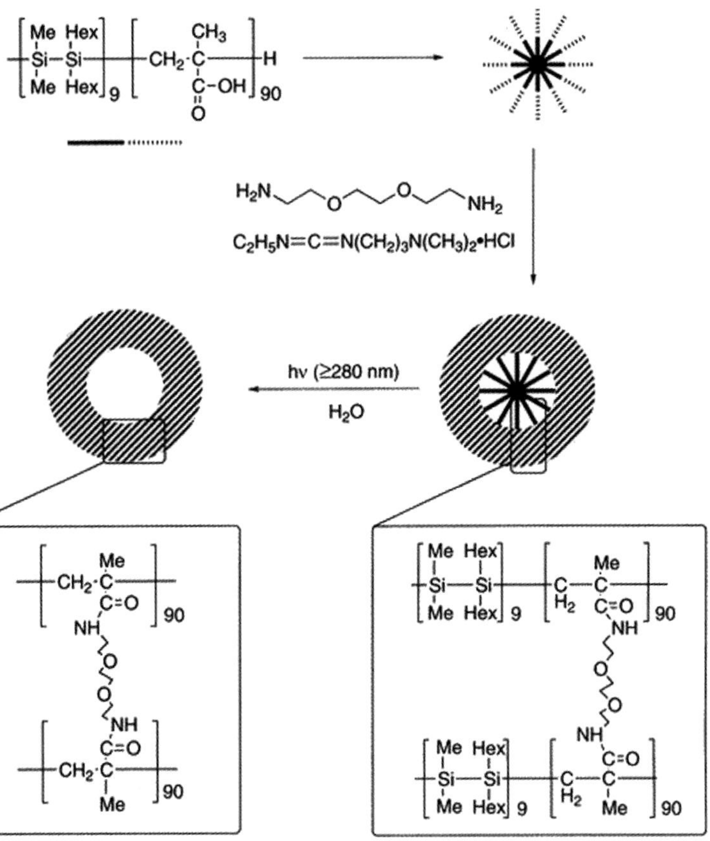

Fig. 24 Schematic illustration of the synthetic pathway for hollow square particles derived from polysilane shell cross-linked micelles templates. Reproduced with permission from [50] Sanji et al. (2000) Macromolecules 33:8524. © American Chemical Society

hollow particles (Fig. 25) [50]. In the UV absorption spectra, a continuous blue shift in the absorption maximum was observed during photoirradiation confirming degradation of the polysilane core. Subsequently 5,6-carboxyfluorescein (CF) was encapsulated into the nanometer-sized hollow particles prepared by a similar procedure; prolonged dialysis led to the release of the dye from the particles. A further application of these SCMs involved the reduction of $HAuCl_4$ with the polysilane core of the micelles [81]. Polysilanes have relatively low oxidation potentials and consequently are able to reduce certain metal ions with the Si-Si bonds undergoing oxidation to Si-O bonds. In this manner gold-nanoparticles of 12±5.7 nm in diameter were produced in the cores of the SCMs with diameters of 25±5.7 nm. The SCM Au nanoparticles were characterized by TEM and UV-vis spectroscopy. In a similar manner the polysilane micelles were used to reduce $PdCl_4^-$ to give SCM Pd nanoparticles of 20±10.7 nm diameter [82]. The SCM-Pd nanoparticles were subsequently shown to be effective catalyst for alkene hydrogenation and in Heck reactions. The polysilane-PMA block copolymers

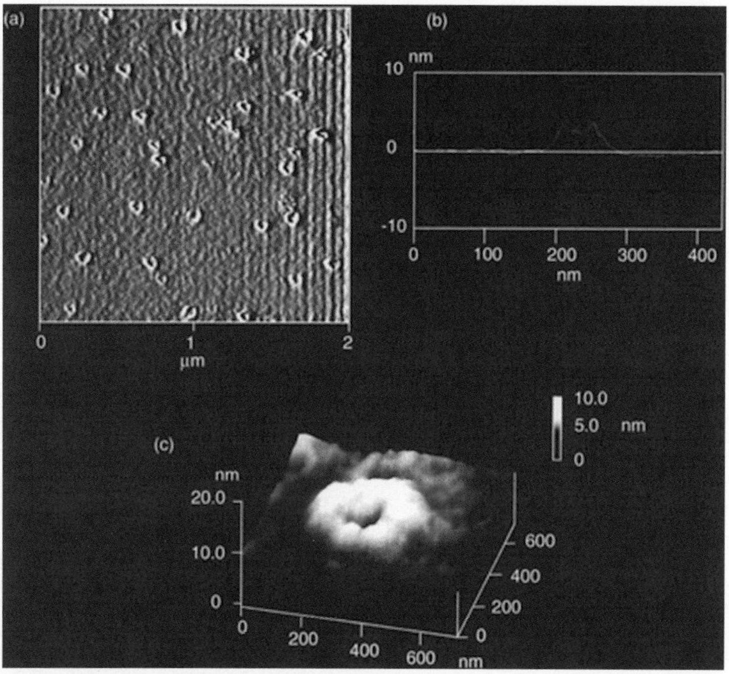

Fig. 25 Hollow particles prepared from SCMs of poly(1,1-dimethyl-2,2-dihexyldisilene)-*b*-poly(methacrylic acid), **(a)** AFM image on Pyrex glass plate with operating in the contact mode, **(b)** vertical profile of the hollow particle shown in part **(a)**, and **(c)** in the tapping mode under THF wet conditions. Reproduced with permission from [50] Sanji et al. (2000) Macromolecules 33:8524. © American Chemical Society

prepared in this study were further utilized in stabilizing silica nanoparticles through chemical attachment to aminopropyl surface groups [85].

To complement the aggregation studies of the multi-block PMPS-PEO polymers, the self-assembly of the amphiphilic triblock copolymers, PHEMA-*b*-PMPS -*b*-PHEMA and POEGMA-*b*-PMPS-*b*-POEGMA were investigated [61]. Series of each of the block copolymers, prepared using the ATRP macroinitiator approach, were studied using TEM and dynamic light scattering. In the case of the PHEMA-*b*-PMPS-*b*-PHEMA copolymers in aqueous dispersion, only micellar aggregates 10–20 nm in diameter were observed (Fig. 26a).

In contrast, aqueous dispersions of POEGMA-*b*-PMPS-*b*-POEGMA contained some large spherical aggregates with diameters of between 300 and 1000 µm among a lot of micellar material of diameter 15–30 nm (Fig. 26b). Silicon having a significantly greater electron capture cross-section than carbon, the dark areas in the centers of the larger aggregates are taken to be silicon-containing regions. In some cases, sheet-like aggregates were formed (Fig. 26c) and electron diffraction patterns (Fig. 26d), revealed these to have a hexagonal close-packed internal structure. PMPS is usually described as being amorphous but has been shown to possess 10%

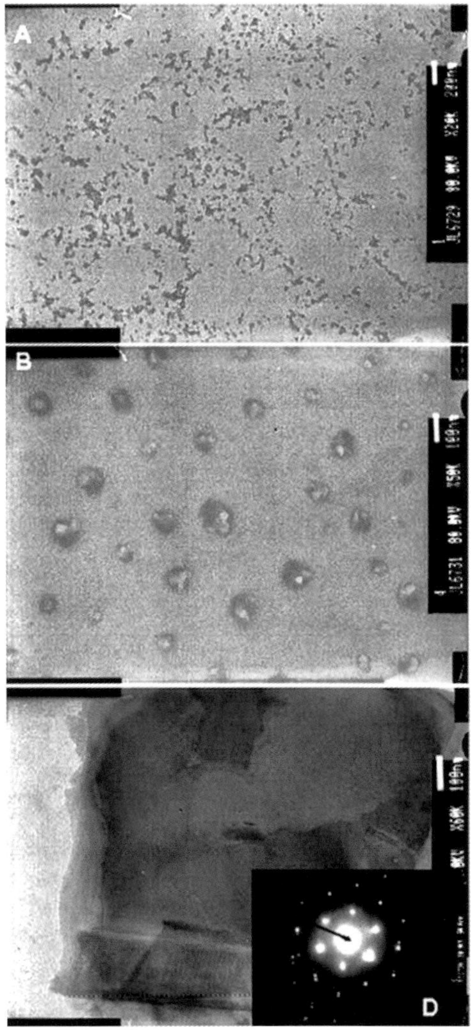

Fig. 26 TEM images of (**A**) PHEMA-*b*-PMPS-*b*-PHEMA micelles, (**B**) POEGMA-*b*-PMPS-*b*-POEGMA large spherical aggregates (**C**) POEGMA-*b*-PMPS-*b*-POEGMA sheet structures with inset (**D**) a diffraction pattern demonstrating hexagonal packing. Reproduced from [61], Holder et al. (2003) J Mater Chem 13:2771. © The Royal Society of Chemistry

crystallinity and to have a diffraction pattern of near hexagonal symmetry for a mesophase [83]. However, the high level of organization within the block copolymer indicates a bilayer with a smectic-like arrangement of the PMPS chains that is not dissimilar to observations of amphiphilic polythiophenes in aqueous dispersion [84]. The degradation of several of the polysilane aggregates upon exposure to UV light (λ_{max} = 254 nm) was demonstrated by monitoring the drop in the intensity of the UV absorption band at 334 nm (Fig. 27). All aggregates were shown to degrade with only slight variations in rate. Such degradation shows that polysilane block copolymer aggregates hold promise as light stimuli-responsive materials. Studies are currently underway in our group to encapsulate hydrophobic materials in the polysilane hydrophobic cores of these micelles and control release of the encapsulated materials through exposure to UV light (Fig. 28).

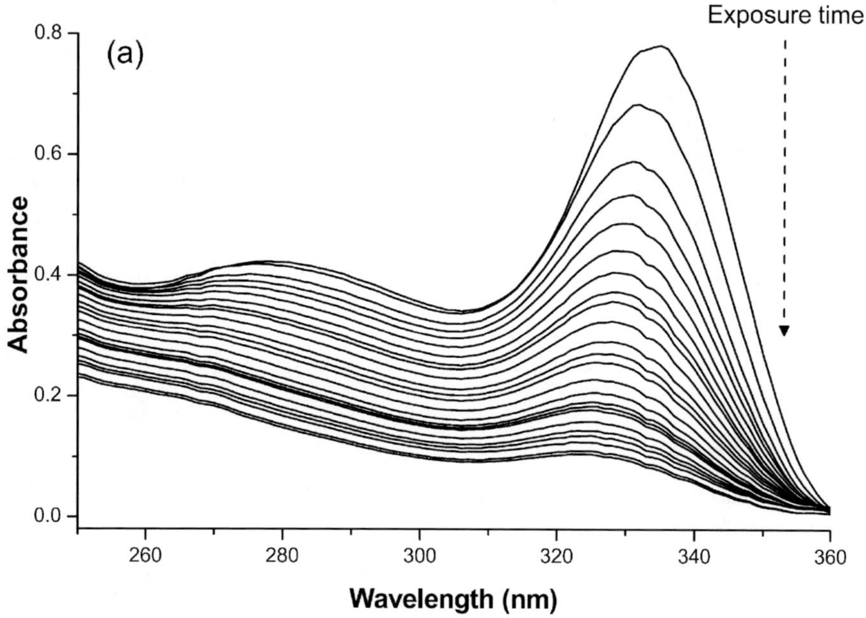

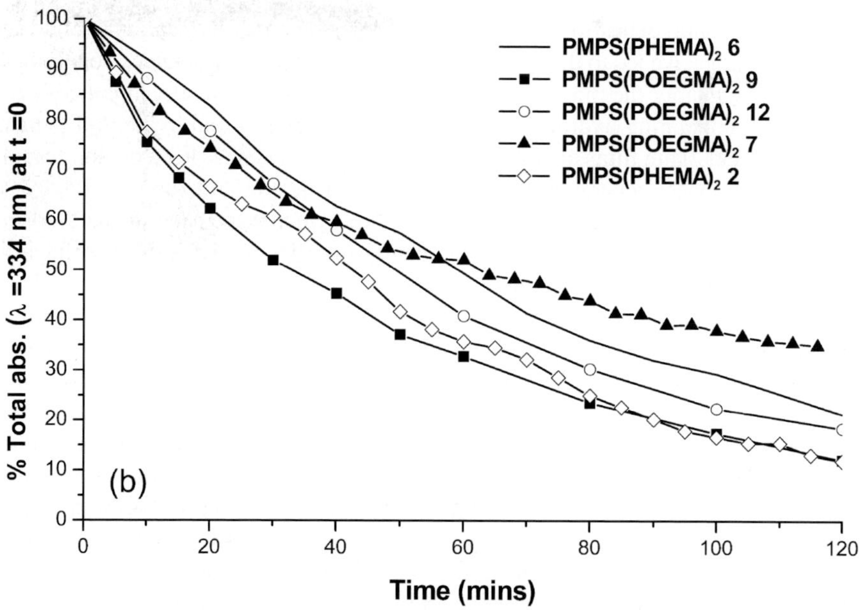

Fig. 27 Irradiation from 180–600 nm: **(a)** UV-Vis spectra recorded at 60 min intervals for the degradation of PHEMA-*b*-PMPS-*b*-PHEMA aggregates; **(b)** a plot of A/A0 × 100% at λmax versus irradiation time for various copolymer aggregates

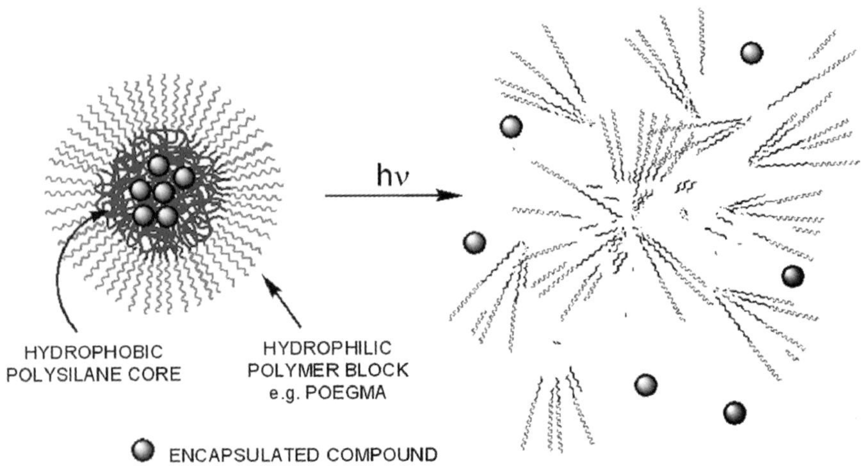

Fig. 28 Schematic illustrating proposed mechanism for the light induced release of encapsulated materials from polysilane micelle cores

4 Conclusions

A number of approaches to the synthesis of polysilane copolymers exist and the most promising have led to the synthesis of a number of polysilane block copolymer structures. Arguably the most intriguing of these are the classic amphiphilic block copolymers containing hydrophobic polysilane components and hydrophilic vinyl polymer blocks. Thin films of polysilane block copolymers have been demonstrated to adopt microphase separated structures with cylindrical or lamellar morphologies predominating. A number of polysilane block copolymers have been shown to form aggregates in aqueous solutions, predominantly micelles but vesicles and bilayers are also evident. The latter structures arising most likely as a result of the rod-like structure of the polysilane components. Despite interest in the optical and electronic properties of polysilanes the most interesting applications of their block copolymers have been in using their photolability as a means to pattern thin films (e.g. for controlled crystallization) or generate unique aggregate structures (e.g. hollow shell-cross-linked micelles). This might not be considered surprising as this is an almost unique feature of polysilane materials; their ability to undergo photodegradation under non-extreme conditions. In contrast many of their optical and electronic properties are matched and/or exceeded by the vast array of π-conjugated polymers and copolymers that have been synthesized and studied over the past two decades. It is likely that the any future applications and interest in polysilane block copolymers will reside in the coupling of this photolability with other properties resulting from the complementary block.

References

1. Miller RD, Michl J (1989) ChemRev 89:1359 and references therein
2. Shieh YT, Hsu TM, Sawan SP (1996) J Appl Polym Sci 62:1723
3. Keppler RG, Zeigler JM, Harrah LA, Kurtz SR (1987) Phys Revs B35:2818
4. Samuel LM, Sanda PN, Miller RD (1989) ChemPhysLett 159:227
5. Suzuki H, Hoshino S, Furukawa K, Ebata K, Yuan CH, Bleyl I (2000) Polym Adv Technol 11:460
6. Hasegawa T, Iwasa Y, Koda T, Kishida H, Tokura Y, Wada S, Tashiro H, Tachibana H, Matsumoto M, Miller RD (1995) Synth Met 71:1679
7. Kishida ZH, Hasegawa T, Iwasa Y, Koda T, Tokura Y, Tachibana H, Matsumoto M,SWada, Lay IT,Tashiro H (1994) Phys Revs B50:7786
8. Yajima S, Hasegawa Y, Hayashi J, Iiinura HJ (1978) J Mater Sci 13:2569
9. Yajima S, Hayashi J, Omori M (1975) Chem Lett 931
10. Peinado C, Alonso A, Catalina F, Schnabel W (2000) Macromol Chem Phys 201:1156
11. Yacgi Y, Schnabel W (1992) Macromol Chem Macromol Symp 60:133
12. Miller RD, Walraff G, Clecal N, Sooriyakurnaran R, Michl J, Karatsu T, McKinley AI, Klingensmith KA, Downing J (1989) J Polym Eng Sci 29:882
13. Hamley IW (1998) The Physics of Block Copolymers, Oxford University Press:Oxford
14. Discher BM, Hammer DA, Bates FS, Discher DE (2000) Curr Opin Colloid Interface Sci 5:125–131
15. Zhang L, Eisenberg A (1995) Science 268:1728–1731
16. Won YY, Davis HT, Bates FS (1999) Science 283:960–963
17. Kataoka K, Harada A, Nagasaki Y (2001)Adv Drug Deliv Rev 47:113–131
18. Allen C, Maysinger D, Eisenberg A (1999) Colloids Surf B:Biointerfaces 16:3–27
19. Kakizawa Y, Kataoka K (2002) Adv Drug Deliv Rev 54:203–222
20. Caruso F, Caruso RA, Mohwald H (1998) Science 282:1111–1114
21. Liu J, Sheina E, Kowalewski T, McCullough RD (2002) Angew Chem Int Ed 41:329–332
22. Jenekhe SA, Chen XL (1999) Science 283:372–375
23. Jenekhe SA, Chen X Science (1998) 279:1903–1907
24. Chen XL, Jenekhe SA (2000) Macromolecules 33:4610–4612
25. de Boer B, Stalmach U, van Hutten PF, Melzer C, Krasnikov VV, Hadziioannou G (2001) Polymer 42:9097–9109
26. van Hutten PF, Krasnikov VV, Hadziioannou G (2001) Synthetic Metals 122:83–86
27. Stalmach U, de Boer B, Videlot C, van Hutten PF, Hadziioannou G (2000) J Am Chem Soc 122:5464–5472
28. Widawski G, Rawiso M, Francois B (1994) Nature 369:387–389
29. Fujiki M (2001) Macromol Rapid Commun 22:539 and references therein
30. Fujiki M, Koe J (2000). In: Jones RG, Ando W, Chojnowski J (eds) Silicon-Based Polymer: The Science and Technology of Their Synthesis and Applications, Kluwer Academic Publishers, Dordrecht
31. Mark JE, Allcock HR, West R (1992) Inorganic Polymers. Prentice-Hall, New Jersey
32. Cypryk M, Chrusciel J, Fossum E, Matyjaszewski K (1993) Makromol Chem Macromol Symp 73:167
33. Sakamoto K, Obata K, Hirata H, Nakajima M, Sakurai H (1989) J Am Chem Soc 111:7641
34. Aitken C, Harrod JF, Samuel E (1985) J Organomet Chem 279:C11
35. Jones RG, Ando W, Chojnowski J (2000) Silicon-Based Polymers: The Science and Technology of their Synthesis and Applications. Kluwer Academic Publishers, Dordrecht
36. (a) Jones RG, Budnick U, Holder SJ, Wong WKC (1996) Macromolecules 29:8036
37. Jones RG, Wong WKC, Holder SJ (1998) Organometallics 17:59
38. Holder SJ, Achilleos M, Jones RG (2005)Macromolecules 38:1633–1639
39. (a) West R, Wolff AR, Peterson DJ (1986) J Radiat Cur 13:35
40. Yucesan D, Hostoygar H, Denizligil S, Yacgi Y (1994) Angew Makromol Chem 221:207

41. Yacgi Y, Krninek I, Schnabel W (1993) Polymer 34:426
42. Wolff A, West R (1987) Appl Organomet Chem 1:7
43. Demoustier-Champagne S, de Mahieu AF, Devaux J, Fayt R, Teyssie Ph (1993) J Polym Sci 31:2009–2014
44. (a) Demoustier-Champagne S, Canivet I, Devaux J, Jerome R (1997) J Polym Sci Part A: Polym Chem 35:1939–1948
45. Schwegler LA, Sheiko SS, Moller M, Fossum E, Matyjaszewski K (1999) Macromolecules 32:5901–5904
46. Lutsen L, Jones RG (1998) Polym Int 46:3–10
47. Hiorns RC, Holder SJ, Schué F, Jones RG (2001) Polym Int 50:1016
48. Sanji T, Kitayama F, Sakurai H (1999) Macromolecules 32:5718
49. Sanji T, Nakatsuka Y, Kitiyama F, Sakurai H (1999) J Chem Soc Chem Commun 2201
50. Sanji T, Nakatsuka Y, Ohnishi S, Sakurai H (2000) Macromolecules 33:8524
51. Sanji T, Takase K, Sakurai H (2001) J Am Chem Soc 123:12690–12691
52. Fossum E, Love JA, Matyjaszewski K (1995) J Organomet Chem 499:253
53. Fossum E, Matyjaszewski K (1995). In: Jones RG (ed) Silicon-Containing Polymers, Royal Society of Chemistry, Cambridge
54. Matyjasewski K, (2000) Controlled/"Living" Radical Polymerization:Progress in ATRP, NMR and RAFT. American Chemical Society, Washington, DC
55. Kato M, Kamigaito M, Sawamoto M, Higashimura T (1995) Macromolecules 28:1721–1723
56. Wang S, Matyjaszewski K (1995) Macromolecules 28:7901–7910
57. Percec V, Barboiu B(1995) Macromolecules 28:7970–7972
58. Anderson RM, Holder SJ, Jones RG, Rossi NAA (2003) Polym Internat 53:465
59. Jones RG, Holder SJ (1997) Macromol Chem Phys 198:3571
60. Rossi NAA, Jones RG, Holder SJ (2003) J Pol Sci PartA:Polym Chem 41:30
61. Holder SJ, Rossi NAA, Yeoh C-T, Durand GG, Boerakker MJ, Sommerdijk NAJM (2003) J Mater Chem 13:2771
62. Hamley IW, The Physics of Block Copolymers, Oxford University Press:Oxford, 1998
63. Bates FS, Frederickson GH (1990) Ann Rev Phys Chem 41:525
64. Thomas EL, Lescanec RL (1994) Philos Trans R Soc London Ser A 348:149–166
65. Klok HA, Lecommandoux S (2001) Adv Mater 13:1217
66. Mansky P, Harrison CK, Chaikin PM, Register RA, Yao N (1996) Appl Phys Lett 68:2586
67. Park M, Harrison CK, Chaikin PM, Register RA, Adamson DA (1997) Science 276:1401
68. Fink Y, Urbas AM, Bawendi MG, Joannopoulos JD, Thomas EL, (1999) J Lightwave Techn 17:1963
69. Urbas AM, FinkY, Thomas EL (**1999**) Macromolecules 32:4748
70. Urbas AM, Sharp R, Fink Y, Thomas EL, Xenidou M, Fetters LJ (2000) Adv Mater 12:812
71. Liu GJ, Ding JF, Hashimoto T, Kimishima K, Winnik FM, Nigam S (1999) Chem Mater 11:2233
72. Nardin C, Winterhalter M, Meier W (2000) Langmuir 16:7708
73. Fossum E, Matyjaszewski K, Sheiko SS, Möller M Macromolecules (1997) 30:1765–1767
74. Hiorns RC, Martinez H (2003) Synth Met 139:463
75. Popescu DC, Rossi NAA, Yeoh C-T, Durand GG, Wouters D, Leclère PELG, Thüne P, Holder SJ, Sommerdijk NAJM (2004) Macromolecules 37:3431–3437
76. Popescu DC, Lems R, Rossi NAA, Yeoh C-T, Holder SJ, Bouten CVC, Sommerdijk NAJM (2005) Adv Mater 17:2324–2329
77. Popescu DC, van Leeuwen ENM, Rossi NAA, Holder SJ, Jansen JA, Sommerdijk NAJM (2006) Angew Chem Int Ed 45:1762–1767
78. Holder SJ, Hiorns RC, Williams SJ, Sommerdijk NAJM, Jones RG, Nolte RJM (1998) Chem Comm 1445
79. Sommerdijk NAJM, Holder SJ, Hiorns RC, Jones RG, Nolte RJM (2000) Macromolecules 33:8289
80. Milling AJ, Richards RW, Hiorns RC, Jones RG (2000) Macromolecules 33:2651–2661

81. Sanji T, Ogawa Y, Nakatsuka Y, Tanaka M, Sakurai H (2003) Chemistry Letters 32:980–981
82. Sakurai H (2006) Proc Jap Acad Ser B: Phys Bio Sci 82:257–269
83. Demoustier-Champagne S, Jonas A, Devaux J (1997) J Poly Sci B:Polym Phys 35:1727–1736
84. Brustolin F, Goldoni F, Meijer EW, Sommerdijk NAJM (2002) Macromolecules 35:1054–1059
85. Sanji T, Nakatsuka Y, Sakurai F (2005) Polym J 37:1–6

Colour Plate Section

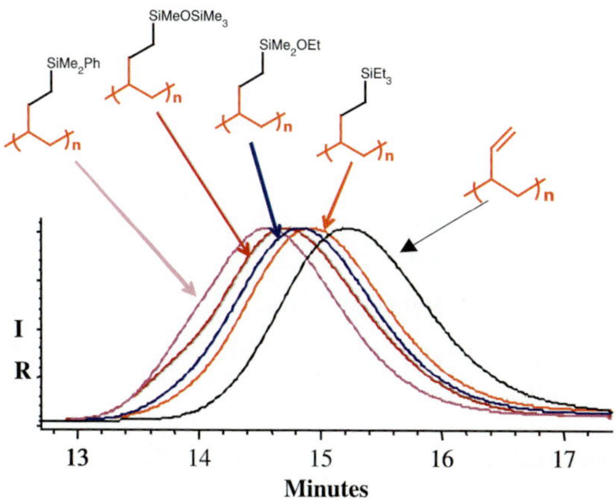

Plate 1 **SEC** traces of the PBD-1 and Polymers **3a**, **3b**, **3c** and **3d**; mobile phase: THF (1mL/min). Taken from Ref [13]. (See also Figure 1 on page 5)

(a) **(b)**

Plate 2 Synthesis of 2c in ethyl acetate/water, (a) reaction mixture at pH 7–8, (b) after acidification at pH 4–5 (See also Figure 3 on page 45)

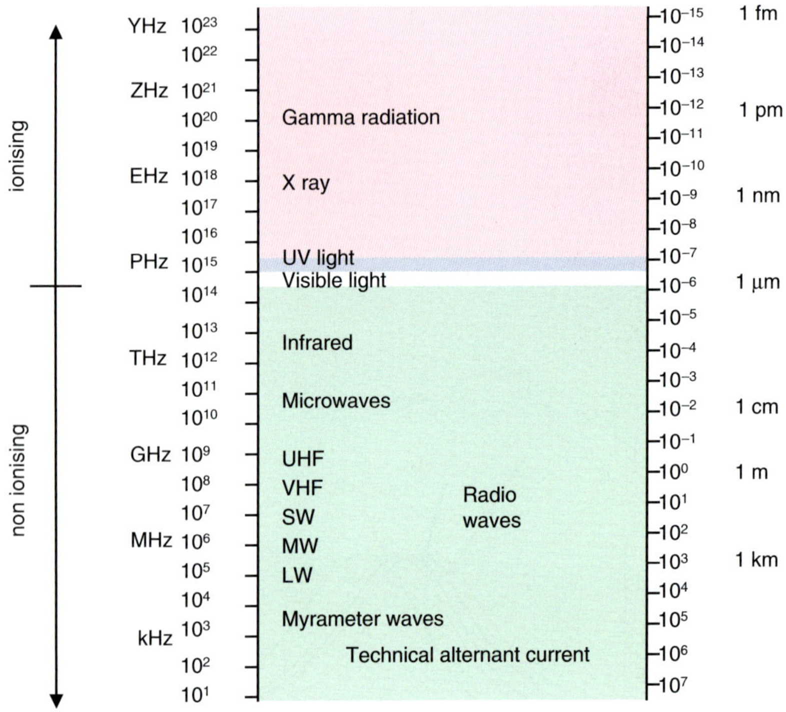

Plate 3 Electromagnetic radiation (See also Figure 1 on page 52)

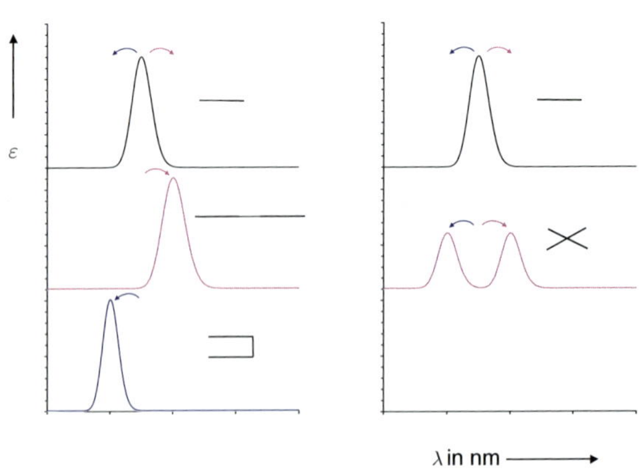

Plate 4 Splitting of UV/Vis absorption (*black line*) caused by various orientation of the transition moments in bichromophores. *Left side*: α orientation (*magenta line*) and β orientation (*blue line*). Magenta line on the *right side*: Skew arrangement (See also Figure 3 on page 54)

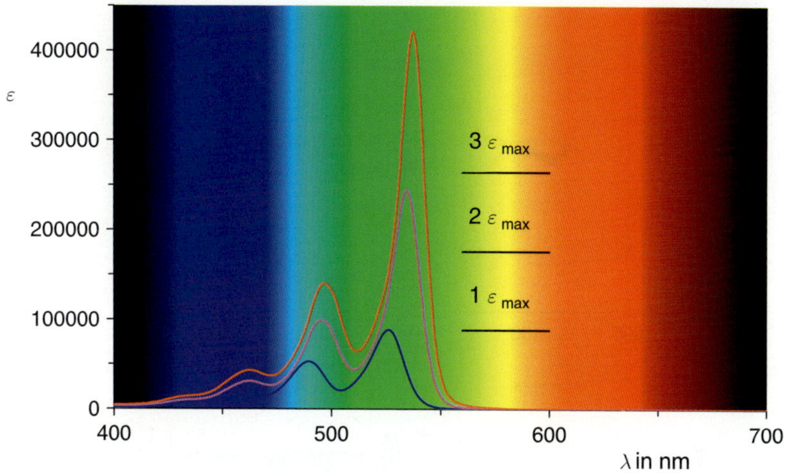

Plate 5 UV/Vis absorption spectra of **1** (*blue*), **4** (*magenta*), and **6** (*red*) in chloroform (See also Figure 4 on page 57)

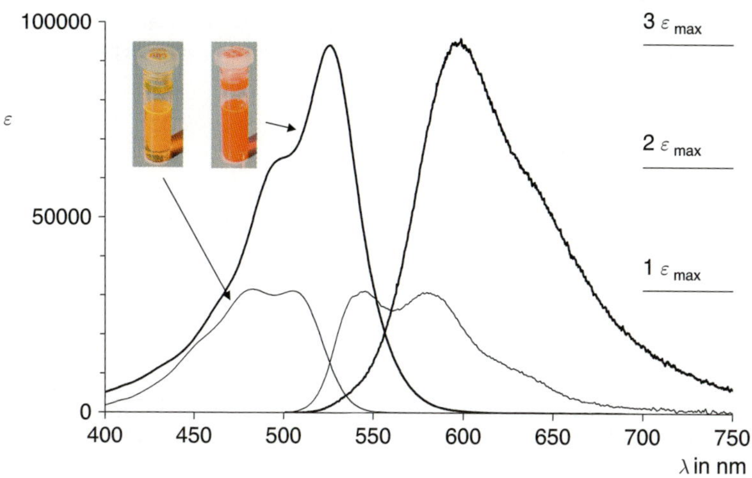

Plate 6 UV/Vis absorption and fluorescence spectra of **7** (*thin lines*) and **8** (*thick lines*) in chloroform (See also Figure 5 on page 57)

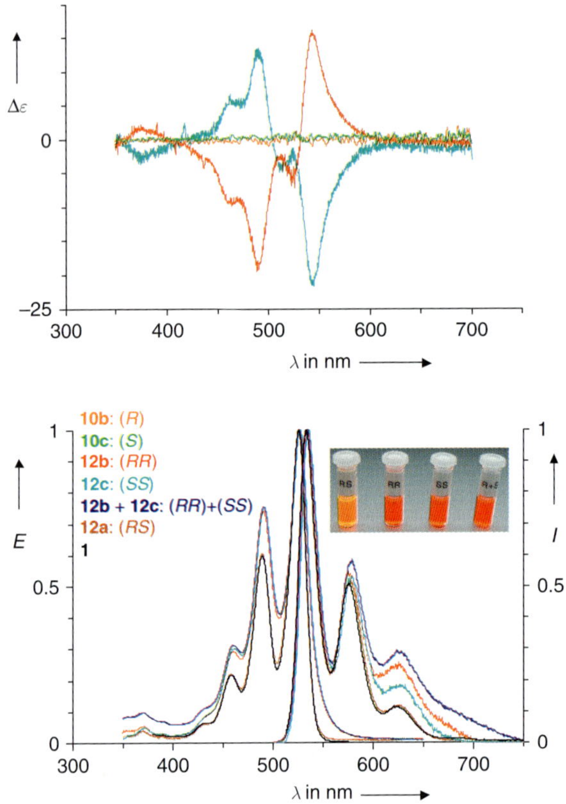

Plate 7 Bottom: UV/Vis absorption (*left*) and fluorescence spectra (*right*) in chloroform. Blue: 1:1-Mixture of **12b** and **12c**, *red:* **12b**, *turquoise*: **12c**, *brown*: **12a**, *black*: reference of standard perylene dyes (**1**). *Top:* CD spectra of **12b** (*red*) and **12c** (*turquoise*) in chloroform and neglible CD effect of the enantiomers **10b** (*orange*) and **10c** (*green*) (See also Figure 7 on page 61)

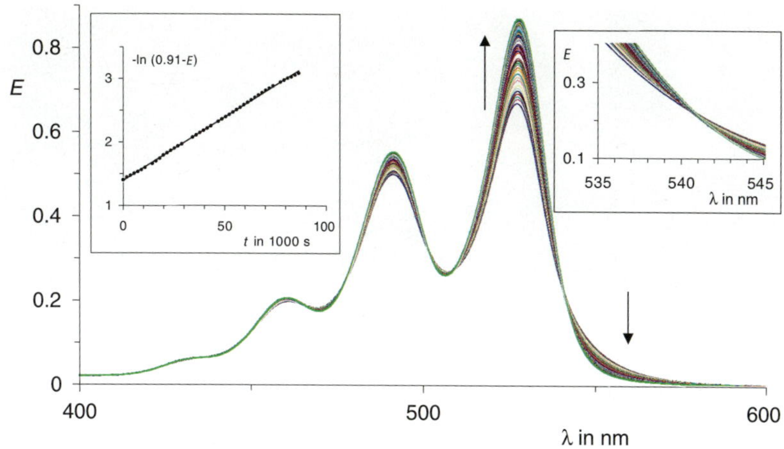

Plate 8 Change of the UV/Vis spectra by the acid-catalysed transformation of a 1:1 mixture of **12b** and **12c** into **12a**. Right insert: amplification of the region of the isosbestic point at 541 nm. Left insert: kinetic analysis of the transformation according to pseudo first order at 527.2 nm (See also Figure 8 on page 62)

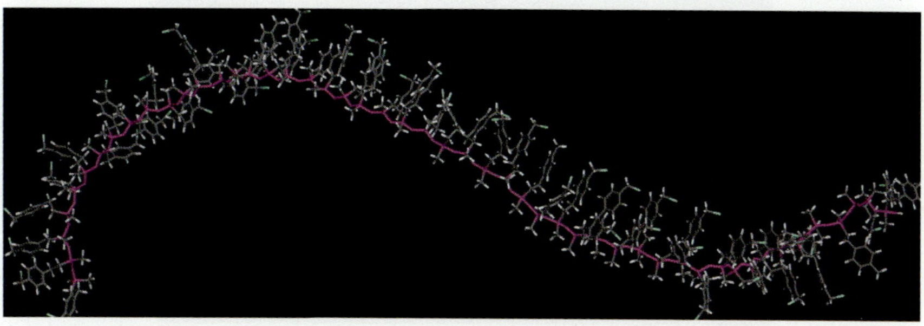

Plate 9 Minimum energy conformation corresponding to the homopolymer containing only B-structural units (See also Figure 3 on page 71)

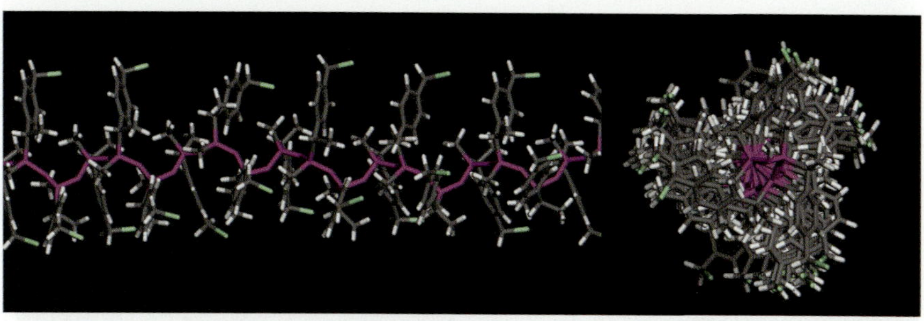

A B

Plate 10 Minimum energy conformation corresponding to the homopolymer containing only α-structural unit: (**A**) longitudinal view; (**B**) axial view (See also Figure 4 on page 72)

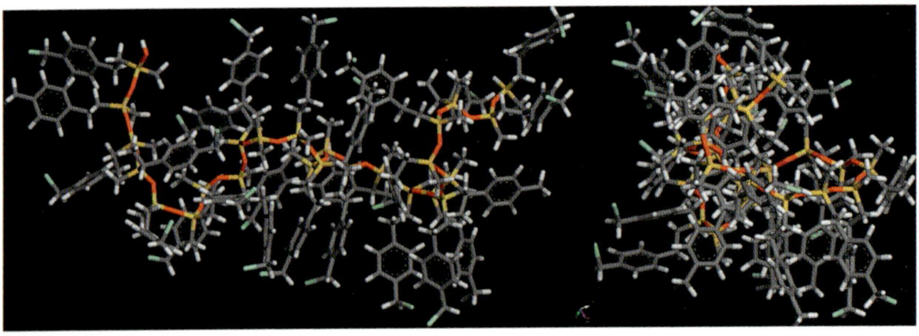

A B

Plate 11 Minimum energy conformation corresponding to the polysiloxane containing α (25%) and β (75%) structural units: (**A**) longitudinal; (**B**) axial view (See also Figure 5 on page 72)

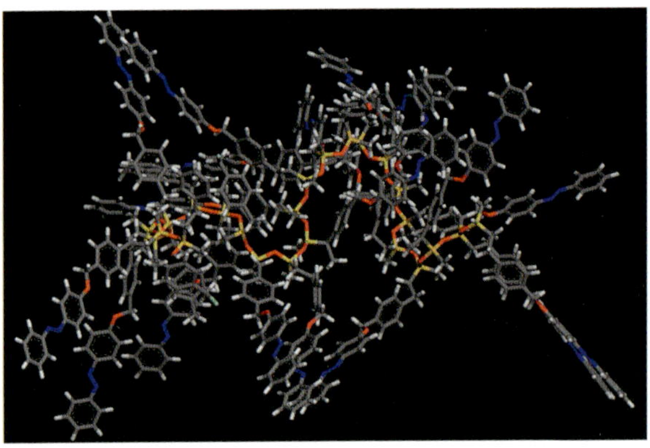

Plate 12 Minimum energy conformation corresponding to a polysiloxane modified with azobenzene (DP = 25) (See also Figure 6 on page 73)

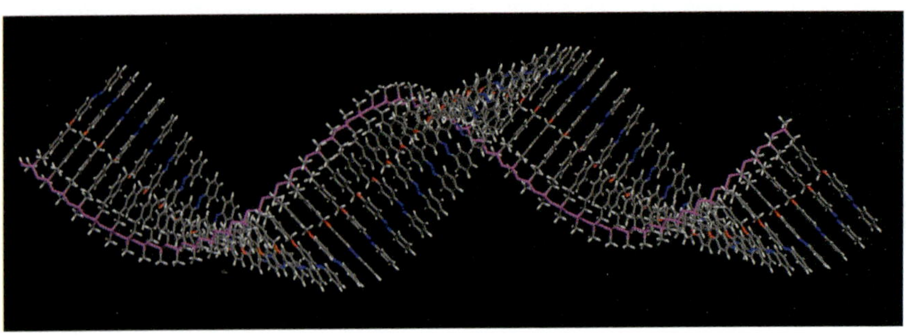

Plate 13 Chain geometry of an azo-polysiloxane based on β – structural unit type (See also Figure 7 on page 73)

Plate 14 9H-carbazole-9-ethanol powder (*left*), oily α,ω-3,5-dinitrobenzoate PDMS (*center*) and final melted-like strong orange complex of both species (*right*) (See also Figure 2 on page 90)

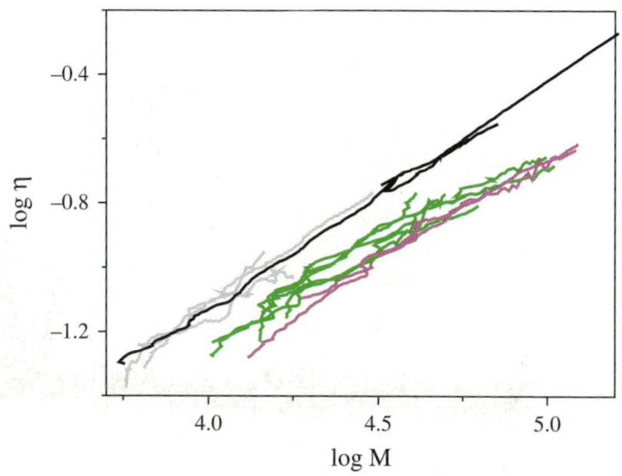

Plate 15 Mark-Houwink-Sakaruda plots for commercial PDMS terminated by $SiMe_3$ (*black*), PDMS synthesized with $B(C_6F_5)_3$ (*grey*), hybrid silicones synthesized with $B(C_6F_5)_3$ (*green*: from precursor **H**/ magenta: from precursor **M**) (See also Figure 12 on page 132)

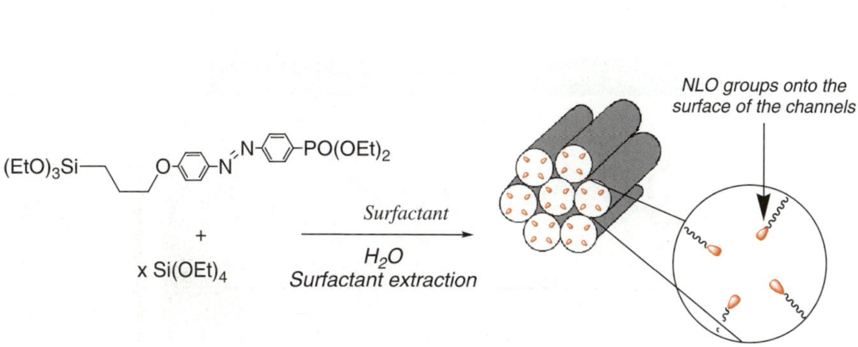

Plate 16 General pathway for the preparation of meoporous bifunctionalized organo-silica thin films (See also Scheme 1 on page 224)

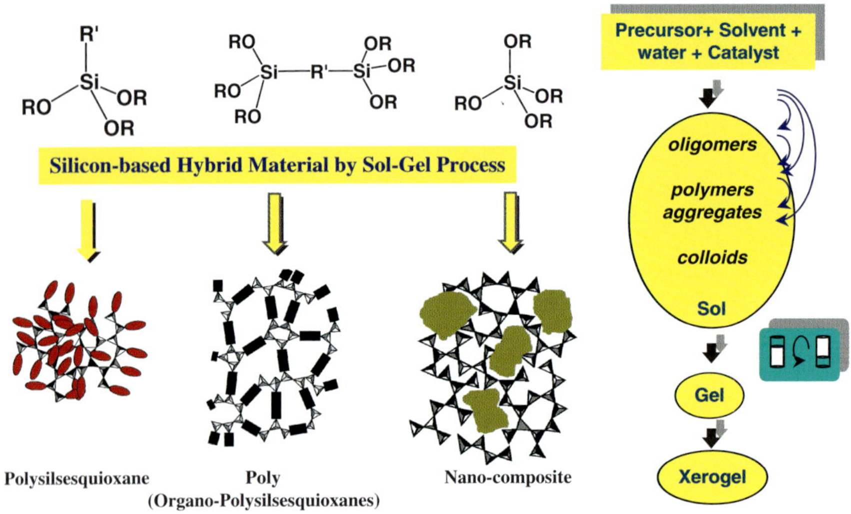

Plate 17 General overview of the different possibilities that offers the Sol-Gel process for the preparation of silicon-based hybrid material (See also Scheme 1 on page 234)

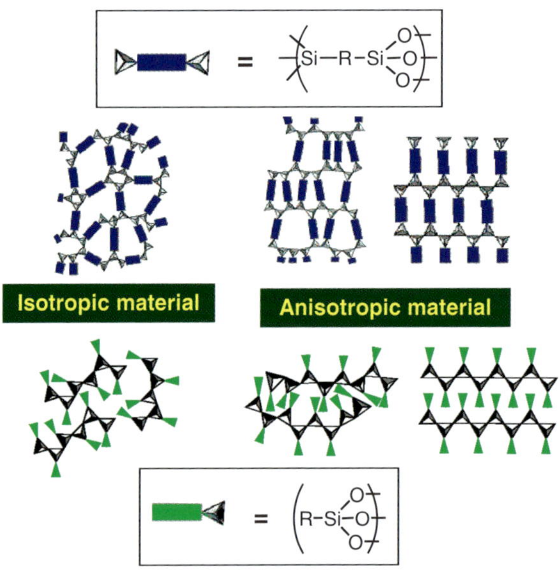

Plate 18 Various possibilities of organisation and structure for silicon-based hybrid materials (See also Scheme 2 on page 235)

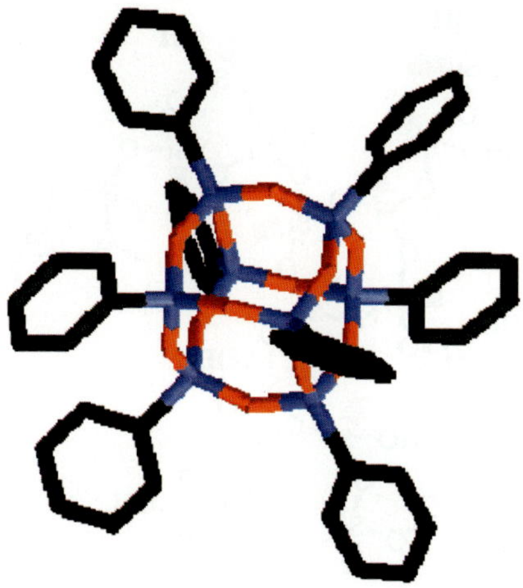

Plate 19 Cage-like structure of T_8 Cube polysilsesquioxane (See also Scheme 5 on page 237)

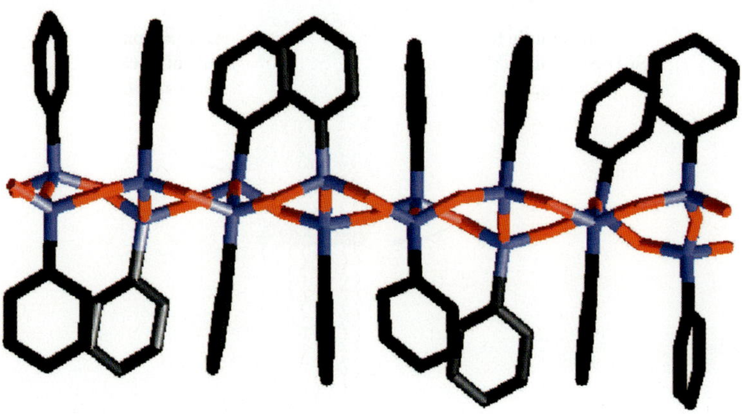

Plate 20 Ladder-like structure of polysilsesquioxane (See also Scheme 6 on page 237)

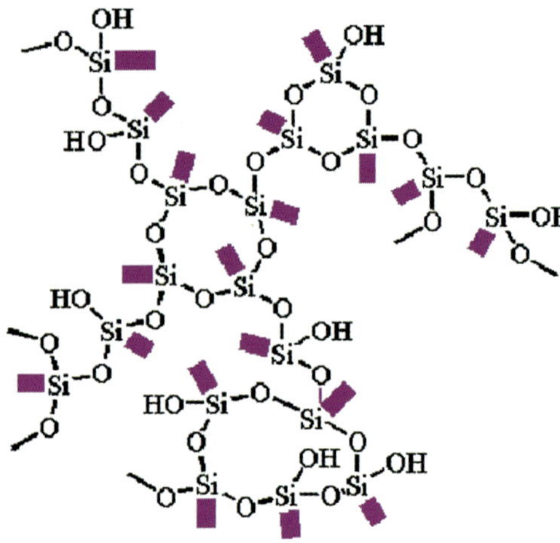

Plate 21 Idealized structure for polysilsesquioxane resin (See also Scheme 7 on page 238)

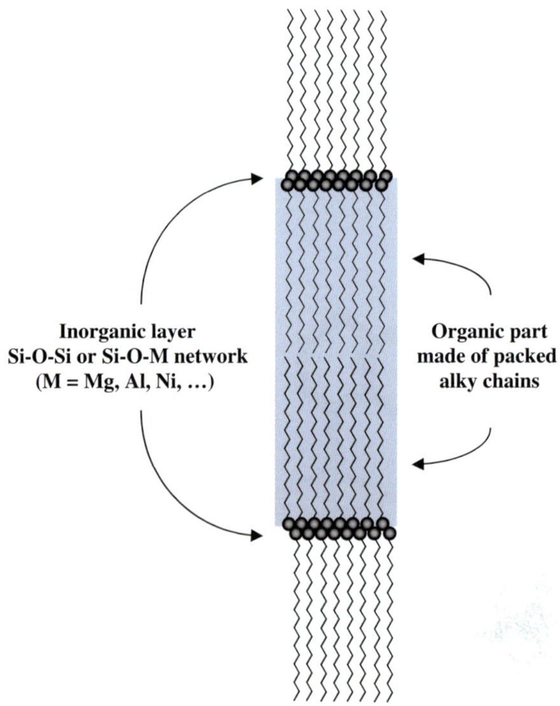

Plate 22 Idealised lamellar structure of polysilsesquioxanes (See also Scheme 8 on page 239)

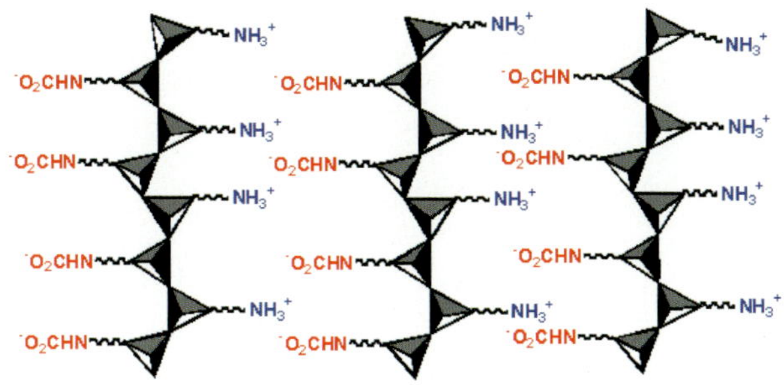

Plate 23 Idealised structure of the assembly of polysilsesquioxanes layers through carbamate/ammonium interactions (See also Scheme 9 on page 240)

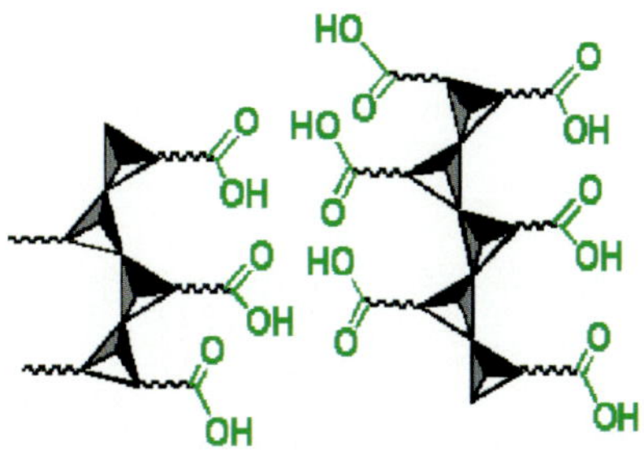

Plate 24 Idealised structure of the assembly of polysilsesquioxanes layers through H-bonds between carboxylic acid groups (See also Scheme 10 on page 241)

Plate 25 Images by microscopy in polarized light of three bridged-polysilsesquioxanes (See also Figure 1 on page 242)

Subject Index

R

S